Innovation and Disruption at the Grid's Edge

Innovation and Disruption at the Grid's Edge

How Distributed Energy Resources are
Disrupting the Utility Business Model

Edited by

Fereidoon P. Sioshansi
Menlo Energy Economics
Walnut Creek, CA, United States

ACADEMIC PRESS
An imprint of Elsevier

Academic Press is an imprint of Elsevier
125 London Wall, London EC2Y 5AS, United Kingdom
525 B Street, Suite 1800, San Diego, CA 92101-4495, United States
50 Hampshire Street, 5th Floor, Cambridge, MA 02139, United States
The Boulevard, Langford Lane, Kidlington, Oxford OX5 1GB, United Kingdom

Notices
Knowledge and best practice in this field are constantly changing. As new research and experience broaden our understanding, changes in research methods, professional practices, or medical treatment may become necessary.

Practitioners and researchers must always rely on their own experience and knowledge in evaluating and using any information, methods, compounds, or experiments described herein. In using such information or methods they should be mindful of their own safety and the safety of others, including parties for whom they have a professional responsibility.

To the fullest extent of the law, neither the Publisher nor the authors, contributors, or editors, assume any liability for any injury and/or damage to persons or property as a matter of products liability, negligence or otherwise, or from any use or operation of any methods, products, instructions, or ideas contained in the material herein.

Library of Congress Cataloging-in-Publication Data
A catalog record for this book is available from the Library of Congress

British Library Cataloguing-in-Publication Data
A catalogue record for this book is available from the British Library

ISBN: 978-0-12-811758-3

For information on all Academic Press publications visit our website at
https://www.elsevier.com/books-and-journals

Working together
to grow libraries in
developing countries

www.elsevier.com • www.bookaid.org

Publishing Director: Joe Hayton
Senior Acquisition Editor: Lisa Reading
Senior Editorial Project Manager: Kattie Washington
Production Project Manager: Kiruthika Govindaraju
Cover Designer: Matthew Limbert

Typeset by Thomson Digital

Contents

Author Biographies ix
Foreword xxi
Preface xxv
Introduction xxix

Part I
Envisioning Alternative Futures   **1**

1. Innovation and Disruption at the Grid's Edge   3
   Fereidoon P. Sioshansi

2. Innovation, Disruption, and the Survival of the Fittest   25
   Stephen Woodhouse and Simon Bradbury

3. The Great Rebalancing: Rattling the Electricity Value Chain from Behind the Meter   41
   Robert Smith and Iain MacGill

4. Beyond Community Solar: Aggregating Local Distributed Resources for Resilience and Sustainability   65
   Kevin B. Jones, Erin C. Bennett, Flora Wenhui Ji and Borna Kazerooni

5. Grid Versus Distributed Solar: What Does Australia's Experience Say About the Competitiveness of Distributed Energy?   83
   Bruce Mountain and Russell Harris

6. Powering the Driverless Electric Car of the Future   101
   Jeremy Webb and Clevo Wilson

7. Regulations, Barriers, and Opportunities to the Growth of DERs in the Spanish Power Sector   123
   Eloy Álvarez Pelegry

8. Quintessential Innovation for Transformation
    of the Power Sector   147
    John Cooper

Part II
Enabling Future Innovations   **165**

9. Bringing DER Into the Mainstream: Regulations,
    Innovation, and Disruption on the Grid's Edge   167
    Jim Baak

10. Public Policy Issues Associated With Feed-In Tariffs
    and Net Metering: An Australian Perspective   187
    Darryl Biggar and Joe Dimasi

11. We Don't Need a New Business Model: "It Ain't Broke
    and It Don't Need Fixin"   207
    Clark W. Gellings

12. Toward Dynamic Network Tariffs: A Proposal for Spain   221
    *Sergio Haro, Vanessa Aragonés, Manuel Martínez, Eduardo Moreda,
    Andrés Morata, Estefanía Arbós and Julián Barquín*

13. Internet of Things and the Economics of Microgrids   241
    Günter Knieps

Part III
Alternative Business Models   **259**

14. Access Rights and Consumer Protections
    in a Distributed Energy System   261
    Fiona Orton, Tim Nelson, Michael Pierce and Tony Chappel

15. The Transformation of the German Electricity Sector
    and the Emergence of New Business Models
    in Distributed Energy Systems   287
    Sabine Löbbe and André Hackbarth

16. Peer-to-Peer Energy Matching: Transparency,
    Choice, and Locational Grid Pricing   319
    James Johnston

17. Virtual Power Plants: Bringing the Flexibility of
     Decentralized Loads and Generation to Power Markets 331

     Helen Steiniger

18. Integrated Community-Based Energy Systems: Aligning
     Technology, Incentives, and Regulations   363

     Binod Koirala and Rudi Hakvoort

19. Solar Grid Parity and its Impact on the Grid   389

     Jeremy Webb, Clevo Wilson, Theodore Steinberg and Wes Stein

Epilogue   409
Index   413

Author Biographies

Eloy Alvarez Pelegry is Director of the Energy Chair of Orkestra, at Deusto University where he manages a group of researchers on energy.

Previously, he held executive positions at Union Fenosa Group, some related to regulation and R&D. He has been Associated Professor at the Technical School for Mining and the Complutense University in Madrid. He is member of the Royal Academy Engineering (RAI) and author of several books and numerous papers.

Eloy has a Bachelor's Degree in Economics and Business from Complutense University, a Diploma in Business Studies from London School of Economics, and PhD in Mining from the Higher Technical School for Mining of Madrid (ETSIMM).

Vanessa Aragonés is an expert on electricity regulation at Endesa where she works on electricity market design and integration of distributed energy resources.

Previously, Vanessa worked for Endesa Latin America developing strategic options with an emphasis on climate change regulation. She has also worked at the Endesa retail division as an electricity and gas analyst during the first stages of the market liberalization in Spain.

Vanessa holds an MSc in Industrial Engineering from the Polytechnical University of Madrid.

Estefanía Arbós is an Energy Policy Analyst at Endesa, focusing on regulated tariffs and prices.

Previous to her current job, she worked as an Operations and Strategy Consultant at Deloitte, as a Research Assistant at BP Chair of Sustainable Development, and as a trainee at the European Commission.

Estefanía is an Industrial Engineer from ICAI and holds a MBA from Collège des Ingénieurs, Paris, France.

Jim Baak is the Program Director for Grid Integration for Vote Solar, a nonprofit advocacy organization. He leads regulatory and policy work for integrating renewable energy into the distribution and bulk power grids. He has testified before the US House of Representatives Natural Resources Committee and state public utilities commissions and legislative bodies on energy issues. He chairs a regional transmission planning group for the Western Electricity Coordinating Council, serves on the Board of Directors for Interwest Energy Alliance, and is an advisor for a multidisciplinary renewable energy research project for the Nevada System of Higher Education.

Previously, he worked for PG&E, Powel Group, Utility.com, Alameda Municipal Power, and ElectriCities of NC.

He has a BS in Economics from the University of South Carolina.

Julián Barquín works as an expert at the Regulatory Affairs Department of Endesa, now part of Enel.

Previously, he was a Professor at Comillas Pontifical University in Madrid, Spain. He has been Visiting Scientist at MIT and Visiting Scholar at the University of Cambridge. He is author or coauthor of several books and more than 100 papers in peer-reviewed journals and a frequent speaker at conferences.

He holds a PhD and Power Engineering degrees from Comillas Pontifical University and a Physics degree from UNED.

Erin C. Bennett is a Research Associate at the Institute for Energy and the Environment and a Juris Doctor candidate at Vermont Law School, where her focus lies in Environmental, Energy, and Natural Resources Law. Prior to her legal studies, Erin earned her Bachelor of Science in Physics from Ursinus College. After a very research-intensive undergraduate career, Erin plans to utilize her scientific knowledge in the field of Energy Law, where science and the environment go hand-in-hand.

Erin is expected to graduate from Vermont Law School in 2018.

Darryl Biggar is the Special Economic Advisor for regulatory matters at the Australian Competition and Consumer Commission and the Australian Energy Regulator where he provides advice and carries out research in the economics of public utility regulation and electricity markets. He has multiple publications on regulation, market power, and investment in electricity networks including a textbook, with M. Hesamzadeh, "Economics of Electricity Markets," published in 2014.

Prior to the ACCC Dr. Biggar worked at the OECD in Paris and for the New Zealand Treasury.

He holds a PhD from Stanford University in Economics and an MA in Mathematics from the University of Cambridge.

Simon Bradbury is a Principal Consultant with over 15 years of experience in the energy sector. His areas of expertise include wholesale electricity trading arrangements, energy market regulation, and market design. Simon has a particular focus on the electricity market design in Europe and in developing markets.

Before joining Pöyry, Simon worked for the GB energy regulator, Ofgem, from 2000–06, where he was latterly a Senior Manager in the Wholesale Markets team.

Simon holds an honors degree in Economics and Business Management from Newcastle University and an MSc (Merit) in Economic Regulation and Competition from City University.

Tony Chappel is Head of Government and Community Relations at AGL Energy Limited. He has previously worked as a Chief of Staff in the NSW Government, and prior to that at the Smith School for Enterprise and the Environment at Oxford University.

Tony holds a Masters degree in Environmental Change and Management from the School of Geography and the Environment at the University of Oxford, and a Masters of Science in Energy Policy from Imperial College London.

Paula Conboy is the Chair of the Australian Energy Regulator, a position she has held since October 2014.

She has over 20 years' experience in public utility regulation in Canada and Australia; first with the Industry Commission and Sydney Water then on to the Ontario Energy Board. She later held senior positions at PowerStream ending up as Vice-President of Regulatory and Government Affairs. In 2010 Paula was appointed to the Ontario Energy Board and oversaw important policy development, adjudicated over 200 applications for cost of service, performance-based regulation, mergers and acquisitions, and leave to construct energy networks.

Paula holds an MSc (Agr) from Guelph University in Canada.

John Cooper is Co-founder and President of Prsenl, the first marketing firm in the new field of Personal Energy, and Founder and CEO of MaaS Energy, a Microgrid-as-a-Service provider of tailored onsite energy solutions for businesses, governments, utilities, and individuals.

John is coauthor of *The Advanced Smart Grid: Edge Power Driving Sustainability*, 2nd Edition; has led numerous innovative projects as a thought leader in the Smart Grid industry and authored numerous white papers, articles, and blog posts.

John has a BA in Government from the University of Texas Austin, as well as an MBA with honors from UT.

Joe Dimasi was appointed Tasmanian Economic Regulator in November 2015 and Senior Commissioner for the Independent Competition and Regulatory Commission (ICRC) of the ACT in 2016. He is a former Commissioner of the Australian Competition and Consumer Commission (ACCC) where he led the Commission's regulatory functions before being appointed Commissioner in 2008. He is also Professorial Fellow at the Monash Business School.

Joe is an applied economist with a long involvement in the area of competition and regulation matters particularly in the regulation and reform of utilities.

He holds Bachelor of Economics and Master of Economics from Monash University in Melbourne.

Clark Gellings is an independent energy consultant specializing in assessing and meeting electricity consumers' needs. He recently retired as a Fellow at the Electric Power Research Institute where he was responsible for technology strategy in energy efficiency, demand response, renewable energy, and clean technologies.

He has an extensive background as an executive in technology management and research; the recipient of numerous awards; has served on various boards and advisory committees; a member of the National Academy of Engineering, an IEEE Life Fellow, a Fellow Emeritus of the Illuminating Engineering Society; and an Honorary Member of CIGRE—the International Council on Large Electric Systems.

Clark has a Bachelor of Science in Electrical Engineering from Newark College of Engineering in New Jersey, a Master of Science in Mechanical Engineering from New Jersey Institute of Technology, and a Master of Management Science from the Stevens Institute of Technology.

André Hackbarth is a researcher and lecturer at Reutlingen University, School of Engineering/Distributed Energy Systems and Energy Efficiency, Germany. His research interest includes business models in the energy sector and consumer attitudes, preferences, and decision-making concerning energy-related behaviors and products.

Prior to his current position, he was Research Associate at the Institute for Future Energy Consumer Needs and Behavior (FCN) at RWTH Aachen University, Germany.

He is PhD candidate at RWTH Aachen University and studied Economics at Heidelberg University.

Rudi A. Hakvoort is an associate professor at the Faculty of Technology, Policy and Management, TU Delft. He is expert in the area of energy network regulation as well as design and regulation of liberalized energy markets.

Previously, Dr. Hakvoort directed the Market and Infrastructure Department of the Dutch Office for Energy Regulation. He has been chairman of the working group on Congestion Management of the Council of European Energy Regulators. Dr. Hakvoort also serves as a consultant for energy utilities worldwide and contributes to training courses in the area of energy policy and regulation in Europe.

Dr. Hakvoort holds a MSc in Applied Physics and a PhD in Materials Science.

Sergio Haro is the recipient of an Endesa grant to analyze advanced tariff structures within the Regulatory Affairs Department.

He has a Power Engineer degree from Zaragoza University and a Master Degree from Pontifical Comillas University. He has also studied at the Bialystock Politechnical University in Poland.

Russell Harris is the Director of Wollemi Consulting, an independent advisor and consultant to the energy industry. Russell's 25 year career has focused on energy productivity in commercial, industrial, and institutional applications and the development of commercial and utility-scale solar power.

He has a Bachelor's Degree in Mechanical Engineering from Swinburne University and postgraduate qualifications in Business (Strategy & Marketing) from Monash University.

Flora W. Ji is a Research Associate at the Institute for Energy and the Environment and currently pursuing her Juris Doctor and Master in Energy Regulation and Law at Vermont Law School. As a member of the Vermont Law Review, she has written about the legal issues in the Clean Power Plan. Previously working for the Associated Press in Beijing, she has done first hand reporting on the disastrous smog China has been suffering from.

James Johnston is the CEO and cofounder of Open Utility, an innovative technology startup based in London, United Kingdom with a mission to democratize energy.

Prior to founding Open Utility, James spent 3 years researching building-integrated direct current microgrids at University of Strathclyde. James is also the founder of Solar Sketch, a design company for the solar industry and worked as a building services engineer at international engineering consultancy Arup.

James holds a BEng and MSc from University of Strathclyde in Mechanical Engineering.

Kevin B. Jones is the Director and Professor of Energy Technology and Policy at the Institute for Energy and the Environment at Vermont Law School.

Previously, he worked at the Long Island Power Authority, Navigant Consulting and as the Director of Energy Policy for the City of New York. He is the coauthor of "A Smarter, Greener Grid: Forging Environmental Progress Through Smart Energy Technologies and Policies" and coauthor of the forthcoming book "The Electric Battery: Charging Forward to A Low Carbon Future."

Kevin has a Doctorate from Rensselaer Polytechnic Institute, a Master of Public Affairs from the LBJ School of Public Affairs, University of Texas at Austin, and a BS from the University of Vermont.

Borna Kazerooni is a Research Associate at the Institute for Energy and the Environment and Juris Doctor Candidate at Vermont Law School. Borna worked as a policy analyst for the Virginia Department of Mines, Minerals and Energy where he provided policy support and analysis for state energy efficiency and renewable energy programs. Prior to working at the state energy office, Borna worked for the Virginia Joint Legislative Audit and Review Commission, a legislative oversight agency for the Commonwealth of Virginia.

Günter Knieps is Professor of Economics at the University of Freiburg, Germany. Prior to joining Freiburg he held a Chair of Microeconomics at the University of Groningen, the Netherlands. He is Member of the Scientific Council of the Federal Ministry of Economics and Energy and the Ministry of Transport and Digital Infrastructure.

Professor Knieps' main research interests include study of network economics, deregulation, competition policy, industrial economics, and sector studies on energy, telecommunications, and transportation. He has published widely in academic and professional journals.

Professor Knieps has diplomas in Economics and Mathematics from the University of Bonn, Germany, and a PhD in Mathematical Economics from the University of Bonn.

Binod P. Koirala is an Erasmus Mundus Joint Doctorate candidate on sustainable energy technologies and strategies at Faculty of Technology, Policy and Management, TU Delft. His current research areas are community energy systems, distributed energy resources, as well as system integration.

Previously, Binod has been working at the autonomous systems and minigrids group of Fraunhofer Institute for Solar Energy Systems in Freiburg, Germany. He was DAAD WISE fellow at Fraunhofer Institute for Wind and Energy Systems in Kassel, Germany.

Binod holds a Bachelor of Technology in Electrical Engineering and Master of Science in Renewable Energy Management.

Sabine Löbbe is a professor and researcher at Reutlingen University, School of Engineering/Distributed Energy Systems and Energy Efficiency, and at Reutlingen Research Institute, Germany and lectures in the Master program at the University of Applied Sciences HTW Chur, Switzerland. Her consulting company advises utilities in strategy and business development and in organizational issues.

Prior to her current position, she was Director for Strategy and Business Development at swb AG, Bremen; Project Manager at Arthur D. Little Inc.; and Project Manager at VSE AG, Saarbrücken.

She holds a Doctorate in Business Administration from the University Saarbrücken and studied Business Administration in Trier, Saarbrücken and EM Lyon/France.

Iain MacGill is an Associate Professor in the School of Electrical Engineering and Telecommunications at the University of New South Wales, and Joint Director for the University's Centre for Energy and Environmental Markets (CEEM). Iain's teaching and research interests include electricity industry restructuring and the Australian National Electricity Market, sustainable energy technologies, and energy and climate policy.

Iain leads CEEM's research in Sustainable Energy Transformation including energy technology assessment, renewable energy integration, and Distributed Energy Systems including smart grids, distributed generation, and demand-side participation.

Dr. MacGill has a Bachelor of Engineering and a Masters of Engineering Science from the University of Melbourne, and a PhD on electricity market modeling from UNSW.

Manuel Martínez works at Endesa's Regulatory Affairs Department where he coordinates tariffs, regulated incomes analysis and proposals, as well as power price statistics. He frequently collaborates with power associations and has been member of several international task forces in network tariffs and electricity price comparison.

Previous to his current job, he was involved in power generation and distribution regulation.

Manuel obtained an Industrial Engineering degree from Universitat Politecnica de Catalunya and an interuniversity MBA from Universitat de Barcelona, Universitat Autonoma de Barcelona, and Universitat Politecnica de Catalunya.

Johannes Mayer is Head of Competition and Regulation at E-Control Austria, the Austrian energy regulator. His main tasks are data analytics, market analysis, market design, and surveillance of wholesale and retail markets in electricity and natural gas.

Previously, he was head of energy policy at the Chamber of Commerce, attache for industrial affairs to the EU, an expert to the Parliamentary Committee on energy policy in Austria, secretary general of the Advisory Council For Economic and Social Affairs. He has authored numerous articles and books on economic policy and energy policy, as well as energy law.

Johannes holds a degree in business administration and studied mathematical economics at the Institute for Advanced Studies in Vienna.

Andrés Morata is Deputy Director of Regulation for Economical Management at Endesa where he leads regulatory analysis and proposals on issues related to tariffs and settlements with special emphasis on regulated activities.

He has worked in the energy sector for more than 20 years; first, in Union Fenosa as International Investment Analyst and from 1997 in Endesa's Regulation Department, carrying out different roles and responsibilities including analysis of the economic impacts of different tariffs, methodologies and remuneration schemes for regulated activities, M&A involving transmission and distribution assets.

Andrés obtained an Industrial Engineering degree from Seville University and MBA from Instituto San Telmo, also in Seville, Spain.

Eduardo Moreda is Deputy Director of Regulation for Generation, Wholesale Power Market and Gas at Endesa. He leads regulatory analysis and proposals on issues related to electricity and gas wholesale markets as well as buying and selling process of generators and suppliers.

Previous to his current job, Eduardo was working in a utility on issues related with generation, distribution, transmission, research, planning, and computing models.

Eduardo obtained a BA in Mechanical Engineering from University of Seville.

Bruce Mountain is the Director of consultancy Carbon and Energy Markets (CME) in Melbourne and the cofounder of retail market data provider, MarkIntell.

As an independent energy economist, he has advised industry, government, associations, investors, lenders, and consumers for the past 25 years on a wide range of issues in the economics of energy and regulation in Australia, Britain, South Africa, and other countries.

He has a PhD in Economics from Victoria University, a Bachelor's and a Master's Degree in Electrical Engineering from the University of Cape Town, and is qualified as a Chartered Management Accountant in England.

Tim Nelson is the Chief Economist at AGL Energy—one of Australia's largest energy utilities—where he is responsible for sustainability strategy; greenhouse reporting; economic research; corporate citizenship program; and greenhouse policy. Tim was previously an economic adviser to the NSW Department of Premier and Cabinet and the Reserve Bank of Australia.

Tim has advised governments and utilities on energy and climate change policy and has published in Australian and international journals. He is an Adjunct Associate Professor at Griffith University.

He holds a PhD in energy economics for which he was awarded the Chancellors Doctoral Research Medal, a degree in economics and is a Chartered Secretary.

Fiona Orton is the Manager of Scenario Planning and Competitor Analysis at AGL Energy, a large Australian energy supplier. She supports the development and execution of corporate strategy through the use of uncertainty analysis and by monitoring critical leading indicators and emerging themes within Australia's energy markets. Previously, Fiona has contributed to AGL's energy policy development, energy market analysis, and economic research and has managed compliance with a range of mandatory and voluntary greenhouse gas programs. Prior to joining AGL she was a climate change and sustainability consultant.

She has authored many policy-related papers that have been influential in the development of Australian energy and greenhouse policy.

Fiona holds an honors degree in chemical engineering, specialized in energy and the environment.

Michael Picker was appointed to the California Public Utilities Commission in January 2014 by Governor Brown and named President in December 2014.

He was Senior Advisor for Renewable Energy to the Governor from 2009 to 2014, Principal at Lincoln Crow Strategic Communications from 2000 to 2009, Deputy Treasurer in the Office of the California State Treasurer from 1998 to 1999, and Chief of Staff to Sacramento Mayor Joe Serna Jr. from 1992 to 1999. He was a member of the Sacramento Municipal Utility District Board of Directors from 2012 to 2014.

He holds an MBA from UC Davis.

Michael Pierce is the Manager Market Analysis at AGL Energy, a large Australian energy supplier. He is responsible for the modeling of wholesale energy markets and advanced customer analytics. Previously, Michael has lead the development of several sophisticated models used by AGL to forecast the impacts of disruptive behind-the-meter technologies and large-scale renewable development on retail and wholesale markets. Prior to joining AGL, he was an energy market consultant.

He is a member of working groups for the Australian Energy Market Operator and has worked on market benefit analysis studies, demand forecast reviews, and asset valuations across several markets.

Michael holds a PhD in Astronomy and an honors degree in Physics.

Fereidoon P. Sioshansi is founder and president of Menlo Energy Economics, a consulting firm, advising clients on energy-related issues. For 25 years, he has been the editor and publisher of *EEnergy Informer*, a monthly newsletter with international circulation.

His prior work experience includes working at Southern California Edison Co., EPRI, NERA, and Global Energy Decisions, acquired by ABB. Since 2006, he has edited nine volumes on different subjects including evolution of global electricity markets, energy efficiency, smart grid, and distributed generation.

He has a BS and MS in Civil and Structural Engineering, an MS and PhD in Economics from Purdue University.

Robert Smith has over 25 years experience working in industry economics, electricity market design, regulation, economic evaluation, energy efficiency, and demand management.

His interests include applied economic analysis and understanding how economics, technology, incentives, regulation, and customers' behavior interact to create change.

Robert has a graduate degree in econometrics, a masters in economics from the University of NSW, and postgraduate qualifications in finance from the Securities Institute of Australia.

Wes Stein is Manager of the Australian Commonwealth Scientific and Research Organization's (CSIRO) National Solar Energy Centre and is the Concentrating Solar Power Stream Leader in CSIRO's Energy Flagship Project.

He is the Australian Solar Thermal Research Initiative's (ASTRI) Principal Investigator for CST projects in high temperature steam, advanced thermal storage, tower air, and supercritical CO_2 Brayton systems and the solarized fuels program.

He has extensive experience in the energy and power industry including thermodynamic cycles.

Theodore Steinberg is a Professor in the School of Chemistry, Physics and Mechanical Engineering at the Queensland University of Technology (QUT) in Brisbane, Australia.

He is the Principal Investigator and Leader of QUT's ASTRI program focused on developing, implementing, and reducing the cost of solar thermal power plants. He has coedited seven books, several book chapters, and authored numerous papers on solar thermal energy and the flammability and sensitivity of materials in oxygen-enriched environments under both normal gravity and reduced gravity conditions with a strong background in materials science and flammability.

Ted has a PhD, MSc, and BSc degrees in Mechanical Engineering from New Mexico State University in Las Cruces, New Mexico.

Helen Steiniger is advisor to the CEOs at Next Kraftwerke where she works on new markets and strategy.

Prior to her current position, she worked in the communications and research department of the same company, being responsible for covering current and future developments in energy markets.

Helen holds an MSc in Environmental Studies and Sustainability Science from Lund University in Sweden and a BSc in Business Administration from the University of Mannheim in Germany.

Jeremy Webb is a visiting researcher at the Queensland University of Technology (QUT) where he is focused on urban transport systems and modal choice. His doctoral thesis examined the historical evolution and the locking in of the automotive mode in urban transport systems. Choice modeling was used to determine the incentives needed to reduce private car usage and commit to public transport. More recently Dr. Webb has collaborated in a number of academic studies and projects focusing on the phenomenon of "peak car" and the likely uptake of shared electric autonomous vehicles in urban environments.

As a former diplomat and head of the Australian Department of Foreign Affairs and Trade's Economic Analysis Unit he has been responsible for a wide range of reporting on global trade and investment issues.

Jeremy holds an MA from University of Hawaii, a Bachelor of Economics with honors from ANU, and PhD in Economics from QUT.

Clevo Wilson is a Professor of Economics at the Queensland University of Technology (QUT). He specializes in environmental, ecological, agricultural, energy, tourism, and development economics with a special interest in using environmental valuation techniques, both revealed and stated for policy decision-making.

He has published widely in diverse topics including energy and water conservation, agriculture, aquaculture, ecotourism, and environmental sustainability.

Clevo has a PhD in Environmental and Resource Economics from St Andrews, Scotland, MSc in Economics from Glasgow University, an MPhil in Environment and Development from Cambridge, and a BA in Economics from the University of Peradeniya, Sri Lanka.

Stephen Woodhouse is a director at Pöyry Management Consulting. He heads Pöyry's Market Design group, which deals with all aspects of energy market policy regulation and design, for private- and public-sector clients. He specializes in electricity market design and the economics of generation, transmission, and interconnection. Stephen leads Pöyry's business development in the interrelated areas of intermittency, market design, and smart energy and is working with clients on customer-centric business models.

Prior to joining Pöyry, Stephen was an Economic Modeler for Ofgem.

Stephen has an MA (Cantab) and a BA (Hons) in Economics from the University of Cambridge.

Audrey Zibelman, was appointed CEO of Australian Energy Market Operator (AEMO) effective March 20, 2017. From 2013 through to her appointment with AEMO she served as the Chair of the New York State Public Service Commission (PSC). During Ms. Zibelman's leadership at the NYPSC, New York Governor Andrew M. Cuomo enacted the "Reforming the Energy Vision" (REV) plan. The REV plan has been internationally recognized for successfully developing and implementing 21st century regulatory reform with a focus on lowering the cost of energy for consumers while building a more resilient and reliable power system.

Her prior experience include senior executive positions at PJM and Xcel Energy, as well as founder, CEO and Chair of Viridity Energy, Inc. Ms. Zibelman has also served on several Board of Directors concerning the reliability and security of the electric system in New York and elsewhere in the United States.

Ms. Zibelman received her BA from Pennsylvania State University and her JD from Hamline University School of Law.

Foreword

One is hard pressed these days to come across an electricity sector that's not undergoing a fundamental change. Policy drivers for decarbonized economies, new business models, and changing technologies are fuelling an accelerating transformation along the entire electricity supply chain, with consumers firmly in the driver's seat.

In Australia, it was the norm to have centralized, synchronous fossil fuel generators transporting electricity over networks in one direction to passive end-use customers. This is now giving way to more decentralized, nonsynchronous, intermittent, and renewable generation sources. Consumers are no longer passive recipients of electricity, but rather actively exercising greater choice and control over when they buy, use, store, and sell their electricity. In particular, Australia has the highest penetration rates of rooftop solar photovoltaic (PV) in the world, with over 1.6 million systems installed.[1] Overall, this equates to more than one in seven Australian households. Some states, such as South Australia and Queensland, record penetration rates as high as 25%, with some suburbs above 50%.

Networks are managing more dynamic and complex energy flows. Stand-alone and microgrids are emerging, particularly at the grid's edge. Improvements in technology, communications, and data availability are providing new opportunities to rethink how existing services are provided and where the boundary between the roles of generators, networks, retailers, and consumers lies.

Late last year the Australian Government committed to reduce the country's emissions by 26%–28% below 2005 levels by 2030. The Australian Governments' focus is to ensure that consumers and industry have access to low-cost, reliable energy, as Australia moves toward a lower-emissions economy. This has given rise to a number of reviews, most recently the Finkel Review,[2] charged with developing a national reform blueprint that will outline national policy, legislative, governance, and rule changes required to maintain the security, reliability, affordability, and sustainability of the National Electricity Market.

---

1. Clean Energy Regulator data as on December 1, 2016. http://www.cleanenergyregulator.gov.au/RET/Forms-and-resources/Postcode-data-for-small-scale-installations
2. https://www.environment.gov.au/energy/national-electricity-market-review

One question being asked at the Australian Energy Regulator is whether the regulatory regime is fit for the future, whatever the future may bring? Can it provide the predictability to attract investment, yet be sufficiently flexible to accommodate a future that is different to what might be predicted? Australia is not alone in this respect. Many others around the world are pondering these exact same issues, as reflected in the Preface and Introduction by regulators in California and New York, respectively, as well as explored in detail throughout the chapters.

Australia is well placed to face these challenges and not new to reform and change. The regulatory regime was developed through the microeconomic reform of the 1990s, during which it saw the structural separation of electricity companies, competitive neutrality between government and private businesses, the removal of regulatory restrictions, and the establishment of the National Electricity Market.[3] The foundational principles developed in the 1990s remain relevant today:

- reliance on competitive markets in those parts of the industry that can sustain competition will lead to lower price, better quality products and services, greater efficiency, and will drive genuine consumer choice and empowerment;
- clear separation between regulated and competitive sectors to allow new energy markets, products, and services to emerge; and
- effective regulation of the monopoly elements to keep costs at efficient levels.

Rising energy prices and plummeting technology costs are prompting customers to choose cheaper, more innovative, and personalized energy products and services. While this disruption and transformation of our energy sector can benefit consumers, it must also be accompanied by appropriate levels of protection. These issues are under active consideration by policymakers and regulators in Australia as further discussed in by Orton et al. and others in this volume.

Many of the current regulatory arrangements and consumer protections in Australia were developed for the traditional energy supply model. Most of the new products and innovative service delivery models were not contemplated when the existing arrangements were put in place. Ensuring consumers have access to the best information to allow them to make informed choices will build their confidence to engage actively in these new markets. It will also help to strike the balance between innovation and supporting customers in an increasingly dynamic and transforming market.

Through a clear separation between regulated and competitive sectors, networks are prevented from discriminating in favor of affiliates offering competitive energy services, such as rooftop solar, smart appliances, and batteries,

---

3. http://www.aemc.gov.au/getattachment/8c426f7d-ea5c-4823-9b86-510dfd4e82dd/The-National-Electricity-Market-A-case-study-in-mi.aspx

to customers. Requiring regulated network businesses to separate their regulated energy services from other unregulated services limits their ability to discriminate and leverage regulated revenues and information asymmetries. Such separation helps to promote a more level playing field for competitive energy service providers, which will ultimately benefit consumers through increased choice, innovative energy products, and service, as well as lower long-term prices. The ring-fencing guidelines will also ensure that network customers do not pay more than necessary by preventing cross-subsidies between regulated and other services. Further, ring-fencing *services*, rather than *assets*, address the potential for storage to generate multiple value streams for multiple players, including consumers, networks, and generators. This work is consistent with New York's plan to Reform the Energy Vision, described in the Introduction, in seeking to promote innovation to deliver greater choice and value for consumers.

Effective ex ante– and incentive-based regulation of network services—the monopoly elements of the electricity supply chain—supports efficient innovation. The Australian Energy Regulator provides networks with a revenue stream that is based on efficient costs and is sufficient to enable the company to operate its network safely and reliably. Once set, the framework encourages the networks to find the most efficient way of delivering the required services and allows for some experimentation, including through demand-side and nonnetwork options. There are also a number of schemes in place to balance cost reductions with reliability and service quality.

Network tariffs have historically overrecovered revenue for off-peak use of the network and underrecovered for peak use. Consumers using most of their electricity at off-peak times have historically been paid more than the cost of network services. Conversely, those consuming electricity at peak times were paying less than the full costs of network services. For example, in 2014 it was estimated that a consumer using a large 5-kW air-conditioner in peak times will cause about $1000 a year in additional network costs compared with a similar consumer without an air-conditioner, but the consumer with the air-conditioner only pays about $300 extra under the most common network prices. The remaining $700 is recovered from all other consumers through higher network charges. Reforms are under way to implement network tariffs that reflect the efficient cost of providing network services, so that consumers can make more informed decisions about their electricity usage and better manage their bills.

The Australian Energy Regulator has supported a measured phase-in of cost-reflective tariffs taking into account the impact of changes in tariffs on consumers. Consumers will be better able to respond to price signals if they can relate their consumption decisions to tariff structures and avoid sudden tariff changes. There is a general agreement that cost-reflective tariffs will help ensure efficient production, investment, and consumption decisions. Network price reform in Australia is further explored in other chapters of this book.

While electricity sectors worldwide are all facing increased innovation and disruption as the transition to alternative futures accelerates, this book provides an excellent review of some of the different mechanisms used to meet these opportunities and challenges. The chapters in this book compile a wide range of expertise and perspectives to promote the sharing of knowledge, ideas, and experience in how some of these challenges might be addressed. I trust you will find it an insightful read.

Paula W. Conboy[4]
Chair
Australian Energy Regulator

---

4. I would like to acknowledge the assistance of Angela Bourke in finalizing this Foreword.

Preface

The editor of this volume has solicited an assortment of contributions from some of the most insightful thinkers advancing the state of knowledge on issues affecting the electric industry at the intersection of the distribution network and customers, referred to as "the grid's edge" in the volume's title. The assembled chapters herein are particularly timely for California and other jurisdictions, such as Hawaii, Australia, and Germany, that are at the forefront of integrating large amounts solar and other distributed energy resources (DERs) interconnecting on the distribution grid, whether on customer premises or in front of the meter. Collectively, these works describe elements of several possible energy futures that may emerge as a result of the transformative trends shaping the industry today.

Two related forces have resulted in the rapid expansion of renewable energy generally, and DERs specifically, in recent years. First, California and other jurisdictions have made serious commitments in recent years to reduce greenhouse gas emissions, improve air quality, and promote renewable energy through a variety of policies and programs. In the electric industry, this calls for mobilizing capital toward new renewable power technologies and away from fossil fuels as the basis of our electric generation. Second, as the policy support has allowed certain clean energy technologies to scale up, performance has improved and costs have fallen rapidly. In the past decade, the costs of wind power and solar PV have fallen by 50% and 75%, respectively, and their costs continue to fall. As a consequence of these forces, the vast majority of the solar and wind capacity connected to electricity grids across the world was installed only in the last decade.

In addition to this shift in the source of power production, we have also seen new technologies resulting in a parallel trend toward decentralization and a new democratization of decision-making within the grid. Like solar and wind technologies, battery storage costs have similarly fallen by approximately half since the mid-2000s. Hundreds of thousands of Californian households and businesses now have experience generating their own power with on-site solar. With the rapid decline in battery storage costs and the advent of smart homes, a small but growing share of customers is gaining experience with *managing* energy usage as well.

As customers take advantage of the additional choices that become available to them, the relationship between customers and utilities will change.

Some experts believe that these trends will, or should, fundamentally alter the nature of electricity markets and the role of distribution utilities. Several visions for the future of the electric industry have been proposed in recent years. As described by proponents, the utility of the future will be characterized by one or more of the following elements: (1) distribution grids are managed by distribution system operators (DSOs), which may or may not be the incumbent utility; (2) DSOs provide a platform for the exchange of energy and other services at the distribution level; (3) transactive energy platforms enable customers to buy and sell energy directly to each other across the distribution system; and (4) microgrids balance load and supply locally to the extent feasible and exchange energy with the transmission grid in response to differing costs or reliability considerations.

While it is important to begin thinking about longer-term implications of low-cost solar and storage, we should be cautious of becoming enamored of innovation for its own sake. Buzzwords in this field have proliferated recently, but we should maintain a healthy skepticism about which investments are most likely to generate real net benefits. I am not convinced that sweeping reorganization of the industry is necessary for DERs to deliver considerably more value to the grid.

For both strategic and practical reasons, I have chosen to focus activity at the CPUC on more tangible tasks that can deliver benefits quickly, rather than questioning the fundamental nature of utility business models. Among the three large investor-owned utilities in California that we regulate, customer-sited solar capacity is approaching 5000 MW. Installed energy storage capacity currently amounts to almost 50 MW, but another 130 MW are in the development pipeline. In addition, at nearly 250,000 electric vehicles, California accounts for half of the United States' fleet. As Californian utilities already have high, and steadily growing, penetration of DERs, we are facing a near-term need to better manage our DERs to avoid detrimental impacts on power quality and reliability.

The overarching philosophy I have followed in pursuit of a more distributed energy future can be described as "Walk, Jog, Run." While deployment of solar and storage adoption continues at a healthy pace, we are not facing an imminent reliability or cost crisis resulting from runaway adoption. We have time to experiment, pilot new technologies, and processes, and get it right before rushing headlong into major new investments or reliance on DERs to provide critical reliability services before these capabilities have been amply demonstrated.

To date, the factors driving the uptake of DERs, whether through technology-oriented programs managed by the CPUC, compensation under net energy metering, or the federal tax incentives, have been relatively blunt, with no differentiation based on locational value. As a result of these incentives, we are failing to derive the full value that DERs are capable of providing. In part this was due to a lack of any agreed-upon process to quantify locational value. After years of fruitless debate, it became clear that new tools and processes were needed to begin deploying DERs more strategically.

The vision we are pursuing is that, over time, DERs will be able to benefit from "stacking" multiple value streams. The FERC recently approved the California Independent System Operator's Distributed Energy Resource Provider tariff, which allows aggregations of DERs to provide energy and other services at the wholesale level. To complement those wholesale revenue streams, we are beginning the process of identifying and quantifying the location-dependent distribution values that DERs are capable of delivering. Once those values have been quantified, the utilities will have mechanisms such as solicitations, incentives, and payments for services to compensate DERs for these services.

To shift deployment of DERs to lower-cost, higher-benefit locations, the CPUC has undertaken three groups of related activities:

- *facilitating* the interconnection of DERs in low-impact areas;
- *targeting* DER deployment to the areas of the grid where they provide the most value; and
- *incentivizing* the utilities to prefer DER alternatives to traditional distribution capital investments where feasible.

To accomplish the first goal, utilities and other stakeholders have been developing a tool that uses highly granular power flow analysis to quantify the DER hosting capacity with a far greater accuracy than the rules of thumb currently in use. Once the initial version of the hosting capacity tool is finalized in early 2017, we will open a proceeding to incorporate the tool into the interconnection process, which should allow a significantly greater DER capacity to interconnect under the expedited "fast-track" process. By making the tool publicly available, DER developers will have better information about where they can expect to interconnect most easily.

Targeting DERs to high-value locations also necessitates development of a tool to highlight areas of the distribution grid where DERs can provide location-specific values, such as distribution capacity deferral and voltage support. With these areas of the grid identified, a mechanism must be established for the utility to recommend specific opportunities for DERs and receive authorization to conduct solicitations or offer incentives in those areas.

Finally, we are piloting incentive mechanisms that encourage utilities to seek DER alternatives to traditional distribution grid investments. As currently envisioned, these incentives will allow the utilities to earn returns on expenditures for DERs designed to compensate utility shareholders by an amount comparable to or greater than the net returns they would expect to receive on the traditional rate-based investment.

While the CPUC is focusing its efforts on more practical near-term concerns, we have been collaborating with our colleagues in New York and at the US Department of Energy to support technical and conceptual research into the longer-term barriers to a cleaner, more resilient, more distributed energy future. The authors who have contributed to this volume are providing critical, cutting-edge thinking to help practitioners begin to grasp how DERs, advanced distribution management systems, price signals, market structures, and utility business models may coalesce to realize that future.

Michael Picker
President
California Public Utilities Commission

Introduction[1]

This is an important compendium. During the recent heyday of books and speeches on the creation of the "Smart Grid", experts frequently repeated the observation that the integrated power system represented one of the great engineering feats of the 20th century. For the 21st century, I believe that regulatory innovation will be cited by experts as the transformational accelerator that gave birth to the electric prosumer, the widespread integration of distributed energy, transactive power, and their chief enabler, the Distributed System Platform (DSP) provider.

In New York's plan to Reform the Energy Vision (REV), the transformation of our electric distribution companies from risk adverse, regulated monopolies into innovative, market-enabling, and increasingly efficient and smart platform operators, DSPs, is a central bet for accelerating the achievement of the power system and network innovations that are discussed throughout this book. Like many transformative actions, the launch of REV was stimulated by a significant event. For New York, it was the devastation the State witnessed by four major storms over a short period, culminating with Superstorm Sandy. Under the leadership of Governor Andrew M. Cuomo, the State determined that combating climate change was no longer optional and, therefore, the manner in which power was produced, delivered, and consumed had to change.

New York has extremely ambitious climate goals. Under Governor Cuomo, New York has mandated that it will reduce emissions by 40% by 2030 and ensure that 50% of the electricity consumed in the Empire State is produced by renewable resources by the same time. To achieve these mandates along with equally important reliability, efficiency, and network security goals, the Department of Public Service and the Public Service Commission, the utility economic regulators in the State, were charged with discovering and then instituting the necessary retail market and regulatory reforms.

Readers of this book likely already know that technological innovations provide us with the ability to achieve essential deep decarbonization, while continuing to improve upon the reliability, resiliency, adaptability, and efficiency of the power networks. The real challenge is to convert these opportunities into a reality in a power system that has heretofore been dominated by markets

---

1. The author wrote the introduction while at the New York Public Service Commission and prior to joining AEMO. The views reflected are those of her own.

designed around status quo preferences for large central station dispatchable generation, monopolistic utilities who earned almost exclusively from investing in expensive and long lived assets, a presumption that consumers are passive and consumption is largely inelastic, and retail regulatory practices based upon 20th century sensibilities. As the state agency that oversees regulated utilities in New York, the intent is to challenge these presumptions; a challenge common to all regulators across the globe as reflected by the Foreword, the Preface, and the Epilogue in this volume by utility regulators from other parts of the world.

As the retail utility regulator in a State that led on competitive restructuring of the wholesale power system and was an early adopter of retail competition, the necessary focus was on the role and function of the regulated distribution utility. New York previously adopted many of the regulatory mechanisms used to ensure that regulated distribution utilities do not actively oppose programs to increase energy efficiency or the adoption of customer-based distributed energy resources. New York's regulatory construct already included forward test years, revenue-decoupling mechanisms, and modest incentives for the privately owned utilities. What was missing, however, was the fact that while utilities were not economically harmed by activities at the edge of the network, they had no business reason, other than regulatory compliance, to actively pursue strategies that would support third-party investments on the customer side of the meter or at the edge of the grid. The regulatory changes put in place under REV are designed to provide that incentive.

The crux of the utility changes contemplated in REV can be summarized into the following five areas, many of which are touched upon by the authors of the following chapters.

1. *The creation of the DSP*: The successful integration of intermittent renewable resources both on the bulk power system and at the edge of the distribution system depend upon the strength of the network business. Transmission and distribution system operators must transform into forward leaning platform businesses that invest in the knowledge, tools, and math that will allow them to function in a world of the millisecond dispatch and a dynamic two-way system, In the modernized networks, reliability and efficiency must be simultaneously managed on the bulk system and at the grid's edge. For the distribution utility, this means that system operators must have the knowledge, confidence, and tools to increasingly rely on third party–owned distributed energy resources to maintain reliable, resilient, and environmentally and economically efficient operations. For the first time, distribution network operators will need to plan for, understand, and then rely on third party–owned resources at the grid's edge, as well as consumer behavior as core elements of a secure and reliable network.

2. *Promoting and encouraging innovation*: Learning to innovate for organizations that pride themselves on low-risk taking and regulatory compliance is a change that requires deliberate and sustained focus. To innovate around

a business model or service delivery requires an opportunity to make small bets that can yield large benefits with minimal regulatory interference. The best entrepreneurs learn to place small bets with rapid deployment followed by quick adaptations based on lessons learned. Integrating that thinking into the mindset of slow-moving and process-oriented utilities and regulators required us to adopt a model that allows—indeed requires—utilities to develop business model demonstration projects that serve this critical role. The result has been the creation and continued creation of multiple demonstrations that are allowing for rapid change in the understanding of how the utility can promote innovative solutions and to create employee-based excitement and enthusiasm for the opportunities that are essential to any transformational activity. While regulators cannot mandate innovation, they should do their best not to obstruct or constrain it. This is particularly important given the fast pace of innovation and disruption taking place at the so-called "grid's edge"—the interface between the distribution network and the consumer, prosumer, or prosumagers—the focus of this volume.

3. *Regulated earnings model*: Peter Bradford, a former Chair of the New York Public Service Commission made the wise observation that all regulation is in fact incentive regulation, it is just what regulators are trying to incent that changes.[2] The broad regulatory change REV is intended to encourage utilities to deliberately move away from regulated capital investing to a business model where significant and sustained earnings can be achieved through information and services that support optimizing the value of distributed energy resources and load management as assets that make the integrated networks more valuable to end-use consumers. Regulators are living in a world where demand for innovation at the edge of the network will continue to grow. The business compliment to this innovation is that regulators have to move away from thinking solely about the volumetric rate for regulated electric services. Rather, regulators need to look at the world from the consumer's perspective of looking at the total bill and value of received services. For consumers the economic and environmental benefit that can be created from a modern grid is to be active participants in an increasingly dynamic and transactive power system dominated by zero-cost energy resources that are made reliable and efficient by a smart, dynamic distribution network.

The utilities could see that change as a threat to their traditional monopoly franchise business, the so-called "death spiral," or a source of an opportunity to innovate and grow in a manner that benefits the end user. In New York, regulators have supplied utilities the latter path. Under REV, they are able to earn from activities as varied as those that reduce peak, increase energy efficiency, reduce transaction expense for DER suppliers, and reduce

---

2. Proceeding on Motion of the Commission in Regard to Reforming the Energy Vision (April 25, 2015).

their own capital spend. Under various formulas, these earnings can more than make up for the earnings they would have achieved under the previous business model. However, unlike New York's current form of regulation, the utility activities in this formula are designed to decrease reliance on spend that finds its way into regulated revenue requirements and thereby result in a system that is smarter, more efficient, and produces an overall bill reduction to consumers. At the same time, New York is fully cognizant of the fact that its utilities remain functioning monopolies that must continually raise considerable capital. Thus, in making these changes, New York will continue to ensure that its regulation balances the needs of shareholders and customers for financially sound companies that are able to attract investment at a relatively low expense.

4. *System information and transactive markets*: Critical to any well-functioning market is the ubiquitous availability of information that innovators and suppliers can use to develop and distribute value-added products and accurate price signals. REV promotes and requires the development of retail markets that will reduce the costs of third parties to participate in retail markets. To that end, New York requires its utilities to develop and file distribution system plans on a regulator basis, develop cost-effective plans for advanced metering infrastructure, pursue advancements in hosting capacity, and develop markets that provide needed load and system data and monetize the value of dynamic load response and energy efficiency.

   The development of accurate prices for dynamic load is a substantial undertaking. Throughout the United States regulators are struggling with the transformation from the convention of net energy metering (NEM) for solar resources, a form of feed-in tariffs, to a price regime that accurately and fully values the ability to manage resources at the grid's edge or the interface with the customer's premise. New York is transforming from its blunt tariff-based system that used conventions, such as demand response payments, NEM, and energy efficiency targets, and measurements, to a transactive system that is nondisruptive, unbundled, and identifies the complete value of these resources. These actions set the foundation for creating competitive statewide distributed energy markets that have the durability and depth needed for third-party investments.

5. *Fair and cost-effective universal access*: The last element of REV that is often overlooked, but critically important, is the maintenance of universal access at a fair and reasonable price. New York has adopted an energy-affordability index that is designed to ensure that no New Yorker is required to spend more than 6% of their income on energy. It is also cognizant of the fact that its economically vulnerable customers are at risk of disenfranchisement from access to energy efficiency and distributed energy without government intervention. Consequently, throughout REV there are numerous initiatives that are designed to ensure that New York avoids the creation of an "energy divide."

This book is notable because it addresses issues that are central to the future governance and functionality of the distribution network, while encouraging technological innovation and disruption to take place at the grid's edge. As a society, regulators have to look to decarbonize our energy systems as a critical component of protecting the natural environment. Economic regulators are on the front lines of achieving this goal in manner that maintains the equally important policy objectives of reliable, cost effective, resilient and secure power. This can only occur if regulators embrace network innovation.

Audrey Zibelman
CEO, Australian Energy Market Operator (AEMO)
and
Former chair, New York Public Service Commission

Part I

Envisioning Alternative Futures

1. Innovation and Disruption
   at the Grid's Edge   3
2. Innovation, Disruption, and
   the Survival of the Fittest   25
3. The Great Rebalancing:
   Rattling the Electricity
   Value Chain From Behind
   the Meter   41
4. Beyond Community Solar:
   Aggregating Local Distributed
   Resources for Resilience and
   Sustainability   65

5. Grid Versus Distributed
   Solar: What Does Australia's
   Experience Say About the
   Competitiveness of
   Distributed Energy?   83
6. Powering the Driverless
   Electric Car of the Future   101
7. Regulations, Barriers, and
   Opportunities to the
   Growth of DERs in the
   Spanish Power Sector   123
8. Quintessential Innovation
   for Transformation of the
   Power Sector   147

Chapter 1

Innovation and Disruption at the Grid's Edge

Fereidoon P. Sioshansi
Menlo Energy Economics, Walnut Creek, CA, United States

1 INTRODUCTION

The unifying message of this book is that innovation and disruption enabled by new technologies—notably information and communication technology (ITC)—are transforming the electric power sector at an unprecedented pace and allowing a growing number of previously passive consumers to become proactive *prosumers*.

These empowered *prosumers*, as further described in chapters in this volume, can reduce their dependence on the services traditionally delivered by the assets and infrastructure upstream of the meter by increasing their reliance on distributed energy resources (DERS), which by definition, are provided, consumed, and possibly stored locally.

With the expected emergence of affordable storage, some *prosumers* can go a step further by becoming *prosumagers*; this they can accomplish by storing the excess generation for use at later times. With zero net energy (ZNE) buildings a virtual reality,[1] it is not far fetched to envisage some *prosumagers* operating virtually independent of the grid for the most part, relying on the network only sporadically, for balancing services and reliability (Fig. 1.1).

Add a host of new intermediaries with sophisticated capabilities who can aggregate flexible loads and distributed generation—which can be effectively bid into wholesale markets—and one can see the power of aggregation enabled by automated machine-to-machine (M2M) communication. Advances in artificial intelligence (AI) are likely to lead to proliferation of services offered by such intermediaries who can provide valuable services to grid operators and distribution networks, while better managing energy consumption and reducing

---

1. California requires all new residential buildings to meet the zero net energy (ZNE) definition by 2020; 2030 for commercial buildings.

Innovation and Disruption at the Grid's Edge. http://dx.doi.org/10.1016/B978-0-12-811758-3.00001-2

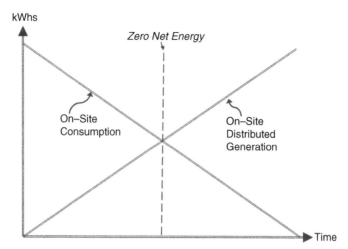

FIGURE 1.1   From consumer to prosumer. Technological innovations, including distributed energy resources (DERs), enable consumers to reach zero net energy (ZNE) by consuming less, while self-generating more as schematically illustrated.

participants' energy service costs. Examples of such aggregators may be found in chapters by Löbbe & Hackbarth, Johnston, and Steiniger, among others.

However, innovation and disruptions don't end there. There is increased interest in transactive energy and peer-to-peer (P2P) trading facilitated by platforms that allow consumers, prosumers, and prosumagers to better manage their consumption, distributed generation, and storage, and not just internally, but with their neighbors and among their peers. While many regulatory obstacles remain to be resolved, the distribution network physically connecting the participants is already in place.

Bitcoin and blockchain technologies[2]—among others—offer new opportunities for such transactions to take place among and between consumers using the existing distribution network and related infrastructure.

Microgrids, another promising emerging technology, offer individual customers and/or a collection of customers to better manage their consumption, distributed generation, and storage, allowing them to operate independent of, or parallel to, the supergrid as described by Knieps.

A combination of these is likely to enable some prosumers to operate virtually independent of the traditional grid, or in the extreme, go off-grid. While this is not a viable option for most, it may be an option for some, especially if the prevailing regulated tariffs are not cost-reflective and/or offer perverse

2. For a timely description refer to Burger, C., Kuhlmann, A., Richard, P., Weinmann, J., 2016. Blockchain in the Energy Transition: A Survey Among Decision-Makers in the German Energy Industry. European School of Management & Technology (ESMT), Berlin.

incentives for consumers to make additional investments that enable them to operate more or less independent of the grid.

As noted in the 2016 MIT report on *Utility of the future*,[3] the perverse economics of grid defection may be among the unintended consequences of regulated tariffs that predate the recent rise of DERs.

This chapter offers the editor's own views on the likely evolution of DERs and their potential impact on the incumbents. Section 2 examines the economics of DERs versus traditional bundled service at regulated tariffs. Section 3 examines the consequences of the bifurcation of customers into consumers and prosumers. Section 4 examines the potential impact of aggregators, intermediaries, and others who are likely to take advantage of new opportunities in the changing business environment. Section 5 speculates how a regulators' role is changing in response to the preceding discussion, followed by a summary of how the book is organized and highlights of each chapter as a quick guide to readers.

2   ECONOMICS OF DERs VERSUS TRADITIONAL BUNDLED SERVICE AT REGULATED TARIFFS

The rise of DERs is for the most part driven by their falling cost: energy efficiency, distributed self generation, and—in the near future—distributed storage are becoming cheaper with the passage of time. By contrast, the cost of buying services from the grid, particularly the kWh component, continues to rise, albeit at a different pace in different places as described, for example, in chapter by Orton et al. for Australia.

Making matters worse for the incumbents in the power sector, in many European countries, Spain and Germany in particular—both featured in this volume—the regulated tariffs are loaded by myriad of levies, taxes, and other surcharges that have virtually nothing to do with the actual cost of generating and delivering electricity to customers. In some cases and for some classes of customers, these surcharges may compromise nearly half of the regulated tariff measured by the cents/kWh. In many countries and parts of the United States, residential retail tariffs are in the 20–30 cents/kWh range, sometimes even higher, such as in Hawaii or Denmark.

By contrast, the costs of DERs, be it energy efficiency in the form of efficient lighting, TVs, refrigeration, HVAC, or services delivered by appliances, or distributed generation in the form of rooftop solar PVs, have been steadily falling.

As many pundits have been saying for years, at some point in some places, the cost of DERs is bound to fall below the cost of buying electrons from the grid—the crossover is often referred to as *price parity*.[4] And it should come as

---

3. Utility of the Future, MIT, 2016.
4. Mountain & Harris examine the relative economics of grid versus self-generation for Victoria.

no surprise to anyone that this has already occurred in some places, and is likely to happen elsewhere over the next decade if not sooner.

And when that happens, consumers are better off to *consume less* (by investing in energy efficiency) and *self-generate more* (by installing solar PVs or other generation options when feasible). By doing so, they avoid paying for more expensive grid-supplied electricity.

This new reality has interesting implications for *developing* countries that currently lack the extensive and expensive grid infrastructure that is already in place in the developed world. In some cases, developing economies may be better off to develop semiautonomous microgrids to serve rural customers who currently do not have access to electricity services, rather than wait for the supergrid to expand to serve them.

The economics of DERs are further fortified by a variety of financial incentives and support schemes such as *net energy metering* (NEM) laws in the United States and *feed-in-tariffs* (FiTs) in many European countries, as further described in chapters that follow.

In the case of NEM, the incentives to self-generate are immensely boosted because consumers not only avoid paying for expensive kWhs, but also can virtually eliminate their entire electric bill by feeding as much into the network as they withdraw from it (Fig. 1.2). As there are little or no fixed fees in many parts of the United States for residential customers, this is not uncommon for many solar households and even some commercial ones.

In states, such as California, where retail tariffs are *tiered*[5]—that is the rate rises at the margin at higher consumption volume—and can be as high as 36 cents/kWh, it is not uncommon to find solar customers who pay virtually little electric bills. Needless to say, such customers—in the absence of reasonable fixed fees or connection charges—do not adequately contribute to the maintenance and upkeep of the network, which they continue to rely on to balance their variable consumption and self-generation.

Aside from the attractiveness of self-generation are major advances in energy efficiency in buildings, lighting and appliances, which makes them ZNE, that is, they consume very little and generate enough to offset their meager usage (Fig. 1.3). The net results are zero or virtually zero electric bills, if tariffs are solely or mostly based on volumetric consumption. California regulators want all *new* residential buildings to meet the ZNE standard by 2020; the same would apply to *new* commercial buildings by 2030.[6]

---

5. Residential rates in California are currently tiered into four blocks rising, as volume of consumption rises. As the top tier is high, heavy users end up paying high monthly bills, which they can substantially reduce or eliminate by feeding excess generation into the network and receiving a credit equal to the prevailing retail tariff.

6. Few cities in California have mandatory solar ordinances that require all new homes to have distributed solar generation, when feasible.

FIGURE 1.2   Consumers feeding the grid. With generous net energy metering (NEM) laws customers are encouraged to become *net* producers by oversizing their rooftop solar installations.

FIGURE 1.3   An experimental ZNE house in West Village, University of California, Davis.

The writing, one might say, is already on the wall. The rise of DERs is likely to reduce net reliance on the grid for kWhs at first, and possibly other services as some consumers become prosumers and possibly, eventually prosumagers.

3 BIFURCATION OF CUSTOMERS

As outlined above, the economics of DERs are already compelling in some places partly due to the fact that existing regulated tariffs are loaded with extra charges—as in many European countries—or because the incentives to invest in energy efficiency and/or self-generate are overwhelming—as in California or Hawaii with high retail tariffs and generous NEM schemes.

Regardless of the specifics, consumers who invest in DERs—or in some cases lease them rather than an outright purchase as in the United States—become prosumers (Fig. 1.4). They buy less from the grid, while injecting their excess generation into the distribution network, and this, as amply described in MIT's *Utility of the Future* report, has a significant impact on the distribution network depending on *when*, *how much*, and *where* the offtake or injection happens. The impact can be positive, negative, or neutral depending on the specifics of when, where, and how much; the details go beyond the scope of this chapter.

The issue of consumers versus prosumers—or solar versus nonsolar customers—has already surfaced as a highly contentions and politically charged topic, especially in the United States, where generous NEM laws are currently in place in around 40 states. Multitudes of studies, for example, allege that solar customers as a class tend to be more affluent and what they save by reducing

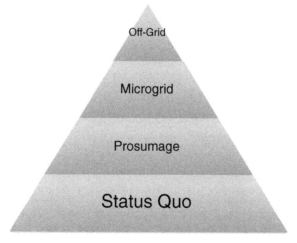

FIGURE 1.4 Consumer pyramid. Depending on the prevailing tariffs and the relative costs of DERs, consumers may opt to remain solely dependent on supply from the existing network or move away from total reliance to varying degrees.

their net consumption from the grid ends up as extra costs that must be borne by the remaining nonsolar customers.[7]

This has naturally led to efforts to introduce fixed fees, connection charges, minimum bills, solar exist fees, and similar regulatory schemes to reduce or eliminate the alleged cost shifting. As described by Baak in this volume, it must be noted that the debate about who is subsidizing whom in the DER space can be highly convoluted because stakeholders with different vested interests have opposing views and present convincing evidence to support their positions.

Setting that aside, it is fair to say that the bifurcation of consumers and prosumers has become contentious, and the regulators, who must find a balance between what is good for the former without jeopardizing the options of the latter—often in the context of broader environmental, societal, and policy objectives—are caught in the cross fire.

There is apparently no easy fix. For example, if the existing NEM laws are modified or voided—as recently happened in the state of Nevada—prosumers who have already invested in solar panels may be tempted to go a step further by investing in distributed storage, improving their ability to store the excess generation for use at later time. Moreover, this will enhance opportunities for energy management while allowing them—potentially—to engage in P2P trade with other customers whose consumption pattern may be different than their own, making their building into a ZNE fiefdom, joining other like-minded prosumers and/or aggregating their consumption and storage in such a way to avoid paying the new fees imposed on them.

In short, if the incentives to sell the excess generation into the grid are reduced or removed, some prosumers may decide to go a step further by becoming prosumagers. Investing or joining semiautonomous microgrids or shared community schemes—as covered in chapters by Knieps and Koirala & Hakvoort—may become even more appealing to such prosumagers.

Future innovations and disruptions to the status quo utility business model cannot be ruled out, as further described in chapter by Jones et al, Löbbe & Hackbarth and Orton et al. This goes to the core of the message in chapter by Smith & MacGill about the need for a "great rebalancing act" and explanations of Woodhouse & Bradbury on how innovation and disruptions will lead to "the survival of the fittest." Likewise, Webb et al. argue that with the continued fall in the cost of distributed solar a new class of customers—large commercial and industrial establishments with much bigger roofs, bigger electricity bills, and more resources—will join residential consumers who have largely been going solar to date. Haro et al. offer more sophisticated *dynamic network tariffs* to address some of these vexing issues.

---

7. Under rate of return regulations still prevalent in most parts of the United States, investor-owned utilities (IOUs) are allowed to raise their retail tariffs to compensate for the fall in revenues due to self-generation.

Moreover, new service options are emerging, offering customers, such as apartment dwellers who cannot physically install solar panels on their roofs, to participate in the DER revolution by becoming virtual solar or renewable customers as described in chapter by Johnston.

This suggests that the bifurcation of customers into those who are and will remain totally reliant on the existing grid for all their electricity service needs and those who may become selectively and sporadically reliant on the grid has only started.

4   AGGREGATORS, INTEGRATORS, AND INTERMEDIARIES

Several chapters in this volume, notably Steiniger, Johnston, and Löbbe & Hackbarth provide a glimpse of the emergence of a host of new businesses and business models that are focused on identifying, developing, and delivering value by aggregating and managing flexible loads and distributed generation in the real or virtual sense.

In many countries—Germany is a prime example—a significant percentage of generation is already distributed and largely produced from variable renewable resources. In California, the investor-owned utilities (IOUs) have already exceeded 5 GW of distributed solar rooftop, as mentioned by Michael Picker in the book's Preface. Australia, with a fraction of the US population, already exceeds 1.5 million solar roofs and counting, roughly the same number as in America.

Making the role of aggregators, integrators, and intermediaries even more urgent is the fact that much of the utility-scale generation is also becoming renewable, which means increasingly variable and nondispatchable as described by Steiniger and Webb et al. In many countries, Denmark, Germany, Spain, Portugal, Ireland, to name a few, the grid operators are already increasingly challenged trying to balance variable generation with demand. It is not just the traditional utility business models that are being challenged, but so are traditional grid-operating and dispatching models, which calls for new perspectives, as explained in chapters by Cooper and Knieps.

Without doubt, this space—the growing role of aggregators, integrators, and intermediaries—is likely to have the biggest impact on the future of the power industry. New entrants, such as Next Kraftwerke[8] and its cohorts, provide a mere glimpse of what is likely to follow.

There will, however, likely to be many others who will find profitable ways to disrupt the traditional ways of doing things at every step of the utility value chain. One example is the recent merger of Tesla with SolarCity. Elon Musk, the CEO of the combined enterprise, has outlined his vision of an integrated energy, mobility, and storage service by combining Tesla's electric cars with SolarCity's

---

8. Refer to chapter by Steiniger.

distributes generation, and its Powerwall distributed storage. Webb & Wilson provide further details on the likely evolution of EVs, including autonomous ones, and their impact on the utility business, both positive and negative.

In a blog posted at Tesla's website describing his vision for a fully integrated company, Musk wrote:

You'd walk into the Tesla store and say: 'I'd like a great solar solution with a battery and an electric car.' And in 5 minutes you're done. It's completely painless, seamless, easy and that's what the customer wants.

Perhaps it will not be quite as simple as Musk claims, but his vision to create an integrated energy services model that is far less complicated than today is not just likely, but definitely on target. Moreover, there are other advantages to integrating energy generation, consumption, storage, trading, mobility, and management. In the case of Tesla, this includes much better use of Tesla's ubiquitous and fancy car showrooms, which can now sell solar PVs and storage batteries under the same roof. One company, one contract, one point of sale, and one integrated bill for a host of integrated services. Clearly Musk is not the only clever entrepreneur trying to disrupt the protected monopoly business model of traditional utilities.

With the approval of the merger, in theory, Musk can now bundle electric mobility, solar PV panels, and storage into an integrated product/service combination that appears to many affluent consumers; those who can afford an upscale EV, may be interested to generate some of their electricity on their roofs, and may wish to store some of the extra generation into a battery for use at another time.

There are many who speculate that integrators, such as Tesla, will not stop there. With their access to valuable data about when their clients use energy, when, and how much they generate, and when and where they store it—including their EVs—newcomers, such as Tesla, can easily manage the entire energy profile of their *portfolio of customers*. Going one step further, they can use this portfolio of distributed load, generation, and storage to sell valuable services to the distribution network operators.

When, for example, part of the distribution network is stressed or needs frequency or voltage support, the aggregators/integrators can reroute some of the generation or stored energy to alleviate the congestion or control frequency or adjust voltage levels.

The projected rapid rise of EVs, described by Webb & Wilson and Alvarez suggests that depending on when, where, and how they are charged and discharged will have significant impact on the future of the distribution network. In their chapter, Orton et al. describe other business plans and models to integrate electricity generation, delivery, and storage services. Many more are likely to emerge.

For the first century of its existence, major innovations and most of the investment were upstream of the meter. The reverse is likely to be the case in the second century, as the rebalancing of the electricity's value chain takes place, as further elaborated by Smith & MacGill. The rising stars of the second century are likely to be the aggregators, integrators, and intermediaries. Their success will increasingly be driven by better management of the data, while offering convenient services, seamlessly and efficiently.

5   EVOLVING THE ROLE OF REGULATORS

Another message that resonates throughout this volume is that the proliferation of new services and service providers in, and on, the periphery of the distribution network—at the grid's edge, the subtitle of this volume—has only begun. It is not going to stop anytime soon. Some pundits believe that the industry is poised to experience massive *disruptions*—Uber and Airbnb are often mentioned as the sorts of disruptions that may happen—in the near future the likes of which has not been seen since Thomas Edison invented the light bulb.

This is likely to put tremendous strain on the regulators around the world who must find a balanced approach to allow—in fact facilitate and encourage—innovation to serve the needs of presumes and prosumagers, while protecting the interests of consumers who may be more or less content with the status quo. Finding the proverbial *level playing filed*, often mentioned in this context, will become more challenging and pressing.

This volume is blessed by having contributions from four prominent regulators from four different parts of the world, who are sharing their perspectives in the Foreword, Preface, Introduction, and Epilogue.

As this editor sees it, two key issues stand out among the many facing the regulators:

- First, how to adjust the existing regulated tariffs for both consumers and prosumers, so that both are contributing their fair share to the upkeep of the network without free riding or being unfairly subsidized, as may be the case today; and
- Second, how to regulate—or not—the access to, and the use of, the distribution network—which is and will likely remain critical to all, except the small minority of customers who, for what ever reason, may decide to go completely off-grid.

Regarding the former, designing fair, and reasonable rates—the fundamental guiding principle for regulators—most experts agree that tariffs must correctly reflect the true cost of service. As described in MIT's *Utility of the Future* report, to get the prices right, tariffs must correctly reflect the four main components of cost of service:

- price of electric energy;
- price for energy-related services, such as operating reserves or firm capacity;

- prices for network-related services, such as the reliability and balancing of load and distributed self-generation; and
- prices to cover the costs of policy-related objectives, such as low-carbon energy mix, taxes, levies, or subsidizing low-income customers.

Currently, these components are bundled in regulated tariffs, usually in highly opaque and nontransparent ways. They are neither obvious to most consumers, nor are they logical from a technical or economic perspective. This increasingly leads to decisions that may make sense to some prosumers or prosumagers, while making little or no sense when viewed from the broader perspective of the society at large.[9]

Moreover, retail tariffs tend to be flat and postage stamp. They do *not* vary by time, location, peak demand, or pattern of usage. This is another major deficiency, highlighted in a number of chapters in this book, including Biggar & Dimasi. Moving to more *granular* tariffs, both on spatial and time dimension, would better capture the actual costs of service, but can make them more complicated than most customers can understand. For example, Baak, in his chapter, describes a number of initiatives in the state of California to move toward more granular and cost-reflective tariffs.

The technology is increasingly available and affordable to design and implement more granular and sophisticated tariffs, but regulators, for the most part, appear reluctant to make major steps in adopting them, favoring a gradual approach.

After examining a host of technical issues, MIT's *Utility of the Future* devotes a lengthy chapter with multitudes of recommendations—30 to be exact—for regulators to consider. Likewise, a recently released manual by the National Association of Regulatory Utility Commissioners (NARUC) offers a rather lengthy discussion on how to treat DERs.[10] The Essential Energy Services, the regulator in the state of Victoria in Australia recently released a report on how to enumerate DERs, focusing—among other things—on the value of distributed solar PVs on the distribution network.[11] Clearly, this already *is* and will likely *remain* a hot area for research, and debate, for some time.

Opinions vary on the topic of how to regulate the use of, and access to, the critical distribution network. As described by Audrey Zibelman in the Introduction, the regulators in the state of New York have embarked on a bold initiative called reforming the energy vision (REV). At its core, REV provides guidelines on who should pay for, who should maintain, and who can use the distribution network. Beyond that, the stakeholders are free to innovate, disrupt, and offer

---

9. Haro et al. offer an interesting proposal for Spain.
10. Refer to Distributed Energy Resources; Rate Design and Compensation, Manual prepared by NARUC staff subcommittee on rate design, November 2016.
11. Refer to ESC at http://www.esc.vic.gov.au/document/energy/36002-distributed-generation-inquiry-final-report-energy-value/

new products and services. The aim in New York, and elsewhere, clearly is to move toward a regulation fit for the emerging distribution network of the future.

Regulators in other states are taking different approaches, depending on the circumstances. In California, for example, it has been suggested that IOUs be allowed a regulated rate of return for investing in DERs, as is currently allowed for approved and prudent investments upstream of the meter. As explained by Baak, while this may sound like a good idea, it clearly makes the regulator's life even more complicated, as they must now decide how much investment in DERs is prudent, necessary, and cost-effective, as they currently do for upstream investments.

In his chapter, Gellings argues that despite all the hyperbole on DERs leading to utility *death spiral* and worse, the industry's traditional rate of return regulation "ain't broke and it don't need fixing," the subtitle of his chapter. His proposal is not to throw the baby out with the bathwater.

Just as IOUs in many states, notably California, have been persuaded to engage in improving the energy efficiency of their customers—a counterintuitive measure under the traditional rate of return regulation—they can in principle be *encouraged* to do the same for DERs, by investing on the customer side of the meter, when it is efficient and cost effective to do so.

Once again, it is easier said than done. An example of the difficulties has surfaced in the debate on who should invest in electric charging infrastructure: IOUs; electric car companies, such as Tesla; municipalities; new private companies; or a combination of all of these? Everyone agrees that this is a classical *chicken and egg problem*: customers will not buy EVs in sufficient numbers unless there is a charging infrastructure in place, while car companies will not invest in EVs unless there is a market for them.[12]

Other pundits who are exploring the future of utility business models have suggested that the power sector is on a path not unlike that of the mobile phone industry. Today, most mobile phone users pay a fixed monthly fee based on a 2- or 3-year contract with a network service provider.

The explanation is simple, while the analogy may not be perfect. Mobile phone service is increasingly about *connectivity* and *access* to the network. It is like a membership in a gym or a private golf course. It is not about volume or frequency of usage.

Customers choose a mobile phone network not only on the basis of cost, but also on the ubiquity of network access—the strength of the signal—bandwidth, and speed. They are rarely charged on a per-call or per-minute basis these days. The cost of service is much better reflected, and collected, through a fixed fee almost regardless of volume of service, say the number of minutes called, number of e-mails sent, etc. Gone are the complicated and lengthy bills that spelled

---

12. This debate has been contentious in California. Incumbent utilities are keen to make massive investments in electric vehicle (EV)–charging stations if they can pass on the costs to all customers, not all of whom are likely to benefit from the investment, at least initially.

out each call number plus time and length of call, whether it was between 9 a.m. and 5 p.m., or in the evenings or weekends. Younger readers probably don't even remember such phone bills.

The same, one can argue, applies to many other services where the bulk of the costs are fixed as in the membership in the gym or garbage collection service. It makes little sense to weigh the garbage being collected each week or charge the customer every time (s)he shows up at the gym.

The electric service, these pundits argue, is moving in the same direction. As much of the new generation is coming from renewable resources, both utility scale and distributed, the cost of electrons—the commodity portion of service—is rapidly falling, eventually approaching zero.[13] When that happens—which is already the case in many places—it makes little sense to charge *any* or *much* for the *energy* or generation component of service, it is often argued.

Moreover, with the advent of ZNE buildings, the volume of consumption is likely to be flat or falling. This is already happening in many advanced economies. The implication is rather clear: tariffs based exclusively or primarily on volumetric consumption are unlikely to deliver sufficient revenues.

Taking this argument a step further, why not charge customers *mostly* for being connected to the network, making the electrons *free* or mostly free? That would make it similar to mobile phone service. After all, the most valuable feature of the grid is the reliability and the balancing services that it provides, for both consumers and prosumers.

This is essentially what one senior California utility insider suggested as the best option going forward. When rhetorically asked what he thought was the simplest way to cover the cost of *electricity service* in the future, he said that if every residential customer paid a fixed $2 per day, he would offer kWhs at 5 cents/kWh in each direction; whether buying from or selling to the grid.

To put this in perspective, current residential rates in California range from 11 to 36 cents/kWh depending on the tiers—hence 5 cents/kWh represents a substantial discount even from the lowest tier. This particular utility has over 5 million residential customers who currently pay virtually no fixed fees. Under his proposal—in the context of a private conversation and a rhetorical question—the utility would collect $300 million per month plus what is applicable from the *net* volumetric consumption at 5 cents/kWh. Presumably, the fixed fees would go toward the fixed component of maintaining the grid, while the volumetric component would cover the variable component of costs, including generation.

Taking a broader context, currently there are 144 million electric meters in the United States, according to the latest annual survey by the Federal Energy

---

13. Wholesale price of electricity in Germany, for example, fell nearly 60% between 2008 and 2015, a trend that is expected to continue with the rising proportion of renewable generation over time.

TABLE 1.1 US Residential Customers by the Numbers

Avg. monthly usage	1000 kWh
Avg. monthly bill	$110/month
Avg. fixed component of bill	$60/month[a]
Fixed charge (as percent of monthly bill)	55

[a]Based on $30/month for distribution services, $10/month for transmission, $19/month for generation assets (not including fuel or actual generation), and $1/month for ancillary and balancing services.
Source: Wood, L., Borlick, R., 2013. Value of the Grid to Distributed Generation Customers, IEE Issue Brief. Institute of The Edison Foundation, Washington DC. Based on the 2011 data from the Energy Information Administration.

Regulatory Commission (FERC).[14] If every meter—for the sake of simplicity—were to pay $2/day for being connected to the "grid," that would amount to over $105 billion per annum, a significant amount toward the upkeep and upgrading of the network.

Ridiculous? Apparently not. The average US residential consumer pays roughly $110 per month for electricity (Table 1.1), of which an estimated 55% is fixed costs.[15] Moreover, as the proportion of renewables in the energy mix rises over time, this percentage will continue to rise. The proposed $60/month fixed charge is roughly consistent with the US national average figures, and would provide network companies with a steady source of revenues that is apparently not out of line with the actual fixed costs of serving typical residential consumers.

A few experts and scholars who have examined the future of network and how to best pay for it would go even further. Some argue that for all but customers, who choose to go totally off-grid, the network can be viewed as a "social good" in the same sense as public libraries or public schools are. The society pays for their upkeep, regardless of how often or how extensively they are used by individuals. For example, singles, unmarried people, childless couples, or those who send their kids to private schools currently pay the same property taxes, which cover the cost of public schools in many parts of the world. The logic is that the society is better off when such services are provided and supported by all, regardless of the level or frequency of usage.

After analyzing the temporal and spatial distribution of costs and benefits and designing sophisticated tariffs that reflect the granularity of costs and

---

14. Assessment of demand response and advance metering, FERC, December 2016, at https://www.ferc.gov/legal/staff-reports/2016/DR-AM-Report2016.pdf
15. Refer to Value of the grid to distributed generation customers, IEE issue brief, Lisa Wood and Robert Borlick, September 2013.

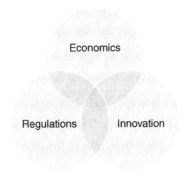

FIGURE 1.5   Outcome of DERs. The evolution of DERs will depend on the confluence of three important factors illustrated.

benefits, why not *socialize* the fixed cost of the distribution network, so that it is supported by all and available to everyone? In such a future:

- consumers would pay an *extra* amount for the services received and the costs imposed on the network; and
- prosumers and prosumagers might *pay* or *receive*, depending on the net costs imposed on, or benefits delivered to, the network.

After carefully examining and contemplating the many insights of the following chapters—the editor was richly rewarded by reading every chapter multiple times during the editing process—the best that can be said is that the longer-term impact of DERs is likely to be influenced—or determined, if you prefer—by the confluence of three critical factors, as illustrated in Fig. 1.5:

- The *economics*, namely the relative costs and benefits of DERs relative to the cost of getting similar services from the grid.
- *Regulations* will determine what is and is not allowed, how much can be charged for it, and who can deliver given services, which is likely to vary from one place to another.
- *Innovation* and *disruptions* are likely to determine what new products and services and new delivery methods are introduced.

The outcome, in other words, is likely to be different in different parts of the world mostly due to differences in regulations, as the other two factors apply more or less equally everywhere.

6   ORGANIZATION OF THE BOOK

This volume consists of 19 chapters and is organized into 3 parts as outlined below.

Part I. Envisioning the Future. This part of the book examines why and how the industry is changing and speculates on the implications of change on incumbent stakeholders and new players.

In Chapter 2, Innovation, Disruption, and the Survival of the Fittest, *Stephen Woodhouse* and *Simon Bradbury* outline the changing nature of the utility business and the struggle for survival—and the relationship with customers—that the incumbent utilities face.

The authors set out a possible model for a customer-centric utility of the future, in which less emphasis is given to ownership of hardware, and customers themselves are treated as core assets. Alternative futures are examined, in which the existing companies are forced upstream by entrants that take over the customer relationship. In that alternative world, there is a further question over whether incumbent companies could extract value from assets and trading, or whether the value will reside with the customer relationship.

The chapter's key contribution is to set up framework to compare alternative future business models, for incumbent utilities and new entrants.

In Chapter 3, The Great Rebalancing: Rattling the Electricity Value Chain From Behind the Meter, *Robert Smith and Iain MacGill* look at how distributed generation is shifting the traditional balance between generation, transmission, and distribution toward customer end uses. It identifies and quantifies the substantial oft-unrecognized investment made by customers in the energy value chain, through more efficient appliances, equipment, wiring, and building design.

The authors consider the extent to which the new business models being developed around DERs, such as the NY REV, are anchored in the value creation behind the meter, yet still reliant on the ongoing operation of the grid.

The chapter's main contribution is to ask if DERs result in value duplication rather than value creation; will this create sunk costs in the grid or see a rebalancing of the grid's functions and charges to reflect new values across the energy chain.

In Chapter 4, Beyond Community Solar: Aggregating Local Distributed Resources for Resilience and Sustainability, *Kevin Jones, Erin C. Bennett, Flora Wenhui Ji*, and *Borna Kazerooni* discuss current developments in community solar. Communities have actively participated in community solar arrays often through group net metering with their utility. While these arrangements have helped community members source more of their energy needs from clean local resources, they fall short of allowing communities to achieve sustainability or energy resilience.

The authors explore alternatives for communities to develop DERs, such as battery storage, demand response, biogas, and microgrids, which increasingly offer opportunities for communities to advance beyond community solar by examining examples, including Marin County, California, and Westchester County, New York to develop a model for community choice aggregation that achieves these goals.

The chapter describes community choice aggregation as means of bringing DERs together for community sustainability and resilience at the grid's edge.

In Chapter 5, Grid Versus Distributed Solar: What Does Australia's Experience Say About the Competitiveness of Distributed Energy? *Bruce Mountain* and *Russell Harris* point out that in South Eastern Australia's 4 regional electricity markets, 21 retailers compete to supply a little over 9 million residential and small customers offering over 5400 different retail packages at any one time.

Using currently available information on distributed generation and storage costs, the authors use currently available offers to assess the economics of distributed generation compared to the grid-supplied alternative.

The chapter's main contribution is to provide insight into the relative economics of grid versus distributed. After all, the relative cost of the two options should dictate the demand for each with significant implications for the future development of retail electricity markets.

In Chapter 6, Powering the Driverless Electric Car of the Future, *Jeremy Webb* and *Clevo Wilson* look at the effect on power demand of likely transformative changes to urban transport. This is being ushered in by the simultaneous emergence and impending fusion of all-electric automotive power trains and autonomous vehicles.

The authors examine the reasons why autonomous vehicles will most likely be all electric (EVs), and largely shared. Provided are projections of the likely rate of uptake of EVs and SEAVs, which indicate a substantial reduction in vehicle numbers. Key determining factors examined are expected rates of technological change in car battery storage capacity and autonomous driving systems, and comparative costs of conventional and all EVs and socioeconomic drivers influencing modal choice.

The chapter highlights that a rapid uptake of EVs and SEAVs will have major implications for electric power utilities in terms of its effect on both the overall demand and distributed power storage capacity.

In Chapter 7, Regulations, Barriers, and Opportunities to the Growth of DERs in the Spanish Power Sector, *Eloy Álvarez Pelegry* explains that generous regulatory support for utility-scale wind and solar investment in Spain encouraged massive investments with serious financial implications. This, in turn, has resulted in a reversal of regulation with implications for the deployment of DERs, including decentralized solar PVs in sunny Iberia.

The author points out that the dramatic drop of photovoltaic costs and the potential reduction of electricity storage prices promises significant opportunities for developing self-consumption and alternative fuels for transport, such as charging EVs.

The chapter examines the Spanish regulatory context taking into account the European objectives for 2020 and 2030 that imply increased reliance on renewables, focusing on the evolution of the regulation and the barriers and opportunities for the development of DERs mostly for self-consumption.

In Chapter 8, Quintessential Innovation for Transformation of the Power Sector, *John Cooper* introduces *the Pace Problem*, a fundamental challenge to business transformation strategies in circulation today. Despite relative advances made by utilities in adapting to changes, more aggressive, speedier change outside the industry leaves utilities falling behind, raising consequent risks to long-term stability in this essential industry.

The author proposes a framework for *iterative innovation* to accelerate the pace of change inside individual utilities and the industry as a whole. Moreover, this framework, in five parts, is tied to a five-part *business transformation maturity model*, to create a standardized platform for change labeled *Quintessential Innovation*.

The chapter's main contribution is to recognize *the Pace Problem*, a crucial barrier that blocks a critical industry from adapting to a complex future, and offer a compelling solution; borrowing successful best practices from other industries to accelerate adaptation to change.

Part II. Enabling Future Innovations. This part of the book examines the role of policy, regulations, and pricing and how the confluence of these influences the ultimate outcome.

In Chapter 9, Bringing DER Into the Mainstream: Regulations, Innovation, and Disruption on the Grid's Edge, *Jim Baak* highlights the challenges regulators, utilities, DER providers, and prosumers face in balancing their conflicting needs and competing interests to enable reliable and cost-effective deployment of DER for the benefit of the grid and consumers.

The author outlines the significant challenges and opportunities of integrating higher levels of DER deployment and the steps necessary to achieve this outcome, which call into question not only conventional grid planning and operations, but also traditional utility business models. At the center of the debate is how will the grid be used and who will pay for it.

By drawing from the experiences of California and New York, the author illustrates different regulatory approaches and offers insights on how best to balance the competing needs of energy users, producers, and regulators, while improving the reliable operation of the grid.

In Chapter 10, Public Policy Issues Associated With Feed-In Tariffs and Net Metering: An Australian Perspective, *Darryl Biggar* and *Joe Dimasi* explore how tariff policies affect pressure for net versus gross metering, embedded networks, and P2P trade.

The authors argue that as long as retail prices are artificially inflated and time averaged, policymakers face an unenviable choice: if embedded generation is paid a different (cost-reflective) price, the end customers have efficient incentives regarding the usage of and investment in embedded generation, but at the same time they have a strong incentive to find ways to offset the output of that generation against their own consumption, and that of their neighbors. On the other hand, if embedded generation is paid the retail price, the incentives to use and to invest in that embedded generation are inefficient.

The chapter's main conclusion is that net metering, and P2P trade will remain controversial as long as retail tariffs are not cost reflective.

In Chapter 11, We Don't Need a New Business Model: "It Ain't Broke and It Don't Need Fixin," *Clark Gellings* takes a contrarian perspective by contending that the traditional regulation that has served the industry well—if not perfectly—is *not* fundamentally broke, and can be modified to serve us in the future.

The author is not convinced that "we" need a new business model for distributional utilities, while acknowledging the dramatic changes in the cost and performance of DERs, energy storage, and hyperefficiency appliances, as well as consumer interest in having control over energy.

The chapter's main point is to argue that there is nothing conceptually wrong with the existing regulatory model based essentially on the rate of return regulation and suggests that, if properly applied, it can continue largely as it was conceived. In fact, it will lead to a preferred basis for the distribution utility of the future.

In Chapter 12, Toward a Dynamic Network Tariffs: A Proposal for Spain, *Sergio Haro, Vanessa Aragonés, Manuel Martínez, Eduardo Moreda, Estefanía Arbós, Andrés Morata*, and *Julián Barquín* explain that new technological developments, and in particular behind-the-meter devices have rendered traditional tariff structures obsolete and induce socially undesirable decision making. On the other hand, they also bring their own solution, in the guise of advanced tariff structures made possible by smart meters.

This chapter proposes a dynamic, "real-time" access tariff methodology motivated by recent developments in Spanish regulations, although the issues are relevant for other systems. The proposal efficiently incentivizes distributed generation and storage deployment, as well as DERs.

The chapter's main contribution is based on an analysis of the smart meter infrastructure currently being deployed, concluding that the tariff is feasible, robust, provides socially desirable incentives, and allows streamlining of an increasingly baroque regulatory system.

In Chapter 13, Internet of Things and the Economics of Microgrids, *Günter Knieps* points out that microgrids consist of two complementary parts: a *physical* low-voltage electricity network; and a *virtual* network consisting of a complementary set of ICT components for two-way communications. The former can be connected to the latter via Internet of Things (IoT) consisting of physical objects embedded with sensors and electronic chips, and connectivity to virtual networks based on All-IP communication infrastructures.

The author analyzes the potentials of ICT for the organization of future microgrids, including the role of standards for home networks, smart metering and sensing networks, the role of multiple virtual networks for cooperation or integration between different microgrids, and pricing strategies within networks.

The chapter's main contribution is a systematic analysis of the ICT innovation potentials and the incentive compatible pricing strategies within microgrids taking into account the welfare improving outside options for microgrids.

Part III. Alternative Business Models. This part of the book examines a number of scenarios for how the developments of preceding chapters may shape the future of DERs on the grid's edge, the theme of the book.

In Chapter 14, Access Rights and Consumer Protections in a Distributed Energy System, *Fiona Orton, Tim Nelson, Tony Chappel,* and *Michael Pierce* note that distributed technologies are delivering energy in ways that were not contemplated when regulations governing consumer rights and protections in the Australian National Electricity Market were developed.

The authors examine the rapid uptake of technologies like air-conditioning and rooftop solar, and the potential for battery storage, EVs, virtual power plants, and other energy management products. They contrast the licensing requirements for traditional and nontraditional energy suppliers, and the processes for approving grid connections for technologies in different jurisdictions.

The chapter concludes that the types of grid services that communities consider to be "essential" may evolve over time, leading to redefined access rights. As the definitions of consumers and suppliers become blurred, consumer protection reforms will need to balance innovation and consumer choice with universal access to essential services.

In Chapter 15, The Transformation of the German Electricity Sector and the Emergence of New Business Models in Distributed Energy Systems, *Sabine Löbbe* and *André Hackbarth* point out to a growing number of concepts, including P2P trading and transactive energy ushering in innovative services and unorthodox business models.

The authors examine a number of such models covering technology, regulatory framework, market potential, customer segments, and the value creation potential of these business models, including energy storage systems and new forms of partnerships, trading, and aggregation.

The chapter's main contribution is to examine the viability and potential impact of such business models that operate in small-scale prosumer networks and envision under what conditions such schemes would prosper and grow and what might be their relevance for future DER developments.

In Chapter 16, Peer-to-Peer Energy Matching: Transparency, Choice, and Locational Grid Pricing, *James Johnston* describes how P2P energy matching could support the transformation of the traditional energy industry into a digital, decentralized, and renewable-powered one.

The chapter introduces the P2P energy-matching concept, using UK start-up Open Utility as an example. It presents the first commercial use case for P2P energy matching: providing customers with transparency and choice over their renewable energy supply.

The author also presents Open Utility's proposed modification to UK Distribution System Operator (DSO) charging regulations to incorporate locational pricing, and concludes with an overview of how P2P energy matching can lead the development of a community-driven and democratized energy industry.

In Chapter 17, Virtual Power Plants: Bringing the Flexibility of Decentralized Loads and Generation to Power Markets, *Helen Steiniger* points out that the rapid rise of variable renewable generation in markets, such as Germany, has reached levels that can no longer be managed through the traditional approaches where different types of thermal plants were dispatched to meet variable demand.

The author argues that in the future, increasing amounts of flexible demand *and* generation must be dispatched in response to variable renewable generation. The chapter describes the concept of a virtual power plant, where the load of numerous consumers with flexible demand and generators with flexible output are aggregated and bid into wholesale markets.

The chapter's main contribution is to illustrate how Next Kraftwerke, a German VPP, has developed a successful business model around the concept and why similar efforts will be needed in other markets where renewables make up a rising share of the generation mix.

In Chapter 18, Integrated Community-Based Energy Systems: Aligning Technology, Incentives, and Regulations, *Binod Koirala* and *Rudi Hakvoort* describe the concept of integrated community energy systems (ICESs), where clusters of residential- and community-level DERs provide a viable alternative to present centralized energy supply system. Although technologies to realize such local energy systems are emerging, the regulatory incentives and institutions to govern them are not.

The authors address a number of critical barriers discouraging further development of ICESs, such as ownership and governance issues, design of local energy exchange, market access of commodities and services, as well as mechanisms for fair allocation of revenues and costs.

The chapter's main contribution is to outline an institutional design of ICESs from a technoeconomic perspective to achieve economic efficiency and fairness for all stakeholders, and to ensure localized sustainability of ICES initiatives, realizing that such institutional settings need to adapt with the changing DER landscape.

In Chapter 19, Solar Grid Parity and its Impact on the Grid, *Jeremy Webb*, *Clevo Wilson*, *Theodore Steinberg*, and *Wes Stein* observe that market forces rather than subsidies will drive a much more diverse mix of PV-based prosumers. Further cost reductions for PV and its locational flexibility means its appeal to commercial and industrial customers is likely to be particularly rapid; the same for urban fringe and remote rural regions where grid connection is uneconomic.

For many such customers, the reliance on the grid will be limited, as daylight PV power generation matches demand. However, as renewables reach a critical proportion of energy generation, flexible concentrated solar power will be needed for backup.

The chapter's key message is that the size and economic function of the greatly expanded class of corporate prosumers will determine the renewable mix, the relationship with the grid, and corporate power purchase agreements (PPAs) with utilities.

In the book's *Epilogue*, *Johannes Mayer* provides an overview of the issues covered in the book and how the insights presented can assist regulators and policymakers to further support innovation and disruption, while maintaining the reliability of the network for those who remain totally or partially dependent on it.

Chapter 2

Innovation, Disruption, and the Survival of the Fittest

Stephen Woodhouse and Simon Bradbury

Pöyry Management Consulting, Oxford, United Kingdom

1 INTRODUCTION

Customers are used to receiving electricity reliably as if by magic, without wishing to see the means of delivery, either in terms of noticing the infrastructure used to supply power or the impact of consumption on their bank accounts.

But a combination of policy objectives and technological advancements is changing the picture significantly.

Focusing first on policy goals, many countries now have clear policy objectives to decarbonize the electricity sector and their economies more generally. Taking Europe as an example, the European Union has an aspiration to reduce greenhouse gas emissions to 80%–95% below 1990 levels by 2050. Within Europe, national objectives have also been defined by some countries. For example, the United Kingdom has set a legally binding target to reduce greenhouse gas emissions by 80% by 2050, Germany has the target to reduce emissions by 80%–95% by 2050, and Denmark has the target of becoming independent of fossil fuels by 2050.

As an engineering challenge, the scale of change is huge, with heavy reliance on emerging technologies and on major IT projects. Public acceptance of new transmission and (some) generation infrastructure is proving a significant barrier, and without better engagement with customers, this will not improve.

The increased need for coordination across the sector is daunting. This applies to both investment planning, as well as actual market operation. Generation is diversifying and shrinking in scale, and demand must respond to generation not just the other way around. But none of this is insurmountable.

Meanwhile, new actors are entering the energy sector, deploying a series of new technologies, which allow them to understand and meet the needs of customers and small producers in ways, which the existing companies have never done. The decentralized energy system, in which customers can be actively engaged, is becoming a practical possibility.

Innovation and Disruption at the Grid's Edge. http://dx.doi.org/10.1016/B978-0-12-811758-3.00002-4

Activity by consumers, network owners, system operators, and in the wholesale markets must be effectively coordinated, and there will be new roles and interactions supported by distributed intelligence and automation. Conflicts will emerge over who has access to customers, who is in ultimate control, and who captures value and holds risk. Ultimately, it is not clear what services will be offered by which actor and who will hold the relationship with customers.

This tension encapsulates the theme of this book and its focus on the effects of innovation and disruption on the energy sector. A number of strands within this chapter are picked up again in others, including chapters by Cooper, which considers business transformation strategies; by Johnston and Sioshansi, which considers transformations needed to allow engagement of proactive consumers; and by Steiniger, which considers the role of virtual power plants.

The emergence of a decentralized industry, with engaged customers and new remote data and control possibilities has opened up an alternative future for the energy sector. Given the scale and nature of the revolution, the underlying question is: what characteristics will a successful energy company of the future have? This chapter considers this question.

This chapter consists of four sections in addition to the Introduction. Section 2 considers the challenges associated with the energy sector transformation. Section 3 considers the characteristics required for the energy company of the future. Section 4 provides an appraisal of what is needed for the energy companies of today to adjust for the future followed by the chapter's conclusions.

2 IS DELIVERING THIS TRANSFORMATION REALLY THAT MUCH OF AN ISSUE?

It is fair to question whether the current energy sector transformation can really be thought of as unprecedented. Similar large-scale transformations have previously been achieved in the energy sector, but only within the context of centrally planned vertically integrated monopolies. The option to go back in that direction exists but it seems to be against the tide of history.

The ability for traditional utility business models, which revolve around the existence of large vertically integrated entities, to drive a digital and customer-centric transformation is doubtful.

In today's market, commercial conditions for large vertically integrated players have deteriorated, signaling a paradigm shift.

The first change is that utilities are suffering from major profitability issues, especially with their thermal generation assets. Running hours and ability to capture revenue are being curtailed as the volume of renewable generation on the system increases. Studies have highlighted further anticipated reductions in load factor for thermal plant, as the penetration of renewables continues to grow. This is illustrated in Fig. 2.1, which presents anticipated falling load

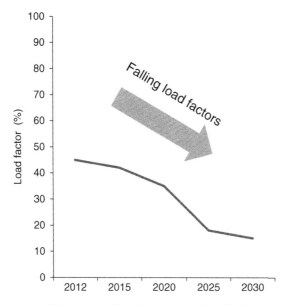

Source: Northern European Wind and Solar Intermittency Study, Pöyry Management Consulting

FIGURE 2.1   Projected load factors for thermal plant.

factors for CCGTs, based on a Pöyry study in relation to the implications of intermittency in Northern Europe.[1]

The second shift relates to the effectiveness of the "vertically integrated" utility model at dealing with both political and market risk. Over recent years, the model has increasingly proved to be poor at dealing with such risks, and its ability to effectively manage risk in a rapidly evolving environment looks unlikely.

The performance of listed utilities' share prices provides an indication of the challenging times that these companies have found themselves in, over recent years, as a result of these shifts. Prior to and during the economic downturn that began in 2008, European utility stocks outperformed the market. All stocks suffered as a result of the downturn, but the recovery of European utility stocks has fallen far short of the wider market. Importantly, in a reverse of the precrash relationship, utilities are now underperforming relative to the market. This provides a clear indicator of the changing fortunes of the utilities. These trends are illustrated in Fig. 2.2.[2]

---

1. The study focused on Austria, Belgium, Czech Republic, Denmark, Finland, France, Germany, Great Britain, the Netherlands, Norway, Poland, Sweden, and Switzerland.
2. Companies included in STOXX Europe 600 detailed at: https://www.stoxx.com/index-details?symbol=SXXGR; companies included in STOXX Europe 600 Utilities detailed at: https://www.stoxx.com/index-details?symbol=SX6p

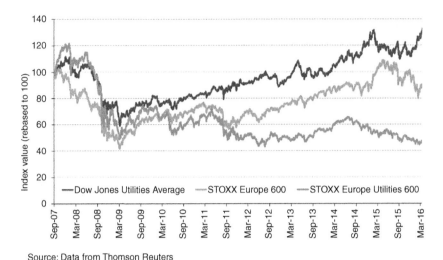

Source: Data from Thomson Reuters

FIGURE 2.2   **Performance of EU utility stocks versus EU stocks more generally.**

The fallout has been pronounced. For example, in 2014 GDF Suez wrote off €14.9 billion as the value of its power plants fell,[3] while both RWE and E.On posted significant losses linked to the decline of their fossil fuel businesses.[4]

Pinpointing the issues and the magnitude of their effect from the perspective of his company, Rolf Martin Schmitz, the COO of RWE, noted that "the massive erosion of wholesale prices caused by the growth of German photovoltaics constitutes a serious problem for RWE which may even affect the company's survival."[5]

The third development is that new technology providers—some of them very well-funded—are entering the energy space and have the potential to undermine the link between existing companies and customers. As an example, while traditional utilities are feeling pain, Eric Schmidt, former CEO of Google, encapsulated the potential but also the uncertainty for the future, commenting that "the Energy Transition in Europe will provide a lot of opportunities for many companies—but nobody knows exactly today where and how to make money in the future." The interest and arrival of new players is changing the competitive landscape of the markets and the mindset of consumers.

There are lessons from other industries that suggest that existing companies with traditional business models are not best placed to adapt to opportunities that new technologies bring.

In telecoms, the invention of a range of mobile technologies forced a major change in the services which traditional players deliver. More recently there has

---

3. https://www.ft.com/content/6005636c-9f8c-11e3-94f3-00144feab7de
4. https://cleantechnica.com/2016/03/21/e-rwe-reporting-big-losses-fossil-fuel-businesses-2015/
5. http://energypost.eu/exclusive-rwe-sheds-old-business-model-embraces-energy-transition/

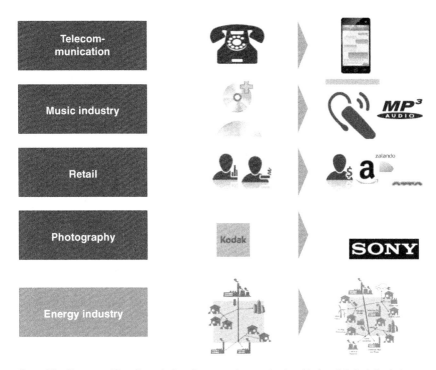

Source: Pöyry Management Consulting, using logos from companies named and graphics from Netanir, shutterstock.com

FIGURE 2.3   Industry transformations.

been a second seismic shift with the introduction of smartphones, in many cases, moving potential revenue away from network operators to the phone hardware providers themselves. The rise of digital downloads has substantially changed the music business value chain, allowing new entrants to go straight to customers. Digital photography left its creator—Kodak—in bankruptcy (Chapter 11), as they failed to ride the crest of the technological wave. Airbnb built a portfolio of 1 million rooms—which it does not own—in 6 years, while it took the Hilton Group 100 years to achieve 700,000.

Fig. 2.3 provides an illustration of the transformations seen in these industries. Such lessons provide a stark warning to those still in traditional utility business model mode—ignore change at your peril.

It is the scale of the ongoing energy sector transformation, the rate of IT advancements, and the difficulties being faced by the traditional utility, in combination, which make the current challenge so significant. If the utilities of today are to survive, successful transformation requires evolution on all fronts—technological, commercial, and cultural. For those that can adapt, there are opportunities. Not adapting looks set to be a short-lived strategy.

In such an environment, what features are needed to create a successful integrated energy company for the future? The alternative is perhaps for new companies to take over the customer relationship, and for the existing companies to be pushed upstream. These themes are explored further by Johnston & Sioshansi, who consider models for engagement of proactive consumers and by Steiniger & Sioshansi, who focus on the role of virtual power plants to aggregate consumers.

3   THE FIVE KEY CHARACTERISTICS OF A FUTURE ENERGY COMPANY

The emergence of a decentralized industry, with engaged customers and new remote data and control possibilities has opened up an alternative possible future for the energy sector. The sections below set out a possible recipe for a future integrated energy company, structured around five key characteristics. These are as follows (and as displayed in Fig. 2.4):

- Characteristic 1: access to a portfolio of generation, storage, and flexible demand will remain important in the future, but with less emphasis on asset ownership.
- Characteristic 2: risk management, optimization, and trading are essential parts of the operation of a utility and will continue to be core business.
- Characteristic 3: control of "Big Data" will give leverage for competitive advantage.
- Characteristic 4: user-friendly applications and automation tools will enhance customer propositions and unlock demand response.
- Characteristic 5: being close to the customers and retaining their trust as their needs change will be important to unlock new sources of value.

3.1   Characteristic 1: Access to a Portfolio of Generation, Storage, and Flexible Demand Will Remain Important in the Future

Any future portfolio will require access to a diverse portfolio. This will need to span generation, demand, and storage or other technologies that can deliver flexibility. Circumstances differ and not everyone needs the same portfolio of generation or the same types of customer. But having access to an appropriate blend of resources will allow for value to be realized through trading activities and for risks to be managed.

This does not necessarily mean continued ownership of the assets by the energy companies, however. In future, a more asset-neutral approach may be expected to emerge.[6] Indeed, this trend is already in evidence, as ownership

---

6. It could be argued that existing energy companies do not own assets, but that the assets own the companies and consume a disproportionate share of management time. The traditional companies are led by asset managers and investors, not innovators or marketing teams, and the primary role of the retail business is one of risk and credit control and IT management rather than capturing consumer value.

FIGURE 2.4   Five characteristics of future integrated energy company.

of generation is diversifying. Initially, this was seen with a wave of independently owned wind farms. More recently, the trend has continued as millions of consumers have invested in solar generation. Looking forward, a similar pattern looks set to emerge with growing installation of domestic scale batteries. Asset ownership is becoming more diversified and decentralized to the extent that, increasingly, each house/business or, with aggregation, cluster of houses/businesses could be considered as an asset.

As an example of a possible model for the future, Statkraft has built a successful business in Germany based on aggregating renewable generation plants from smaller developers, and offering trading and risk management services. They have over 9.5 GW of wind and solar energy from more than 1300 sites under contract, which dwarfs its small portfolio of hydro and gas generation in Germany. As a result, Statkraft is the largest direct selling company in Germany and one of the biggest traders in the German day-ahead and intraday markets, without owning any wind assets there. It is notable that their "asset-light" approach is concentrated on portfolio management and trading and risk management, rather than a full focus on retail and the needs of household customers.

Ongoing market design changes in Europe are increasing the potential for this model to be applied more broadly. Now that renewable energy has a significant role in many markets, the emphasis is increasingly on integrating renewables into the market rather than insulating them from it. In the European Commission's recent Clean Energy Package,[7] for example, we are seeing proposals for balance responsibility and, prospectively, removal of priority dispatch for all but the smallest renewable installations. In addition, there is emphasis on future support regimes for renewables structuring payments around their revenues from the wholesale market. Entering into arrangements with an aggregator to provide trading and risk management services in the vein described above is one option available for dealing with greater market exposure.

Therefore, having access to a portfolio of resource across generation and demand, integrating commercial arrangements with the assets of others into this portfolio, will be an important feature of the future energy company, as well as a valuable service to others in the market.

3.2 Characteristic 2: Risk Management, Optimization, and Trading are Essential Parts in the Operation of a Utility

Electricity is the world's most volatile traded commodity. The true strength of traditional energy companies is their ability to invest in assets with a 20–40 year lifetime on the back of a series of short-term contracts. The series of trading, risk management, optimization, and financial and collateral management tools, which support this is world class. The complexity and volatility of the wholesale market is a barrier to entry for new players, and existing energy companies will retain a competitive advantage over new players for some time to come in this regard.

But the nature of risks within the market is changing. The growth of renewables has increased price risk but has also brought volume risk to the fore. Thermal plants now have real uncertainty over when and how much they will operate, as running patterns flex around renewable generation output with near-zero short run marginal costs. The challenge lies in that the industry broadly continues to trade the same products as 10 years ago, which suited the risks faced at that time, but are not geared toward the risks of the future.

For example, a decade ago, most CCGTs would be able to predict reasonably well their likely running patterns for the coming weeks, months, and even beyond, based on expectations of typical demand patterns and fuel prices relative to coal generators. But today, a CCGT's output is dependent on wind and solar levels, which are only known in the nearer-term, and so it does not know its likely hourly production profile significantly ahead of real time. Unlike the situation 10 years ago, the CCGT now has volume risk, as well as price risk, changing the basis for forward energy trading. The spark spread relates to a

---

7. https://ec.europa.eu/energy/en/news/commission-proposes-new-rules-consumer-centred-clean-energy-transition

profile of generation that has not been seen in most of Europe for 5 years. A negative spread just means that the market is trading the wrong products.

New trading instruments are available, though, in the form of options. This is a natural product to complement trading of firm energy. Generation capacity can essentially be considered an "option" to produce energy. While all production capacity represents such an option, some resources offer greater flexibility than others in how the option can be exercised and how frequently. In an increasingly decarbonized system with high levels of low carbon generation, energy market volatility could be significantly greater, increasing the value of such options.

The appropriate response is to ensure that market participants are able to trade energy "options" in the form of a sufficiently broad range of products mirroring the various types of services the market requires. Multiple option variants could be developed to reflect different requirements, such as deliverability and flexibility, and to suit the capabilities of different types of capacity. Our own suggested approach develops the concept of capacity as an energy option and proposes a bilateral model for the trading of energy "options."[8] Under a bilateral approach, options for use of a resource can be sold by the provider to other parties who may require flexibility to balance their own positions (e.g., as a complement to variable generation resources or to help manage retail positions).

Moving forward, traders need to find means of packaging risk and managing complexity in ways that will continue to bring in capital for new projects. The new risks need new market instruments, such as the energy options discussed above, to manage volume risk and complement the markets for delivered energy. With a system of energy options, parties with flexible assets could sell options for access to their flexibility to others, such as retailers who are then able to exercise the options when needed to access energy and help manage price volatility. This gives:

1. the flexible assets revenue that is not tied solely to production and
2. the counterparty with a type of insurance that helps to protect them against price fluctuations.

As an alternative to trading hedges, another possible route for risk mitigation is offered through physical hedges. Tesla provides an example of an integrated model, which now spans electric vehicle, solar PV panels (since the acquisition of SolarCity[9]), and electricity storage through its Powerwall product. Through its Supercharger network, Tesla (for now) offers free fast charging for its EV owners; and the addition of solar and storage to the mix offers the possibility for customers to break links with traditional utilities and

8. http://www.poyry.com/sites/default/files/media/related_material/revealing_the_value_of_flexibility_public_report_v1_0.pdf
9. https://www.bloomberg.com/news/articles/2016-11-17/tesla-seals-2-billion-solarcity-deal-set-to-test-musk-s-vision

instead rely on Tesla for all their energy needs. Importantly, this is all under the banner of a well-recognized and established brand. Customers buying into the Tesla product range will have a long-term relationship with the company, given the lifetime of the underlying products themselves. Tesla has the ability to define these relationships such that it can access the flexibility offered by the fleet of EV charging activities, storage, and generation assets of its customers. It also has a way in to become the energy supplier for its customers and their Tesla products. Therefore, through its brand and products, Tesla hopes to make big waves in the energy market.[10]

3.3   Characteristic 3: Control of "Big Data" Will Give Leverage for Competitive Advantage

Put simplistically, utilities have tended to know their customers as deemed profiles, only obtaining details on actual consumption in aggregate across a period of months significantly after the event and with very little understanding of their needs or desires. The advent of smart meters and other data processing technologies/businesses offer the opportunity to change this. The dawn of readily available "Big Data" (i.e., the ability to integrate with data from other sources, such as social media) provides access to much more granular and timely information concerning consumer behavior and the underlying reasons.

This is a powerful new resource. It allows utilities to obtain a much clearer understanding of consumer preferences and behavioral patterns and, therefore, the needs and requirements of their customers. To make the most of this opportunity, utilities need the ability to effectively access, analyze, and process the new sources of data to harness the valuable information that they contain concerning consumer behavior. Based on this, utilities will have the ability to develop propositions that are suited to the needs of the consumer and so capture value. This, in turn, may open opportunities to bundle energy with other services. This theme is picked up further as part of characteristic 5 below.

Energyworx provides an example of a company harnessing big data for use in utility sector applications.[11] Energyworx converts data delivered through smart meters and sensors into propositions that its clients can act on. Utility clients benefit from Energyworx's high granularity information about individual consumers' energy consumption through faster and more accurate billing to consumers, more precise demand forecasting for planning generation, or better targeted pricing and marketing depending on their customers' installed energy equipment. Equally, prosumers benefit, for instance, by being better able to judge whether they should consume or redistribute their energy or by being able to compare their energy efficiency against other users from the community.

---

10. https://www.bloomberg.com/news/articles/2016-06-29/-tesla-solar-wants-to-be-the-apple-store-for-electricity
11. https://www.energyworx.com/

Big Data methodologies and enhanced connectivity of devices will also revolutionize day-to-day operation of the system. With the increased complexity of a decentralized system, the challenge of efficient operation will multiply. Efficient optimization requires better data on the state of different devices and their ability/willingness to respond. Additionally, more powerful tools are needed to allow data to be turned into intelligence and control. But in many regards, data are not available today in timescales that support such optimization. For example, it is remarkable that the settlement timetable for domestic customers has barely changed since the time when urgency in data transfer meant using first class instead of second class postage.

The modernization process in energy has begun. Domestic "smart metering" is gathering pace. However, it is rudimentary and far from being universally applicable. There are beacons of positive development, though. For example:

- Estonia has rolled out a centralized data hub which supports applications from third-party developers. Eesti Energia in its role as a retailer now has around one-third of its domestic customers facing hourly spot pass through tariffs.
- Next Kraftwerke—further described in chapter by Steiniger in this volume—bundles around 3000 small- to medium-scale generators and consumers into a virtual power plant, using a digital platform to connect and optimize flows between producers and consumers.[12]
- Open Utility—further described in chapter by Johnston in this volume—uses its peer-to-peer energy matching platform to allow consumers to select their own supply source based on current supply and demand from measured smart meter data.[13]

The full "Internet of Things" vision may be overkill for the energy system for now, as electricity costs for most applications are low and transaction costs are material in that context. However, it is clear that existing information gathering and monitoring tools in many markets are not yet fit for the decentralized future. Capitalizing upon the functionality offered by rapid advances in IT and communications technologies offers the chance to forge a model for the future focused around the customer.

3.4   Characteristic 4: User-Friendly Applications and Automation Tools Will Enhance Customer Propositions and Unlock Demand Response

Many customers now expect instant information and feedback via smartphone or Internet. Customer service in most industries has moved away from call centers to sophisticated applications and Internet-based solutions. Energy

---

12. https://www.next-kraftwerke.com/
13. https://www.openutility.com/

companies are adapting to this service model and there has been a proliferation of smart apps in recent years.

However, for the demand side to become truly active by changing consumption patterns in response to system conditions, users will not want continuous active involvement. For most domestic consumers, the potential saving even from very sophisticated decision-making is low. Typically, households in Europe pay around €2 per day for electricity in total, and the saving from optimal consumption patterns is a fraction of that. With (flexible) electric heating or vehicle charging the value increases. For example, to charge a 30 kWh electric vehicle battery, the saving based on typical day–night tariffs in United Kingdom could be €5 or more per charge. But these decisions will still need to be automated if users are expected to take optimal decisions everyday. This requires energy companies to offer excellent, user-friendly, and robust automation tools, which allow users to retain control.

Therefore, decentralized decision-making must become automated, and the customers' experience should not be affected. In this vein, in the United Kingdom, npower has partnered with Nest Labs,[14] who specialize in smart thermostats and smoke/CO detectors.[15] The Nest Thermostat learns from the user and programs itself based on this, while also allowing remote control via smartphone. This provides the consumers with energy for heating based on learning from observed patterns of behavior, while also offering control where desired.

German company beegy[16] offers another example where consumers enjoy automated integration of a suite of smart energy technologies behind a streamlined interface. Their service consists of managing a home's PV panels, Tesla Powerwalls, and EV charging points through a single beegyHUB, whose activities the homeowner can track through a web portal or smartphone app. The company consolidates all beegyHUBs Germany-wide into a single beegy Community. Algorithms allow users to store their energy or feed it into the Community when their consumption dips below their generation capacity. Alternatively, when their consumption exceeds their domestic generation and stored energy, users can also draw on the Community for electricity—at a flat rate that beegy locks in for 20 years. The company also offers its clients constant monitoring of their equipment, providing a 20-year warrantee for all products. Alongside this specific German example, the chapter by Löbbe & Hackbarth describes examples of innovative companies in Germany.

Heating represents another area where smart technologies could enable demand-side response. Glen Dimplex, the world's largest manufacturer of electrical heating equipment, has been focusing since 2015 on this issue through the RealValue project, in collaboration with partners from Ireland, Germany, and Latvia. This 3-year project, funded through the EU's Horizon 2020 program,

---

14. https://nest.com/uk/
15. https://nest.com/uk/energy-partners/npower/
16. http://www.beegy.com/

aims to demonstrate how Smart Electrical Thermal Storage (SETS) can, along with a management platform and interface for end users, meet domestic heating needs while also acting as energy storage capacity for the electricity industry.

The trend of twinning automation and user-friendly apps with energy applications is clearly on the rise, with early movers seeing and benefiting from the access to and engagement with consumers that it offers.

3.5 Characteristic 5: Being Close to the Customers as Their Demands Change

The needs of customers do not always align with the plans of vertically integrated energy companies. The most obvious conflict is over incentives for energy efficiency, which causes customers to consume less energy and, hence, companies to produce and sell less of their basic product. This dilemma sheds light on how a truly customer-focused energy company might approach its customer offering.

Ultimately, most successful mass-market business models involve a combination of high volume and low margin sales accompanied by opportunistic niche sales with higher value. To be sustainable, the higher value sales need to be based on predicting and then meeting customer wishes. The bundling of services may be within the energy sector or extend beyond. For example, cars or appliances may come with an energy supply agreement.

Again, Tesla is an example here. Its branding is an essential part of its proposition and its customers feel that they are buying into the brand. Strong brands from the nontraditional, newer entrants into the energy sector may hasten the demise of traditional integrated energy companies, whose own brands have, in some cases, been tarnished by concerns regarding misselling to or profiteering at the expense of their customers. Trust in conventional utilities has been negatively affected, as a result.[17]

A successful energy business must build a strong relationship with its customers, including trust on both sides, and has to build a good understanding of its customers' needs. It must come to think of its relationship with customers as part of its asset base, noting that, in a long-term business, assets must be carefully managed to not be exploited.

4 THE NEW ENERGY COMPANY

Taking the above characteristics into consideration, who in the energy sector has this combination of skills today? Arguably, no one.

The existing energy companies excel at characteristics 1 and 2: portfolio management and trading/risk management. But how do they defend their

---

17. https://press.which.co.uk/whichpressreleases/consumer-trust-in-the-energy-industry-hits-new-low/

businesses from left-field entrants already skilled in Big Data, user-friendly applications, and automation tools? They must focus on their existing strengths while covering the gaps—and improve their relationships with their customers while doing so.

What appears to be central to a successful strategy for the future is to focus on the consumer and to place the customer at the heart of the offering. While the sector has always focused on providing energy to meet consumers' needs, the traditional approach has been for consumers to be passive and for there to be a one-way flow down the supply chain from utilities to consumers. Energy needs have been met, but the consumers and their wider preferences have been largely extraneous to the workings of the markets.

With Big Data providing insights into consumer behavior and needs, IT providing scope for greater consumer interaction and increased ownership of energy assets at smaller scale, the scope for more mutually beneficial engagement between energy company and consumer is vastly increased. In this context, companies which dwell on their existing skills and focus on their own technologies and assets at the expense of their customers run a high risk of failure, especially in a fast-changing industry.

Some of Europe's utilities are seeking to reinvent themselves. RWE has restructured, choosing to bundle renewables, networks, and retail into a new subsidiary, Innogy, separate from the conventional generation business (although still under the RWE banner[18]). E.On has also restructured, but has elected to retain renewables, networks, and retail under the E.On brand, with the conventional generation business now operating as Uniper.[19]

Whether such steps will prove effective in improving their fortunes remains to be seen. But it seems unlikely that the newer potential entrants to the energy sector, such as Tesla or Google, will withdraw, having made effective forays and gained traction. The drive toward decentralized energy, with a growing role for demand-side response, aggregation, and engaged customers has also irrevocably changed the game.

In the current environment, our advice to the utilities contains the following messages:

- hesitate before taking big decisions on conventional generation investments;
- invest in trading and risk management tools, such as options, and trading capabilities that are appropriate for the future market context;
- understand the customer base and the requirements and perspectives of the consumer; and
- take a strategic view on customer propositions that will work for your customers and on the branding and partnership models that will allow you to achieve them at scale.

---

18. https://www.rwe.com/web/cms/en/113648/rwe/press-news/press-release/?pmid=4014358
19. http://www.eon.com/en/media/news/press-releases/2016/1/4/separation-of-eon-business-operations-completed-on-january-1-uniper-launched-on-schedule.html

5   CONCLUSIONS

The energy industry is ripe for revolution toward a system which uses new technologies to put its customers first. The set of characteristics above set out one possible recipe for a future integrated energy company, which retains its connection with customers and enshrines these relationships as the crux of its business. Other futures are available, including ones in which new companies take over the relationship with customers and force the energy companies to retreat upstream toward production and trading.

Utilities need to find new ways to engage with their customers to drive value propositions through the combination of smart apps, appliances, and Big Data capabilities. In short, focusing on the customer side of the business will not be an alternative business model—it will be the lead.

Changes are being seen in the utility–customer relationship and new propositions are being developed for electricity consumers. For example, in Australia, Sumo Power is borrowing from concepts in other sectors by offering an "all you can eat" energy contracts,[20] which gives the customer cost certainty and avoids the prospect of unexpectedly high bills. But there is scope for further developments and innovation in this relationship.

Electricity will not be going out of fashion anytime soon. As a whole, its value to society is growing as electrification increases. The question is whether the existing companies will retain their customer-facing role, or become someone else's outsourcing partner for top-up kWh. To remain viable, companies within the sector today need to adapt to keep ahead. That means changing the way business is conducted and focusing on the needs and wishes of customers before someone else does.

This theme is the focus of many of the accompanying chapters, which explore specific examples of innovation and disruption on future business models.

---

20. https://www.sumopower.com.au/all-you-can-eat

Chapter 3

The Great Rebalancing: Rattling the Electricity Value Chain from Behind the Meter

Robert Smith* and Iain MacGill**
*East Economics, Sydney, NSW, Australia; **Centre for Energy and Environmental Markets (CEEM) and School of Electrical Engineering and Telecommunications, University of New South Wales, Sydney, NSW, Australia

1   INTRODUCTION

Change is afoot for the electricity sector. As reflected in other chapters in this volume, the exact end point remains contentious but there is broad consensus on the direction and the drivers. Distributed energy resources and renewables are in; centralized generation and fossil fuels are on the way out. Driven by climate change and new technology, the grid is being reshaped and a recalculation and redistribution of the electricity value chain is under way. Innovation and disruption at the grid's edge is happening on a scale not seen for over 100 years. And within this metamorphosis the links in the value chain facing the greatest uncertainty appear to be the bits in the middle, the regulated monopoly transmission and distribution networks.

The electricity sector in transition is a study in contrasts. Generation is still dominated by large-scale coal, hydro, and nuclear generation who's designs, and in some cases actual plants, are 50 or more years old. Cables, transformers, poles and wires, and the simple accumulation meter have changed little in the past 100 years. However, the past decade has seen extraordinary growth in renewables, particularly wind and solar. Over 15% of Australian households have rooftop PV installed as further described by Mountain & Harris, Biggar & Dimasi, and Webb et al. Yet even as ubiquitous connection of end-use appliances foreshadows an "Internet of everything," described by Knieps, many key end-use technologies, like electrical resistance storage hot water systems, space heaters, and cooktops, remain similar to they were in Edison's days.

With the changes underway the traditional value chain and business model seems a poor fit for the 21st century grid. The traditional model is based on

Innovation and Disruption at the Grid's Edge. http://dx.doi.org/10.1016/B978-0-12-811758-3.00003-6

monopoly regulation that guarantees returns on regulated assets deemed to have a "used and useful" life of 40 years and funded through volumetric tariffs. Guaranteeing returns out 40 years is problematic when auguring the future of the grid out even 5 years has become increasingly challenging. One way to see beyond the uncertainty, to separate what is possible and practicable from what is fanciful, is to follow the value chain.[1]

The value of electricity to its original end-use customers was immense. At the start, in its first key role of lighting, it was considerably more expensive than existing gas lighting but still offered greater value—particularly in terms of safety, simplicity, illumination, and cleanliness.[2] From its Pearl Street New York roots the original electricity value chain quickly grew to serve an amalgam of diverse and disparate end users linked together by transmission and distribution.[3] A value chain anchored to remote centralized generation at one end and local customers at the other.

Now new technologies in local energy creation and storage means that, for increasing numbers of end users, the value chain has slipped its anchor. Even the need for the "poles and wires" is being reassessed as with DER generation value finds a haven closer to home embedded among energy users.

This chapter examines the disruption and innovation of the value chain at the grid's edge, looking: first, in Section 2 at what made the traditional value chain so effective; then in Section 3 at new visions for the value chain and complexity; next in Section 4 at the value of the tariff cost stack compared to the full value chain; and finally in Sections 5 and 6 at DER's role in the full value chain and drawing conclusions.

2 GREATER COMFORT AND CONVENIENCE

As the widespread consensus has shifted in favor of a disruptive future for the grid it is worth revisiting and remembering the strength of certainty about the centralized value chain which existed in the past.

The image the industry sought to project is shown in Fig. 3.1 from a utility public relations comic of the 1950s. While this is utility propaganda from the postwar golden age for electricity it also reflects the view largely supported by regulators, investors, and customers into the start of the 21st century.

---

1. The value chain often describes the journey of a product through a chain of activities that each add value; Porter's corporate strategy approach with the firm as the unit of analysis. In this chapter we use the term as an industry value chain or supply chain, like an input output model, and a societal approach based on utility to end users.
2. See Fouquet and Pearson (2006) on light value, Shiman (1993) on early competition between electricity and gas while diversity still exists across Australian, with gas having a large role in Victoria and wood significant in Tasmania.
3. See Yakubovich et al. (2005) Sydney (a late starter on electrification) had a major coal-fired power station (now the Powerhouse Museum) operating within 1.5 km of the town hall till 1959.

FIGURE 3.1   Electricity service for your greater comfort and convenience.

The advantages of the traditional centralized utility network model emerged after Westinghouse, with Tesla in his corner, defeated Edison in the "current wars" and alternating current became the standard for electricity distribution, largely because of its superior ability to carry power over long distances. Thomas Insull at Consolidated Edison in Chicago then shaped the business model, institutions, and value chain to support utilities as regulated monopolies.[4] Since Insull's day, regulation, deregulation, and reregulation has fractured and recombined the institutional and financial elements of the value chain within different

4. See Smith and MacGill (2014) and also Yakubovich for discussion of the mid-1890s debate on the Wright and Barstow pricing systems.

jurisdictions. However, the physical elements of the chain in generation, transmission, distribution, and retail remained relatively stable through the 20th century and the utility model also proved remarkably resilient.

This stability was based on delivering value. Three things combined to drive the centralized electricity grid's value proposition: economies of scale, continuous load growth, and diversity of use.

The scale economies come from the engineering properties of equipment, bigger is better value per unit of output, and from customer density, the more customers at a location the cheaper to serve each. Continuous growth ensures that you can capture the benefits of scale economies. You could never build too big only too soon, as eventually demand grew to absorb the extra capacity obtained cheaply though scale economies. Indeed, scale economies, supported by load density from geographic proximity and continuous growth, created enough surplus value from urban, commercial, and industrial supply that it could often be used to cross-subsidize remote, rural, and disadvantaged users.

Less obvious, particular once the industry matured, were the critical advantages of load diversity. The recently developed concepts of the share economy, shareconomy,[5] the connected economy, peer economy, platform economy, access economy, and collaborative consumption are based around increased utilization of assets through customer's joint usage. Under this new value chain, the business models of companies like Uber and Airbnb are built around facilitating consumers to share the underutilized assets they own with others. Diversity of load in the electricity grid, however, also ensures asset sharing that dramatically cuts the cost of capacity and of supplying reliable and resilient energy services to customers. Customers are interconnected, value is shared, and consumption is collaborative, but assets are owned by regulated monopoly utilities.

Diversity works like magic to get extra use out of a given capacity. As customers' individual loads are not perfectly correlated, the average customer load at time of system peak is substantially less than the sum of each customer's individual peak load. Australian customers' data have shown that while an individual residential customer's 90% Probability of Exceedance (POE) peak demand may average 11 kVA, for a group of 32 customers this more than halves reducing to 4.6 kVA and then for 1000 customers reduces further to 2.6 kVA, and for 1 million customers then shrinks to 2 kVA.[6] Where a standalone electricity system might therefore have to provide 11 kVA capacity to serve its single customer, a network serving 1 million similar customers only needs to be able to provide 2 kVA per consumer.

Economists hold the "there is no such thing as a free lunch" but when scale favors larger generation and distribution, growth ensures scale advantages are

---

5. Terms like "shareconomy," while in vogue, are problematic as they are applied across a range of activities involving new commercial arrangements between consumers, rather than sharing as it is traditionally understood.

6. Calculations are from Davies (2015) using data from the Smart Grid Smart City project.

realized and diversity spreads capacity usage across customers the marginal cost of grid capacity becomes extremely low. These scale and diversity benefits tend to be largest at the generation level and smaller closer to the customer. And this has implications for the economics of DER that aren't well appreciated in much of the discussion.

While DER is becoming increasingly attractive compared to averaged volumetric tariffs it is easy to forget that the traditional electricity grid benefits which come at the aggregated system level (and underpinned low average kWh tariffs) are potentially lost or unable to be realized at the customer level for DER. Therefore, the network benefits of individual customer's efforts to manage their own peak loads are not fully reflected up the value chain as the impact of individual load profiles on the system peak has already been reduced due to diversity, and the cost reduced due to scale economies.

The attraction of the old grid model was not lost on investors. Benjamin Graham in his 1949 seminal work on value investing, *The Intelligent Investor*, described the "comfortable and inviting situation" of utilities for conservative investors:

> *The position of utilities as regulated monopolies is assuredly more of an advantage than a disadvantage for the conservative investor. Under law they are entitled to charge rates sufficiently remunerative to attract the capital they need for their continuous expansion, and this implies adequate offsets for inflated costs. While the process of regulation has often been cumbersome and perhaps dilatory, it has not prevented utilities from earning a fair return on their rising invested capital over many decades.[7]*

Over 75 years later, Grahams' protégé and major proponent, Warren Buffet, has continuing confidence in the business model for electricity utilities investments:

> *Our confidence is justified both by our past experience and by the knowledge that society will forever need huge investments in both transportation and energy. It is in the self-interest of governments to treat capital providers in a manner that will ensure the continued flow of funds to essential projects. It is concomitantly in our self-interest to conduct our operations in a way that earns the approval of our regulators and the people they represent.*

But notes that this confidence is being challenged:

> *In its electric utility business, our Berkshire Hathaway Energy ("BHE") operates within a changing economic model. Historically, the survival of a local electric company did not depend on its efficiency. In fact, a "sloppy" operation could do just fine financially. That's because utilities were usually the sole supplier of a needed product and were allowed to price at a level that gave them a prescribed*

---

7. Graham, p. 356.

return upon the capital they employed. The joke in the industry was that a utility was the only business that would automatically earn more money by redecorating the boss's office. And some CEOs ran things accordingly. That's all changing....... tax credits, or other government-mandated help for renewables, may eventually erode the economics of the incumbent utility, particularly if it is a high-cost operator. BHE's long-established emphasis on efficiency—even when the company didn't need it to attain authorized earnings—leaves us particularly competitive in today's market (and, more important, in tomorrow's as well).[8]

However even Warren Buffet, the world's most successful investor, may be underestimating the changes planned by some. BHE's long-established efficiency in the old business model may not be enough if the shape of the value chain is radically altered by cost-effective DER and reconfigured regulation. While Gellings in his chapter doubts the need for a new business model, most other contributing authors doubt the continued viability of the old model.

3   NEW VISIONS OF THE VALUE CHAIN: RHETORIC, REALITY, REGULATION, AND THE REV

Despite early concerns, it is now apparent that customers with existing grid supply and DER will not to go off-grid in large numbers, at least in the near term.[9]

In essence, while volumetric tariffs price the grid mainly on kWh usage, a grid connection provides a bundle of services which are independent of energy volumes and which are still highly valued by customers. Due to these services the new value chain and business models being constructed still hold a role for the "poles and wires" networks. A continuing role for the grid is uncontentious. What is highly contentious is what the role will be, how it is paid for, and by whom. The uncertainty created by DER, particularly for transmission and distribution, has spawned a spate of new models with snappy labels and snazzy infographics but few insights and fewer answers.[10]

Looking at PwC's analysis as one of the better examples, Fig. 3.2 shows 10 areas interconnected in a "new market paradigm" which is so dense with links that the real interest becomes why some areas are not interconnected. With 35 of 45 possible links shown why should new entrants not have a connection to retail? Or distributed generation lacks a connection to storage? The picture can be replaced by a simple statement; in the new value chain you need to understand how everything is connected to everything else.

PwC's attempt to encapsulate the value chain and different parameters into eight possible business models is shown in Fig. 3.3 but where to disentangle the

---

8. Buffet (2015).
9. See Rocky Mountain Institute (2013, 2014) as an example.
10. See PwC (including their caveats), Deloitte (2016), ENA and CSIRO (2016), and Bain & Company (2015).

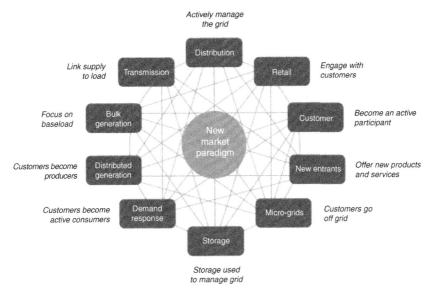

FIGURE 3.2   PwC's new market paradigm.

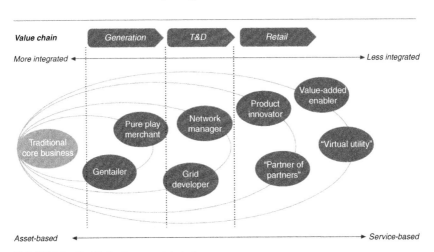

FIGURE 3.3   PwC's business model choices.

value which underpins the business model labels is left to imagination.[11] Knowing that things are connected and that value is involved is not enough to enable you to map how it is connected and can be restructured.

Regulators in major jurisdictions, as Conboy, Picker, and Zibelman state in the opening of the book, are repositioning the rules and incentives to respond to

---

11. PwC acknowledges the overwhelming uncertainty and similar approaches with similar issues that appear elsewhere.

DER but also to rattle and rearrange the traditional value chain. New rules are needed to better align the underlying economics of DER with tariffs and costs. And some rule makers are proving highly proactive. For example, Ofgem's RIIO (Revenue = Incentives + Innovation + Outputs) approach in the United Kingdom is attempting to embed climate change, innovation and stakeholder's interests into the regulatory frameworks for gas and electricity prices. It involved radical change and is designed to accommodate innovation and DER, yet at its core it is still cost-to-serve monopoly regulation.[12]

The most ambitious, certainly most impressive sounding, attempt to rearrange the regulated links of the value chain, however, comes from the birth place of the grid, New York. Reframing the Energy Vision (REV) represents one of the most radical attempts to reshape regulation of the grid in the past 100 years.

The REV has its genesis in 2012 after Superstorm Sandy when the New York regulator, as discussed by Zibelman, took the approach that traditional cost-to-serve ratemaking was untenable for 21st century.[13] The response was the REV a "modern regulatory model that challenges utilities … by better aligning shareholder financial incentives with consumer interest."[14]

The REV framework proposes radical changes to support and incentivize third-party involvement and a Distributed System Platform (DSP) which mimics other sectors like Telcos and smartphone apps where:

> *the traditional provider's role has evolved to a platform service that enables a multi-sided market in which buyers and sellers interact. The platform collects a fee for this critical market-making service, while the bulk of the capital risk is undertaken by third parties.[15]*

The new REV role for the networks is as market-maker not as asset owner and manager; no longer an owner of physical links in the value chain but a linker of chain components which will "reorient the electric industry and the ratemaking paradigm toward a consumer-centred approach that harnesses technology and markets."[16]

This effectively aims to turn the traditional electricity business model on its head. It assumes new digital technology can solve major problems by creating efficient signals through market prices. New price signals that will vaporize

---

12. Cambridge Economic Policy Associates (2016) and Biggar (2011) for an excellent summary of the history of regulation which highlights the considerable issues that have always been present with the traditional cost to serve model, even with deregulation and RPI-X incentives.
13. NYPSC Chairman Richard Kauffman's speeches share Buffet's fundamental dislike of the inefficiencies of the old business model, and favors a platform approach based on perceptions of what is possible from high tech sectors.
14. New York Public Service Commission (2015).
15. Ibid, also check web for the hype on the "platform economy."
16. NYPSC, Case 15-M-0252.

the old problems of monopoly market power, information asymmetry, the sunk costs in large and lumpy capital, scale economies, instantaneous flow, socially sanctioned and accepted cross-subsidies, and the value of diversity and trade; problems that have exercised regulators minds for over a century. The success or failure of radical visions like the REV depends on whether these problems are actually addressed rather than merely shuffling them along the value chain. Whether this is rhetoric or reality depends on whether the underlying problem of DER innovation and disruption is complex or merely complicated.

3.1 Complicated or Complex?

Understanding the technical issues of voltages, storage losses, two-way flows, efficient dispatch, and establishing efficient prices is important for imaging the future of the grid but is unlikely to be sufficient. This is because the issues of innovation and disruption at the grid's edge are complex rather than complicated.[17]

Complicated problems are ones where rules and relationships are numerous and may not be known, but are known to exist. They differ from a simple problem mainly in scale. Big computers and big data can crunch and crush these types of problems. Yet while the management of well-established industries like the grid tends to be complicated, industry transformations are complex. A complex problem is one where rules are not known and not knowable, and relationships are not clear or changeable. A complicated problem is likely to involve estimates of performance and risk which are measurable or have probabilistic boundaries; a complex problem is likely to involve uncertainty, which by definition is not amenable to measurement.

Designing an electric car, the battery and power delivery system that drive it is complicated but now seem poised for mass market adoption.[18] By comparison, forecasting the adoption rate of electric vehicles is complex, more so now that EVs are naturally twinned to the emergence of autonomous vehicles as further described by Webb & Wilson. Not only are forecasts required of uncertain EV technology progress there are also uncertainties about policy responses, autonomous vehicles, conventional fuels and fuel prices, geopolitics, climate change, and customers' preferences and behaviors.[19] And the uncertainty that prevails for EVs has parallels across other products, technologies, sectors of the economy, and customer segments at the grid's edge.

A view that the value chain is complicated rather than complex, and hence amenable to organized rearrangement, underpins much of the rhetoric of current attempts at grid value chain redesign and business model creation.

---

17. See Colander and Kupers (2015); Kay (2010).
18. LeVine (2016) traces the path of recent battery progress.
19. Productivity Commission (2016).

This perception underestimates the complexities of interactions across the value chain and, in particular, the understanding required of customer value.

Speaking at a World Future Society function in 1977 Ken Olsen, founder of the then technology giant Digital Equipment Corporation, said "There is no reason anyone would want a computer in their home." This quote went on to be listed as one of the worst predications of all time.[20] Yet it was taken out of context. Olsen's speech actually debunked the popular and persistent postwar notion of the fully automated home where a computer could, and would, control the key aspects of our lives from the lights and appliances to diet and exercise. In its original context it has proven remarkably prescient for over 40 years, a time span similar to predicting from now to 2057. Technical complication has not stopped the automated home developing beyond the grid's edge, it is the complexity of aligning technology and regulations with what customers' value.

What was missing in 1977, and still needs to be found, is a place within the existing full electricity value chain to fit a new customer value proposition, and around which new business models that incorporate innovation and disruption can be built.

4 THE TARIFF COST STACK, THE MYSTERY BEYOND THE METER AND THE FULL ELECTRICITY VALUE CHAIN

In the early days of the electricity sector, utilities were focused on customer value and what went on beyond the meter. Utilities had appliance stores, promoted extra plug points into houses, had salesmen, actively competed with gas, and saw themselves as providing a new, modern, and cutting edge technology. With regulation, rate cases, deregulation, and reregulation, together with the inevitable torpor that comes with technology and industry maturity, this customer focus was lost.

As the industry matured the customer became a number, a meter identifier, a billing code, or at best a stream of data points from a meter. Volumetric tariffs, based on averages and therefore economically inefficient,[21] raised enough revenue and worked well enough in a world of continued growth (see Haro et al.). The industry took for granted that it delivered an "instant, continuous, low-cost electricity service for your greater comfort and convenience." The utilities link to end users weakened and the incentives for them to see the world from beyond the meter, to understand what customers' value, grew fainter.

So, while the value of the utilities regulated assets which are built into tariffs was weighed and measured, debated and routinely dragged through the courts, what customers do, including the value of customers' assets, became a mystery. The electricity value chain behind the meter has not been thought of as

---

20. Chaline (2011).
21. See discussion in Biggar & Dimasi, Orton et al.

either part of the larger electricity system or its own household microgrid but as individual elements. Appliance purchases are made individually, as one-off decisions by consumers for a fridge, a phone, or a George Foreman grill and as part of the value chain of product manufacturers and retailers not the electricity sector.

Once considered as a whole it becomes clear that the majority of the investment in the grid, broadly defined, is done by it users and the largest part of the value chain is behind the meter. The traditional grid value chain, as it is commonly calculated and depicted, is better viewed as a "wholesale + network + retail" tariff cost stack. The tariff cost stack is typically driven by generation pricing (with competitive wholesale markets in some jurisdictions), monopoly network cost-plus and rate of return regulation (either with or without added incentives), and a retail operation and profit margin placed on top to create a price, the tariff, that is seen by customers in the market.

The asset value of the traditional grid value chain, the "tariff cost stack" elements, is simultaneously both simple to measure and complex to understand. This is because some parts (the networks) have published regulated assets values, while others have revenues set by market values and no official asset values. The prices which drive the regulated revenues are based on estimated asset values which in a competitive market would be the basis for setting asset values.[22] Regulated distribution and transmission asset values are therefore well reported but artificial.

In Australia pool prices and long-term contracts should allow calculation of true generation asset values but estimating these is increasingly confabulated and confused by decisions about energy policy. Estimates are unstable as, for example, current high pool prices reflect underlying costs but also incentives for renewables and carbon pricing, recent closures of gas fired and brown coal generation, uncertainty about interconnectors, national and international climate change policy, and geopolitics more generally.

An official view at a time of relative stability, c.2010, estimated the value of traditional grid assets in the Australian National Electricity Market (NEM) as over $100 billion comprising of: $40 billion generation (for the major generation assets operating in 2008 and not including new renewables); $17 billion for transmission; and $46 billion for distribution.[23] This can be viewed as the tariff cost stack or value chain to the meter. More recent regulated asset values, as at 2015, have transmission assets as similar at $18 billion but distribution increasing to $64 billion[24] and this suggests a total tariff cost stack value of around $125 billion (Table 3.1).

---

22. This is the cornerstone of Benjamin Graham's approach to value investing.
23. Productivity Commission (2011).
24. Missing from this traditional NEM view is around $10 billion for Western Australian assets not in the NEM and new renewable generation assets.

TABLE 3.1 The Electricity Tariff Cost Stack, Australian Grid Asset Values 2015

	$ Billion
Powerlink	6.6
TransGrid	5.8
AusNet	2.5
ElectraNet	2.0
TasNetworks	1.2
NEM Transmission	18.2
Energex	10.9
Ergon Energy	9.0
Ausgrid	14.6
Endeavour	5.7
Essential	6.9
ActewAGL	0.8
Powercor	3.1
AusNet Services	3.2
United Energy	1.9
CitiPower	1.7
Jemena	1.1
SA Power Networks	3.6
TasNetworks	1.5
NEM Distribution	64.1
Western Power	10.0
Total	74.1

Source: www.aer.gov.au

The traditional value chain of the electricity grid is shown in Fig. 3.4. Regulation of the industry, and the tariff structures that come with it, have shaped the way the electricity system is seen, operated, and people relate and respond to the grid. It has a linear flow that follows the flow of electrons through the system. The grid boundary, in people's minds, in regulation and in legislation stops at the utility meter. Its end point is seen as the value at the meter where the retail tariff is measured and applied.

FIGURE 3.4   The traditional electricity grid value chain.

Even before recent changes this view was limited. Missing from the tariff cost stack view of the electricity value chain is the activity on the customer side of the meter.

The meter and the retail tariff are better seen not as the end point but as the midpoint of the value chain for electricity customers. Beyond the meter customers have always made substantial investments to derive value from electricity. The value of electricity is not just the tariff cost but the cost of everything that consumers themselves do and pay for on their side of the meter to get benefits from electricity as shown in Fig. 3.5.

FIGURE 3.5   The customer-side value chain.

Most obviously the true end point includes the cost of appliances and equipment - the whitegoods, TVs, phones, and gadgets for households and equipment, machinery, and computers for businesses. Beyond this are the customer internal electricity infrastructure, fixtures and fittings, wiring, lighting, building design and maintenance, as well as investments in energy efficiency that support customers in deriving benefit from grid electricity. But even adding this customer cost stack on top of the tariff cost stack does not create the true top of the value chain.

The true top of the value chain is the well-being, quality of life, or satisfaction that is felt by end consumers. The gap between costs and value in use is consumer surplus[25]: the difference between what a customer actually pays for all the components required for what electricity delivers, and what they would pay if they were made to pay the maximum they could. This component of the value chain is also missing from most of the new business models' understanding of value.

When broadly defined and considered as a whole it becomes clear that the majority of the value chain is on the users' side of the meter. However, customers' holdings of electricity assets are challenging to calculate. There is no centralized accounting as is done for regulated assets but rough estimates show that meter-side assets greatly exceed in value the grid-side assets that make their way into tariffs.

Variations in the value chain exist for particular customers, locations, circumstances, and needs—such as off-grid customer, cogeneration, and controlled loads—but for most of the 20th century and the majority of electricity customers, the customer cost stack has been remarkably stable and linear.

---

25. The concept of consumer surplus originates with Alfred Marshall's "Principles." Michael Porter's popularization of the value chain approach to corporate strategy focused on value capture of companies with value assumed at market prices.

FIGURE 3.6 DER additions to the value chain.

Now, as DER is becoming widespread it is stepping across and blurring the boundary of what value belongs to the grid (Fig. 3.6). These "new" components of the value chain have been possible for decades, but struggled to compete with the value for money of centralized grid electricity. As their kWh costs have fallen compared to conventional industry tariffs they are now poised to increase investment behind the meter and tip the existing imbalance further toward customer assets.

Importantly, the new DER additions to the value chain is not simply tacked on to the end of the tariff cost stack, it needs to be seen as a value wedge positioned between the meter and customers' own existing value add and consumer surplus.

4.1 Customer Assets Beyond the Meter

A full estimate of end users' asset holdings within the electric value chain is a separate, larger topic. However, preliminary estimates from various data sources available for Australia confirm that the investment by customers, before considering new technologies in DER, considerably exceeds the roughly $125 billion value of grid-side asset investment.

One simple perspective of households electricity asset holdings comes from household expenditures shown in Table 3.2.

Across expenditure groups, from the lowest to the highest quartile, more is spent each week on appliances and supporting assets needed to get utility from electricity than on electricity tariffs. On average households spent 72% more to get value out of electricity than they did to have it supplied by the grid. Across all households, c.2010, this spending was around $19 billion a year which, if assets have a life of 10 years, would exceed the $125 billion total value of all grid assets. And this is only the value of basic household applications before considering the full range of residential assets or any nonresidential assets.

The asset imbalance would be still more stark if to this simple view the following are added: household wiring, hot water systems, installation cost, lighting, design features, insulation cost, fixtures and fittings, and the increased uptake of newer technologies since 2010. Moreover, electricity tariffs include not only asset costs (the deprecation of longer lived regulatory assets and a WACC) but also operating costs as well as environmental levies and concession payments.

This imbalance between the grid side and customer side of the meter would be even greater but for the continuing efficiency gap across the meter due to the emergence of new technology and real price deflation. The benefits of technology have mostly accrued behind the meter in appliance efficiency and function,

TABLE 3.2 Australian Household Expenditures (Australian Bureau of Statistics, 2011)

Household expenditure on goods and services $ per week 2009-10

	Lowest	Second	Third	Fourth	Highest	All house-holds
Electricity (selected dwelling)	**16.83**	**21.00**	**23.28**	**27.45**	**32.60**	**24.23**
Whitegoods and other electrical appliances (excluding stoves and related)	5.47	10.14	9.53	10.47	15.64	10.25
Cooking stoves, ovens, microwaves, hot plates and ranges	0.98	*1.68	1.69	*3.45	5.17	2.60
Audio-visual equipment and parts	7.28	10.74	12.29	18.08	23.46	14.37
Home computer equipment (including pre-packaged software)	1.58	5.15	7.87	9.21	13.16	7.39
Blank and pre-recorded media (excluding pre-packaged computer software)	2.31	4.33	6.09	9.54	13.42	7.14
Appliances (selected & media)	**17.62**	**32.04**	**37.47**	**50.75**	**70.85**	**41.75**
Appliance spendi as a % of Electricity spend	**105%**	**153%**	**161%**	**185%**	**217%**	**172%**

The latest 2015–16 survey is completed but unpublished however air conditioning, smartphone, and mobile computing usage has increased along with rooftop PV. Simultaneously there had been a bubble in distribution and transmission asset spending in Australia, as well as renewable generation and closures of fossil fuel plants.

particularly in ICT. The grid side has lagged both because of its long lived assets (which are slow to turnover and incorporate new technologies) but also as, by their nature, grid assets are predominantly electromechanical whereas technology improvement has been greatest in electronics. This change is reflected in the price changes customers have seen for electricity and appliance as shown in Fig. 3.7.

Balancing this lower cost of appliances is the increased number of new appliances. Household electric appliance and equipment numbers grew from an average of 46 in 2000 to 67 in 2010[26] and will be larger now. This trend has also seen a shift to high value but lower energy consumption appliances, such as smartphones, LED lights, digital TVs, and computers. These newer devices typically have lives shorter than traditional whitegoods and much shorter than the 40-year deemed life of regulated assets.

---

26. Energy Efficient Strategies (2008).

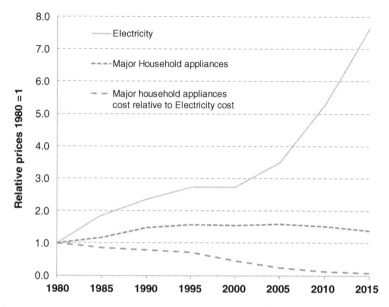

FIGURE 3.7   Relative prices of electricity and household appliances, 1980–2015 (Australian Bureau of Statistics, 2016). Prices are shown at 5-year intervals. Customer anger over electricity price rises has flowed from both historic expectation of real price deflation and "just price" concepts, which date back to St. Thomas Aquinas.

With little change in real prices, Australian retail sales of fridges and freezers have slowed to an average of around $1 billion a year, while washing machines, dishwashers, and clothes dryers combined also average about another $1 billion.[27] Sales of these five major whitegoods over the last 20 years amount to around $32 billion, equivalent to a current optimized depreciated replacement cost of around $18 billion (approximately the value of all transmission assets). And, this has happened while the relative cost of appliances has been steady in real terms and the average fridge annual kWh usage has fallen over 40% due to energy efficiency.

Another perspective of the value chain is possible by taking an industry view, using IBISWorld[28] estimates of revenue in selected industry sectors. This puts the total "tariff cost stack" value, based on annual electrical utility retail revenue, in 2016 as $46 billion, which is around $10 billion less than the revenues for just electrical equipment wholesaling.

For the retailing and service sectors where there is a clearly identifiable link to customers' electricity use behind the meter, revenues top $50 billion per annum, as shown in Table 3.3. This is without being able to identify other

27. ABS Retail sales.
28. IBISWorld revenues are for all chain components so cannot be seen as the value add of a particular activity.

TABLE 3.3 IBIS World Revenue Estimates, 2016–17 ($ Billion)

Selected Wholesaling	$B
Telecommunications and Other Electrical Goods	30
Computer and Computer Peripheral	18
Household Appliance	8
Electrical Wholesale	56
Selected Retail and services	
Electrical Services	19
Domestic Appliance Retailing	14
Air Conditioning and Heating Services	7
Computer and Software Retailing	6
Electrical and Lighting Stores	2
Insulation Services	1
Elevator Installation and Maintenance	1
Domestic Appliance Repair and Maintenance	1
	51
Electricity Grid	
Fossil Fuel Electricity Generation	14
Hydro generation	2
Wind and Other Electricity Generation	2
Electricity Transmission	3
Electricity Distribution	17
Generation transmission and distribution	38
Electricity Retailing	46

Source: https://www.ibisworld.com.au/

substantial commercial, industrial, and transport uses. A comprehensive list of grid connected assets would uncover larger swaths of economic value. For example, it overlooks the billions invested in electric transport in trams and trains (not to mention elevators and lifts), all of which is outside of the traditional "tariff cost stack" accounting.

Taking a different asset stock view, Australian national accounts estimates of net capital stock do not separate electricity equipment assets from other types of assets. Fortunately, the information technology capital stock (a part of the full electricity value chain) is estimated separately. On this measure alone, shown in Table 3.4, nonutility industry ICT assets exceed the value of all utility assets (including water and gas assets along with electricity).

Difficult as meter-side assets are to identify and value, consumer surplus is more problematic. The value of customer reliability (VCR) provides an imperfect proxy of consumer surplus, in terms of the costs associated with not having

TABLE 3.4 Australian System of National Accounts Information Technology Net Capital Stock, Selected Items by Industry

	$ Billions			
	Computers and peripherals	Electrical and electronic equipment	Computer software	Total
Electricity, gas, water and waste services	1.4	15.1	2.6	19.1
Information media and telecommunications	0.7	9.7	4.0	14.5
Financial and insurance services	3.0	1.6	9.2	13.8
Public administration and safety	2.8	1.8	5.4	10.0
Transport, postal and warehousing	0.8	3.3	4.6	8.7
Professional, scientific and technical services	1.6	0.9	5.4	7.9
Rental, hiring and real estate services	2.3	4.0	1.1	7.4
Retail trade	1.3	2.6	2.7	6.7
Wholesale trade	1.1	2.6	2.7	6.4
Manufacturing	1.0	1.9	3.3	6.2
Health care and social assistance	1.0	2.1	2.3	5.4
Construction	1.1	2.8	1.2	5.1
Education and training	1.0	1.0	3.0	5.0
Arts and recreation services	0.4	2.2	1.0	3.6
Mining	0.2	1.6	1.7	3.5
Accommodation and food services	0.3	2.4	0.3	3.0
Other services	0.3	2.1	0.5	2.9
Administrative and support services	0.6	0.3	1.6	2.5
Agriculture, forestry and fishing	0.2	1.5	0.4	2.0
Total	21.0	59.6	52.9	133.4
Total nonutility	19.6	44.4	50.3	114.3

supply when it is expected. VCR estimates in Australia are typically two orders of magnitude greater than the cost of electricity.[29]

Even if an accurate estimate of the VCR was available the full customer surplus from electricity is still problematic. This is because in a real sense there is no standalone value chain for electricity. Electricity is always an input into another value chain, whether of cold beer, hot showers, or aluminum smelting.

---

29. See Productivity Commission (2013) for international comparisons.

Electricity customers, now being labeled "prosumers" and "prosumagers," have always used electricity to produce their version of value or utility in their businesses or homes. The formal objective of the Australian National Energy Market is "the long term interest of customers" but customers are only interested in electricity as an input into their household and business production functions for other outputs and outcomes.

The traditional electricity value chain is only one part of the overall economy and an increasingly smaller part as energy intensity declines. Yet the digital economy, climate change policy, and EVs make the overall economy more tightly coupled and bundled with reliable electricity supply. This can be seen in the stock of software assets in Table 3.4 which are worth more than ICT hardware assets. As electricity's relative resource use falls and energy intensity declines its role in underpinning broader economy wide benefits is increasing. This highlights the importance of maximizing benefits, not just minimizing and allocating costs, as the way to increase value.

As the meter is clearly not the end point of the full electricity value chain the utility industry's heavy focus on smart meters, tariffs, and the meter as the focal point for managing customer value is likely misplaced. The business models that will emerge need to look beyond the tariff and the meter to where the largest share of assets and greatest volume of value sits within households and businesses. And DER, wedged in just behind the meter, needs to find a place and value within the true value chain.

5   THE DER DILEMMA FOR THE TRUE ELECTRICITY VALUE CHAIN

DER costs, benefits, and grid impacts need to be seen in the context of the true value chain, including the consumer surplus rather than just the tariff cost stack view. This reflects Amory Lovin's "hot shower and cold beer" approach, and also Lancaster characteristics theory's that all goods and services can be viewed as the characteristics they deliver for the customer.

Household expenditure on distributed energy has been significant in Australia over the past decade. There is already around 5 GW of household PV systems involving an overall investment well over $10 billion. This has been a win for customers with PV but the former generous feed-in tariffs and non cost reflective kWh pricing has led to challenges elsewhere in the value chain.

Accommodating DER in the traditional business model was not an issue when investment was small, just one of the many manageable distortions and cross-subsidies ever present in the electricity sector. But when DER becomes large the kWh tariffs that create individual value at individual cost shifts too much of the shared network costs onto other consumers. In particular, DER may reduce network revenue far more than it reduces network costs creating an imbalance.

The dilemma for DER on the value side is that much of what it can offer is already included as part of the bundled service provided by grid electricity, and DER still requires some of the services of the grid to realize its value. Parts of the grids value proposition remain unchallenged and are valued even when other components are achieved with DER. In particular, if a grid connection is valued for reliability, emergency use, DER backup, or facilitating trading electricity, grid access to obtain any of these values means customers have access to all the others that come bundled with it. For example, if you have DER but retain a grid connection for reliability, the cost imposed on the transmission and distribution part of the value chain may be little different from that of a consumer without DER.

The prospect of adding DER for demand management benefits that reduce network costs, particularly transmission and distribution network augmentation, has diminished in Australia and many other jurisdictions, falling along with peak demand and energy usage. Even in the world of high growth this demand management benefit was liable to be locational and time dependent so this channel for DER value add is limited.[30]

Put another way, DER's problem is that a substantial part of its value creation is more appropriately considered value duplication. It is in competition with the value proposition of grid electricity that is not only relatively low cost due to economies of scale, diversity, and sunk costs but comes as a bundle where the major cost of transmission and distribution for a grid connection, available for reliability, energy export, and backup is little different than for full grid energy supply. This is acceptable if customers choose based on an equitable balancing of costs and benefits among customers.

The original move to electrification involved a rebalancing. In the industrial revolution factories that were tied to wind but particularly water mills first escaped the constraint of place with onsite steam engines. Then Edison designed a system with centralized power generation and distributed use. Power moved offsite, became centralized, and as homes and business' organized their assets around remote electricity generation people's investment on their side of the meter became separated from the grid. But this was done with trial and error in a social construct for sharing of value, not templates for value chain restructuring and business model paradigm shifts. New technologies tend not to replace older ones, they reposition them. Prior to electricity, shafts and belts delivered first wind and water then steam power from a factory's single power source to worker's stations in an arrangement like that in Fig. 3.8. Initially electrification replaced the single factory power source and only over time did wires replaced shafts and individual motors appeared at workspaces.[31] Rebalancing after innovation and disruption took time and required understanding, accommodating, and adapting rather than predicting and proscribing.

---

30. The NY REV make much of the Brooklyn Queens Demand Management (BQDM) Project, but it predates the REV framework and the latest 2016 Distributed System Implementation Plan (DSIP) filings show limited scope for DM compared to total grid costs.
31. Devine (1983).

FIGURE 3.8 Factory line shaft power drive. *(Source: http://www.singen-hegau-archiv.ch/ singen-industrie.html.)*

Now with DER we face another rebalancing. Generation is moving back onsite and transmission and distribution are changing from a monopoly on electricity supply into a monopoly for electricity exchange. Where electrification allowed expansion and economies of scale to cause tariffs to fall, now DER holds the prospect of deelectrification and with it the possibility of increasing tariffs undermining the viability of the utilities' old business model.

The handsome returns utilities make, on an asset base that now looks outdated, are being questioned. However, while returns were made handsome to attract capital, as Graham and Buffet note, there is also limited upside for utilities compared to other investments. A utility would never reap the rewards of an Apple, an Amazon, or an Uber so it never had the risks placed on it of value loss.

New business models need to recognize that the old model bundled a range of highly valuable services and also managed the risks to consumers of underinvestment by rewarding conservative and reliable infrastructure choices. If the cost of this risk reduction is not covered by tariffs it will invariably appear somewhere else in the value chain.

Rearranging value capture in a world of increased DER will require revisiting the old largely volumetric supply tariffs levied at the meter, and providing more economically efficient prices to increasingly empowered and enabled consumers, as further explored by Steiniger and Johnston, among others, in this volume. This is not the same as assuming, as for example the REV seems to imply, that the monopoly elements no longer have a value or that their value can now be delivered efficiently at market prices by third parties through market mechanisms. Underpinning the monopoly elements of the grid is its bundle of services that still have a very high value as inputs to the needs of customers.

A value so high that it can support different tariff and pricing models for grid elements, accommodate cross-subsidies, and still allow for the large investments customers make beyond the grid.

6 CONCLUSIONS

To make the business case for anything, someone must want something and must be willing to pay more for it than it costs, or else someone needs to find a way to charge less for something than people are already paying for it. You need to add value or reduce cost. This is the conundrum for DER.

Generation can be replaced by DER, as long as reliability and system stability can be maintained. But for effective value delivery from DER the transmission and distribution infrastructure is still required and their costs are effectively fixed over the short and medium term. In particular, adding DER cost behind the meter doesn't necessarily let you take equivalent network cost away in front to the meter. In many cases, a large part of the value that DER creates is a duplication of the bundled value already provided by the grid. In such cases, DER may end up increasing costs and eroding the overall customer surplus that end users might enjoy. Like the move from shafts to wires with electrification, creating real value with DER will be a complex process, likely to involve extended trial and error and a deeper understanding of what customers' value and where both value and costs can be shifted across the new and broader value chain.

As the digital economy increases, as the connected economy and the share economy increase, as EVs and autonomous vehicles increase, dependence on a reliable and resilient grid also increases. Wishing for similar new business models to materialize for electricity and assuming that the value chain for the grid can delink and reconfigure to suit, as a range of the proposed business models appear to assume, will not work.

Solving complex problems in the grid and setting prices to recover costs with DER innovation and disruption requires an understanding of what customers' value and how value is created and bundled. The costs of the grid can only be replaced and the benefits augmented by understanding and addressing them and not by downplaying them and assuming them away. The first complex step to this is seeing a full view of the electricity utility value chain and this means being wary of any approach that is not firmly anchored in understanding the true top of the chain, utility as seen by the consumer.

REFERENCES

Australian Bureau of Statistics, 2011. 6530.0 Household Expenditure Survey, Australia: Detailed Expenditure Items, 2009–10.
Australian Bureau of Statistics, 2016. 6401.0 Consumer Price Index, September Quarter 2016.
Bain & Company, 2015. World Economic Forum in collaboration with Bain & Company. The Future of Electricity-Attracting Investment to Build Tomorrow's Electricity Sector.

Biggar, D., 2011. The fifty most important papers in the economics of regulation. Working Paper No. 3, ACCC/AER Working Paper Series.

Buffet, W., 2015. Berkshire Hathaway Inc. Shareholders letter. Available from: http://www.berkshirehathaway.com/letters/2015ltr.pdf

Cambridge Economic Policy Associates, 2016. Future regulatory options for electricity, Networks Energy Networks Association (ENA) and Commonwealth Scientific and Industrial Research Organisation (CSIRO).

Chaline, E., 2011. History's Worst Predications and the People Who Made Them. Chartwell Books, New York, NY.

Colander, D., Kupers, R., 2015. Complexity and the Art of Public Policy: Solving Society's Problems From the Bottom Up. Princeton Press, Princeton, NJ.

Davies, M., 2015. Tariffs, What are the Issues (Load Factor, Diversity & Density). Available from: https://www.atse.org.au/Documents/Events/intelligent-grids/8-Electricity-Tariffs-Mervyn-Davies.pdf

Deloitte, 2016. The future of the global power sector, preparing for emerging opportunities and threats. Available from: https://www2.deloitte.com/global/en/pages/energy-and-resources/articles/future-of-global-power-sector.html

Devine, Jr., W.D., 1983. From shafts to wires: historical perspective on electrification. J. Econ. His. 43 (2), 347–372.

Energy Efficient Strategies, 2008. Energy use in the Australian residential sector 1986–2020. Report for the Department of the Environment, Water, Heritage and the Arts.

Energy Networks Association, CSIRO, 2016. Electricity network transformation roadmap. Unlocking value: microgrids and stand alone systems roles and incentives for microgrids and stand alone power systems.

Fouquet, R., Pearson, P.J.G., 2006. Seven centuries of energy services: the price and use of light in the United Kingdom (1300–2000). Energy J. 27, 139–177.

Kay, J., 2010. Obliquity: Why Our Goals Are Best Achieved Indirectly. Profile Books, London.

LeVine, S., 2016. The Powerhouse: America, China, and the Great Battery War. Penguin Books, New York.

New York Public Service Commission, 2015. White Paper on Ratemaking and Utility Business Models.

Productivity Commission, 2013. Electricity Network Regulatory Frameworks, Report No. 62, Canberra.

Productivity Commission, 2016. Digital disruption: what do governments need to do? Canberra.

PwC Global Power & Utilities, 2014. The road ahead. Gaining momentum from energy transformation. Available from: https://www.pwc.com/gx/en/utilities/publications/assets/pwc-the-road-ahead.pdf

Rocky Mountain Institute, 2013. E-Lab, New Business Models for the Distribution Edge: Transition From Value Chain to Value Constellation.

Rocky Mountain Institute, 2014. The Economics of Grid Defection: When and Where Distributed Solar Generation Plus Storage Competes With Traditional Utility Service.

Shiman, D.R., 1993. Explaining the collapse of the British electrical supply industry in the 1880s: gas versus electric lighting prices. Econ. Bus. Hist. 23 (1), fall 1993.

Smith, R., MacGill, I., 2014. Revolution, evolution or back to the future? Lessons from the electricity supply industry's formative days. In: Sioshansi, F.P. (Ed.), Distribute Generation and Its Implication for the Utility Industry. Academic Press, Oxford, (Chapter 24).

Yakubovich, V., Granovetter, M., McGuire, P., 2005. Electric charges: the social construction of rate systems. Theory Soc. 34, 579–612.

FURTHER READING

Allis-Chalmers, 1955. The Ghost Town that Came to Life. Available from: http://www.misterkitty.org/extras/stupidcovers/stupidcomics490.html

Australian Bureau of Statistics, 2015. 5204.0 Australian System of National Accounts.

Australian Energy Market Operator (AEMO), 2014, Value of Customer Reliability—Final Report. Available from: http://www.aemo.com.au/-/media/Files/PDF/VCR-final-report--PDF-update-27-Nov-14.pdf

Australian Energy Regulator (AER), 2015. State of the Energy Market 2015. Available from: https://www.aer.gov.au/publications/state-of-the-energy-market-reports/state-of-the-energy-market-2015

Australian PV Institute, 2016. National Survey Report of PV Power Applications in Australia 2015. Available from: http://apvi.org.au/pv-in-australia-2015/

Brox, J., 2010. Brilliant: The Evolution of Artificial Light. Houghton Mifflin Harcourt, Boston, MA.

Consolidated Edison, 2016. Distributed System Implementation Plan (DSIP self-assessment and five-year view of the integration of Distributed Energy Resources into Planning, Operations, and Administration). Consolidated Edison, New York, NY.

Energy Efficient Strategies for the Commonwealth of Australia, 2011. Third Survey of Residential Standby Power Consumption of Australian Homes—2010.

Granovetter, M., McGuire, P., 1998. The making of an industry: electricity in the United States. In: Callon, M. (Ed.), Sociological Review Monograph Series: The Laws of the Markets, vol. 46, issue S1. Wiley-Blackwell, Oxford, pp. 147–173.

Hawken, P., Lovins, A.B., Lovins, H., 2000. Natural Capitalism: Creating the Next Industrial Revolution. Back Bay Books; Routledge, Boston, MA; London.

Jonnes, J., 2003. Empires of Light: Edison, Tesla, Westinghouse, and the Race to Electrify the World. Random House, New York, NY.

Ledovskikh, A., 2016. Electricity Retailing in Australia, IBISWorld Industry Report D2640 and other reports. Available from: http://www.ibisworld.com.au/

Marshall, A., 1895. Principles of Economics. Available from: https://archive.org/stream/principlesofecon01marsrich/principlesofecon01marsrich_djvu.txt

McDonald, F., 1962. Insull: The Rise and Fall of a Billionaire Utility Tycoon. University of Chicago Press, Chicago, IL.

New York Mayor's Office of Long-term Planning and Sustainability, 2013. A Stronger, More Resilient New York.

New York Public Service Commission, 2014a. Reforming The Energy Vision Staff, Report and Proposal.

New York Public Service Commission, 2014b. Symposium Technical Conference—An Energy Agenda for the Future.

New York Public Service Commission, 2014c. Order Establishing Brooklyn/Queens Demand Management Program.

Nordhaus, W., 1998. Do real-output and real-wage measures capture reality? The history of lighting suggests not. Crowles Foundation Paper No. 975, Yale University.

Porter, M.E., 1985. The Competitive Advantage: Creating and Sustaining Superior Performance. Free Press, New York, NY.

Schlesinger, H., 2010. The Battery: How Portable Power Sparked a Technological Revolution. Smithsonian Books, Washington, DC.

Waide, P., Tanishima, S., 2006. Light's Labour's Lost, Policies for Energy-Efficient Lighting, In Support of the G8 Plan of Action. International Energy Agency, Paris.

Wasik, J.F., 2006. The Merchant of Power: Sam Insull, Thomas Edison, and the Creation of the Modern Metropolis. Palgrave Macmillan, New York, NY.

Chapter 4

Beyond Community Solar: Aggregating Local Distributed Resources for Resilience and Sustainability

Kevin B. Jones, Erin C. Bennett, Flora Wenhui Ji and Borna Kazerooni
Institute for Energy and the Environment, Vermont Law School, South Royalton, VT, United States

1 INTRODUCTION

Communities across the globe have become increasingly interested in strategies to reduce their local carbon footprint and increase their energy resilience to severe weather events. Distributed technologies, such as solar photovoltaic (PV), demand response, battery storage, and electric vehicle (EV)–charging infrastructure have provided real alternatives for communities to invest in more clean and resilient energy resources. Aggregating these distributed resources in a manner that produces sufficient economies of scale to promote affordability for the consumer remains one of the primary challenges facing many communities. Both residential and small commercial consumers face technological and cost hurdles if they pursue it alone, compared to larger commercial and industrial customers, and thus aggregating these customers together on a community basis, is an important strategy for bringing these options to the residential and small commercial customer.

This book is about the transformational change entering to the electric utility sector, the technological change that is driving this transformation and the business models that are emerging. If communities in the United States and across the globe want to be more than spectators in this process of change and harness some of these forces for the local communities benefits, then they must engage in this process. To provide some insight into how communities can actively participate in the transformation of the grid, this chapter explores opportunities for local communities to be directly involved in the process.

In Sections 2 and 3, this chapter examines the successes and challenges of community solar as a means to meet community energy goals. Section 4 then

Innovation and Disruption at the Grid's Edge. http://dx.doi.org/10.1016/B978-0-12-811758-3.00004-8

examines community choice aggregation (CCA) as an alternative means to allow communities to go beyond what community solar can achieve to offer a roadmap for a more comprehensive model for local sustainability. In an effort to better inform local communities, the next three sections of the chapter look closely at three diverse examples of CCA in different stages of their development in Marin County, California; Lowell, Massachusetts; and Westchester County, New York. In exploring these three programs, the chapter examines how the programs are developed, their ability to encourage the adoption of renewable energy, and their adaptability toward improving the resilience of their community's energy systems. Through this analysis, the final section explores whether CCA is a viable means for going beyond community solar and providing both low-carbon resources and higher energy resilience for communities as they seek cost-effective ways to both help mitigate and adapt to our climate challenge.

2 THE GROWTH OF COMMUNITY SOLAR

Community solar development is currently one of the most popular trends within the realm of distributed energy resources (DERs). According to NREL, half of the population in the United States cannot participate in solar net metering because these customers live in areas that cannot support solar arrays (NREL). Rental agreements and improper roof orientation serve as just two of the reasons why hosting an array in these areas is challenging.

Community solar has functioned as an opportunity to provide solar access to those unable to host their own arrays. The Department of Energy defines community shared solar as "a solar-electric system that provides power and/or financial benefit to multiple community members" (USDOE). NREL estimates that shared solar could represent "32-49% of the distributed market in 2020," an additional 5.5–11 GW of generating capacity worth $8.2–16.3 billion of investment (NREL). Other attributes of community solar are: (1) it is located offsite for, at least, a portion of the customers; and (2) it requires appropriate state rules authorizing group (or virtual) net metering. Under group net metering, customers sign up for a percentage of the output of an offsite solar facility and receive financial credits on their utility bills for their pro rata share of the output.[1] Like traditional net metering, the community solar facility's production and the community solar customers usage are netted on a monthly basis. While community solar is projected to grow rapidly, it is currently less than 1% of the solar PV capacity in the United States, with 108 projects serving 70,000 customers with 110 MW of capacity (O'Shaughnessy, 2016).

Community solar lowers the solar energy's financial barriers, both, in hard and soft costs. It can provide affordable renewable power and financial benefits to more consumers, including renters and low-income households that

---

1. For customers to truly go solar from a community solar facility, those customers must also retain the title to the renewable energy credits associated with their net metered output.

face participation barriers. Community solar has the potential to significantly increase renewable energy production in the United States, which would assist communities and individuals to meet their energy needs with clean, resilient, locally owned energy production. Each state has different laws, regulations, and incentives; and each utility has policies that promote or discourage community solar. These policies exist in a constant state of flux due to ongoing legislative and regulatory changes (Jones and James, 2017).

3   COMMUNITY SOLAR CHALLENGES

While community solar offers residential and small commercial customers a cost effective opportunity to share in the benefits of solar energy, there are limits to a community's solar ability to fully reduce a community's carbon footprint and provide energy resilience. Scaling up community solar through a net metering regime to allow it to fully serve a local area's energy supply can present challenges, given both the intermittent nature of solar energy and the limitations of net metering as a billing arrangement, rather than a more complete distributed energy system.

For example, while group net metering allows a community solar customer to net their electric bill on a monthly basis, on any given day an individual customer may be overproducing solar (based on its pro rata share of output) and exporting energy to the grid; or, alternatively, consuming significantly more energy than is produced by the solar array and thus consuming energy provided by the grid. As customer adoption of net metered solar grows, so do the utility concerns about cost shifting, given that community solar customers still rely on the grid for some of their purchases, but may avoid paying an appropriate share of the system's fixed costs (Blansfield and Jones, 2014).

In addition, while community solar arrays may be distributed generation that is independent from the grid, when the local grid is impacted by a regional power outage or even a local disturbance to the distribution system, the community solar facility is designed to disconnect from the grid and shut down its energy output. As a result, community solar on its own does not contribute meaningfully to local community resilience. When the grid goes down so does the community solar facility, and the solar customer's energy supply is no more resilient than other grid-supported customers. Similar to Koirala & Hakvoort, conclusions in regards to Europe, in the United States, a comprehensive and integrated approach where communities can take control and capture the integrated benefits of distributed technologies is still lacking.

4   COMMUNITY CHOICE AGGREGATION: TAKING STEPS BEYOND COMMUNITY SOLAR

An increasingly popular option to insure more integrated control for local communities that is being explored across the United States is CCA. CCA is a municipal energy procurement model that is typically authorized by the state government. It allows bulk procurement of electricity and other energy services through the municipality. Under CCA, the municipal government is authorized to negotiate the purchase of energy services on behalf of their residential and business customers. While the municipality contracts with a third party for energy services, those services are still delivered to customers through the local utilities distribution network. CCA while typically allowed in states that offer customers to choose their electric supplier through retail choice programs, could also be offered on a voluntary basis in states that have not fully opened up their retail markets to competition.

Six states have passed laws allowing CCA, including California, Illinois, New Jersey, Ohio, and Rhode Island (USDOE, 2016). In addition, the New York Public Service Commission has allowed CCA, as a pilot under its Reforming the Energy Vision proceeding (Hales, 2016). The underlying policy reason for the CCA legislation seems to vary by state with no clear trends apparent for those states that have authorized CCA compared to those who have not.

Under CCA, communities can competitively procure their electricity from third-party energy service companies (ESCOs) based on lowest cost or other factors, such as the renewable content of the electricity. Cost savings can accrue through the aggregation of electric demand and the reduction of overheads, such as customer acquisition costs, which have historically reduced ESCOs' willingness to serve residential and small commercial customers. CCA is typically implemented as an opt-out program, which means that all customers are included in the program on a default basis, but are allowed to return to the utility default service option if they opt to do so (The Solar Foundation, Community Choice Aggregation). Utilizing an opt-out arrangement has resulted in CCAs being much more successful in enticing customers to enroll in renewable energy choice programs, with the lowest participation for CCA programs being approximately 75%, and the highest voluntary utility green power programs only achieving enrollment rates in the low 20s (USDOE).

In the following sections, we will take a closer look at CCA, as it has been implemented in California, Massachusetts, and New York. These three programs have been chosen to compare models in three states that have been innovators in grid modernization and leaders in clean-energy policies. Our discussion begins with California, which has been an early leader promoting opportunities for CCA.

5 CASE STUDY: MARIN CLEAN ENERGY

After California's brief, but tumultuous, experiment with "deregulating electricity markets," the state legislature in 2002 authorized cities and counties to establish CCA Programs (Stepanicich, 2013). According to the California Public Utilities Commission (CPUC), there are five active CCA programs. Communities can establish CCAs to procure and market energy services. Although the CPUC does not regulate CCAs in California, the statute establishes minimum reporting requirements and requires an initial approval by the commission for the CCA to commence operation.

Marin County in California is located on the San Francisco Bay, and according to the 2010 census, the population was 252,409. Marin Clean Energy (MCE), formerly known as the Marin Energy Authority, was California's first CCA program, which started in 2008 (MCE, 2015). In the beginning, MCE only had eight jurisdictions as part of its joint powers agreement, but by mid-2016, the number of participating jurisdictions grew to 24 (Stepanicich, 2013; MCE, 2015). More recently MCE has expanded to include both Marin and Napa Counties, plus the cities of Richmond, Benicia, El Cerrito, San Pablo, Walnut Creek, and Lafayette.

MCE serves over 170,000 customers. Customers are automatically enrolled unless they provide a written notice opting out. Customers who remain part of the CCA have three energy procurement options that are either provided through bundled or unbundled renewable energy certificates or limited local solar generation. Although customers are enrolled automatically, customers may withdraw from the program at any time.

5.1 Governance and Structure

CCA formation and operation is governed by the California Public Utilities Code (§366.2). Under the law, localities can establish community choice aggregators to "aggregate" customer loads to procure electricity on their customers' behalf. Any locality can establish or join a CCA program, as long as the locality is not in an area currently served by a municipal utility. The law establishes a limited role for the CPUC in reviewing, approving, and overseeing the CCA. The CPUC does not regulate CCA's specific rates or contracts; instead, the CCA's own governing body is responsible for these tasks.

To form a CCA, a locality needs to pass an enabling ordinance. If two or more jurisdictions want to participate jointly, they can by agreement establish a joint powers agency. This agreement must specify that the members will not be liable for the obligations of the agency to limit the liability to members of a joint powers agreement. A "liability shield" in a joint powers agreement protects a local government agency's operating budget from the contractual obligations of the joint agency (Stepanicich, 2013).

Prior to operation, the jurisdiction or public agency responsible for establishing the CCA needs to prepare and submit to the CPUC a statement of intent, an implementation plan, and other information the CPUC requests, and submit it to the CPUC. The implementation plan must include a description of the organizational structure of the CCA. This includes issues ranging from how the CCA operates and how it is funded to how to terminate the program (Cal. Pub. Util. Code §366.2 (c) 3 and 4). The CCA must provide universal access, reliability and "equitable treatment of all classes of customers" (Cal. Pub. Util. Code §366.2 (c) 4).

5.2 The Formation of MCE

MCE was formed initially in 2008 as a joint powers authority. MCE is an independent entity and governed by a governing board consisting of "one elected official appointed by the governing body of each member" (Stepanicich, 2013). MCE has a very broad authority to function as a separate legal entity, including, but not limited to, entering contracts, borrowing money, and hiring personnel to the extent allowable under the law (Stepanicich, 2013). Two major issues that MCE needed to address early in its formation were liability and voting.

Although the law and the joint powers agreement protect individual joint powers agency members from contractual obligations, this protection does not extend to tort liability. As a result, MCE has three layers of protection. First, the joint powers agreement provides that MCE indemnify the member parties. Second, Section 8.3 of the joint powers agreement provides that MCE will need to acquire indemnity insurance to protect the parties to the agreement. Third, any third parties contracting with MCE will need to agree that they will "have no legal rights or remedies against the individual members" (Stepanicich, 2013).

MCE also needed to have a voting structure that would address the needs of all the members. When MCE first started, there was a tension between larger members and smaller members. For decision-making purposes, larger members preferred a voting scheme according to load, while smaller members worried about the dominance of larger members. As a result, MCE created a two-tiered voting system. Under this system, major decisions require a majority vote of the electrical load, as well as the majority vote of the members. Some matters, such as termination or amending the agreement, require a two-third vote (Stepanicich, 2013).

5.3 Energy Services

MCE offers three energy procurement options for its retail customers:

1. standard: minimum 50% renewable energy,
2. opt-in 100% renewable energy option, known as the Deep Green program, or
3. opt-in 100% local solar option (currently limited to 600 customers).

Under the first two options, renewable energy is procured through bundled and unbundled Green-e certified RECs through contracts between MCE and energy providers. Under the third option, MCE purchases electricity directly from locally developed solar facilities (MCE, 2015). Although MCE has a limited local production capacity, MCE is actively seeking to expand their portfolio of local solar generation. In addition to contracts with energy marketers, other bundled contracts, and two community solar projects, MCE also acquires energy from a feed-in tariff and a net metering programs. Beginning in 2016, MCE is working to reduce its unbundled (REC only) purchases to no more than

3% of its retail load, and overall MCE's energy mix at over 51% renewable includes one of the highest percentages of renewable energy of any California utility (MCE, 2015).

MCE's feed-in tariff program allows customers to sell their power output for a period of 20 years at a fixed rate. The program is limited to smaller facilities under 1 MW. The feed-in tariff is connected to a power purchase agreement that provides approximately $0.14/kWh for a period of 20 years for a solar project (slightly less for wind or biomass), with a declining rate as total FIT capacity reaches certain thresholds. Any environmental attributes of the energy, including RECs are transferred to MCE. As of October 2015, MCE's annual IRP update reported that the program includes approximately 5.7 MW of renewable energy aggregate capacity from existing and proposed feed-in tariff projects (MCE, 2015).

MCE also offers a net metering program available to all customers for systems smaller than 1 MW. There are a number of attractive features to the MCE's net metering program. First, MCE offers a $0.01/kWh premium adder to MCE's generation rate for every kWh produced in excess of the customer's consumption. Third, if the credits accrued exceed $100, the customer will have the opportunity to "cash out" annually every April. As of May 2015, MCE has over 5300 customers who participate in the net metering program, who provide over 35 MW of the renewable energy capacity. MCE hopes to increase such generating capacity to 47 MW by 2021 (MCE, 2015). MCE's website contains technical resources for customers interested in installing solar on their property, including information about "free" solar assessments and financing arrangements.

5.4 Grid Resilience and Storage

As of today, there are a few opportunities for major energy storage or grid resilience benefits in MCE. To move toward resilience and grid independence, MCE will need to work with the incumbent utility Pacific Gas and Electric Company (PG&E) to update infrastructure and install energy storage.

The CPUC in 2013 set energy storage targets for CCAs in compliance with California's Energy Storage law (AB2514) enacted in September 2010. This target requires CCAs to meet an "energy storage procurement target" equal to 1% of their 2020 forecasted peak load. MCE's October 2015 IRP estimates that, given the current forecasts, MCE will need to complete installation of 3 MW of storage options by 2024. Starting on January 1, 2016 and every 2 years thereafter, MCE must report on MCE's progress and plans for complying with the requirement. In February 2016, MCE outlined plans to increase "behind the meter" storage options.

MCE has already made progress with this goal through a partial funding of a Tesla battery project on the College of Marin's campuses (MCE, 2016). According to Renewable Energy World, "the College of Marin demonstration project consists of a 2.4-MW system (five lithium-ion battery units delivering

480 kW of power each) on the college's Kentfield campus and a 1.44 MW system (three 480 kW units) on its Indian Valley campus" (Bloom, 2016). The Kentfield system stores power generated by a carport- and rooftop-mounted PV system that was originally installed in 2008 (Bloom, 2016). The Tesla project is expected to reduce the College of Marin's utility demand charges by about $150,000/year by utilizing the battery storage to plateau the College's demand peaks. The reported $5.3 million in costs was fully paid for by PG&E, MCE, and Tesla financial incentives (Bloom, 2016).

6 CASE STUDY: LOWELL, MASSACHUSETTS COMMUNITY CHOICE POWER PLAN

The City of Lowell (Lowell) is located in the Commonwealth of Massachusetts. Lowell is located on the Merrimack River, about 25 miles northwest of Boston. Once serving as the forefront of the American Industrial Revolution, Lowell continues to thrive, indulging in a plethora of artistic exhibitions and performances, taking place at its myriad of venues. Additionally, Lowell is home to multiple large-scale festivals, including the Lowell Film Festival and the Lowell Folk Festival, a festival that draws 250,000 spectators every summer. As of 2013, the population for Lowell, Massachusetts was 108,861.

On December 6, 2012, Lowell filed a petition with the Massachusetts Department of Public Utilities (DPU) for an approval of a municipal power plan pursuant to a Massachusetts Municipal Aggregation Statute (G.L. c. 164, §134). Following multiple revisions, the DPU accepted the plan, approving Lowell's CCA Program (The Commonwealth of Massachusetts Department of Public Utilities, 2014). The detailed Lowell Aggregation plan explains everything from the plan's basic purpose to exemplifying the rates for a customer with a monthly energy usage of 500 kWh (City of Lowell Community Choice Power Supply Program: Aggregation Plan, Section 6.3).

The plan's purpose is to represent consumer interests in competitive markets for electricity and to aggregate consumers in the city to negotiate rates for power supply. It brings together the buying power of almost 40,000 consumers allowing the city to take control of their energy prices. Participation is voluntary and eligible consumers have the opportunity to opt out of the plan's service and to choose their own competitive supplier. Based on enrolment figures from the previous community aggregations, it is anticipated that 97% of the eligible consumers will participate (City of Lowell Community Choice Power Supply Program: Aggregation Plan).

6.1 Governance and Structure

The Community Choice Program (Program) is authorized by Massachusetts' statute (G.L. c. 164, §134), which allows municipalities to "aggregate the electrical load of interested electricity consumers within its boundaries...." Under subsection (a), a town may launch a Program in Massachusetts by a majority vote in town council or with the approval of the mayor. Upon an affirmative vote or mayoral approval, "a municipality or group of municipalities establishing load aggregation pursuant to this section shall, in consultation with the department of energy resources... develop a plan, for review by its citizens, detailing the process and consequences of aggregation. Any municipal load aggregation plan established pursuant to this section shall provide for universal access, reliability, and equitable treatment of all classes of customers and shall meet any requirements established by law or the department concerning aggregated service." (G.L. c. 164, §134).

The Lowell Program's organizational structure operates under five levels: (1) consumers, (2) city council, (3) city manager, (4) consultant, and (5) competitive suppliers. Lowell regards each level of its organizational structure as crucial and necessary, assigning explicit duties to be upheld by each structural participant. The unique CCA participants to the structure are the city's consultant who serves as the City's agent while conducting and carrying out the Program's day-to-day business ventures. Under contract agreement, the Consultants serves as the City's procurement agent, "utilizing its existing staff to solicit services as requested by [Lowell]." In addition, the competitive suppliers contract through the city manager and are monitored for compliance by the Consultant. A competitive supplier upholds certain responsibilities agreed to contractually by itself and the City, which can be found within the Electric Service Agreement between the competitive supplier and Lowell (City of Lowell Community Choice Power Supply Program: Aggregation Plan, Section 2.2).

6.2 Goals

The goals of the Lowell Program are: (1) to provide the basis for aggregation of eligible consumers on a nondiscriminatory basis; (2) acquire a market rate for power supply and transparent pricing; (3) provide equal sharing of economic savings based on current electric rates; (4) allow those eligible consumers who choose not to participate to opt-out; (5) provide full public accountability to participating consumers; and (6) utilize municipal and other powers and authorities that constitute basic consumer protection to achieve these goals (City of Lowell Community Choice Power Supply Program: Aggregation Plan, Section 2.3).

The Massachusetts Department of Energy Resources' (DOER) Guide to Municipal Electric Aggregation defines the term universal access as "electric services sufficient for basic needs available to virtually all members of the population regardless of income." The DOER Guide dictates that a city's municipality aggregation plan, such as Lowell's, constitutes universal access "by giving all consumers within its boundaries the opportunity to participate, whether they are currently on Basic Service or the supply service of a Competitive Supplier." Lowell's largest affordability goal is to "[p]rovide the basis for aggregation of eligible consumers on a non-discriminatory basis." (City of Lowell Community

Choice Power Supply Program: Aggregation Plan, Section 2.3). This includes low-income consumers.

6.3 Energy Procurement

As a first step, Lowell chose Colonial Power Group, Inc., a Massachusetts energy consulting company, in a competitive process to design, implement, and administer the Community Choice Power Supply program on behalf of the city. In working with its consultant, Lowell initially signed a contract with Dominion Retail, which provided 100 percent carbon neutral energy (City of Lowell Becomes First Massachusetts Municipality to Achieve Carbon Neutral Electrical Consumption for the Entire Community; USDOE, 2016). The term carbon neutral specifies that "greenhouse gasses were not generated in producing the energy." Lowell's agreement with Dominion Retail differed from other Massachusetts municipalities, as well as other states because, while municipalities have the option to purchase energy in bulk to save residents' money, Lowell is one of the few communities nationally to fully purchase carbon-free energy. In Lowell, the energy received by customers is derived from multiple renewable resources within the state of Massachusetts, such as solar, wind, and hydropower. Moreover, "[the energy also] includes conventional fossil-based power from outside Massachusetts for which power producers bought renewable energy credits—certificates to help promote renewable energy—to make up for the carbon generated." (internal quotations omitted) (Sato, 2016).

Following its initial agreement with Dominion Retail, Lowell switched energy suppliers and now contracts with Hampshire Power for the Community Choice Power Supply Program. Hampshire Power is a Massachusetts-based nonprofit energy supplier that is part of the Hampshire Council of Government. With the use of carbon-neutral energy, Hampshire Power and Lowell hope to reduce the community's carbon footprint, thus reducing their impacts on global climate change (Hampshire Energy Renewable Energy Co-Operative; Lowell Community Choice Power Supply Program Rates).

7   CASE STUDY: WESTCHESTER, NEW YORK

Westchester County is just north of New York City and occupies 450 mile^2 in the Hudson Valley. Nearly 1 million people reside in the still mostly rural area (About Westchester). In May 2016, the county launched New York State's first CCA program. Twenty municipalities in the county signed up for the program and named it the "Westchester Power." The program is intended to lower the electricity cost through community-based bulk energy purchasing and also increase use of renewable energy in the county. Approximately 110,000 residents are in the program during its initial launch in May (Westchester Power).

In February 2014, a group of local officials and clean-energy advocates in Westchester tried to push a bill in New York State Senate to authorize municipalities participating in a CCA program, to "coordinate efforts to procure electric and/or gas supply services on behalf of its residents." The bill would have established New York's pilot CCA program in Westchester County, which is served by New York State Electric & Gas Corporation (NYSEG) and Consolidated Edison of New York (ConEdison). Although Governor Andrew Cuomo supported the idea of CCA, he vetoed the bill because of the restrictions on the Public Service Commission's (PSC) ability to intervene and the lack of protections for third-party access to consumer data. Following the veto, Governor Cuomo directed the PSC to implement CCA programs under its existing authority (Giamusso, 2015).

7.1 Governance and Structure

Unlike other states where CCAs were established by enabling legislation, in December 2014, Sustainable Westchester, a nonprofit advocate group of energy advocates, residents, and local governments petitioned to the PSC to allow local communities to purchase their own power, instead of continuing with the traditional utility model. Following 3 months, the Commission approved the Westchester CCA program, and the Commission Chair described this project as "innovative," furthering Governor Cuomo's Reforming Energy Vision (REV) strategy.[2]

7.2 Structure of the Program

Participating towns and cities have adopted local ordinances to initiate the program, which was formally launched in May 2016 (Department of Public Service, 2015). Westchester Power is a partnership program of Sustainable Westchester and the 20 participating municipalities. Sustainable Westchester is the manager and administrator on behalf of the communities. Participating communities are in two utility territories: ConEdison and NYSEG. Like other CCA programs, Westchester Power purchases electricity in bulk from third-party ESCOs.

7.3 Energy Procurement

For Westchester, the contracts are fixed priced from a single supplier for each jurisdiction in the program. In March 2016, Sustainable Westchester entered into a $150 million 2-year agreement with ConEdison Solution, a Consolidated Edison subsidiary to serve the 17 municipalities in the ConEdison territory. This contract will cover roughly 90,000 residents and small business owners.

---

2. Reforming Energy Vision (REV), Governor Cuomo's energy strategy for the state, to build a "clear, resilient, and more affordable energy system for all New Yorkers." The 2030 goals are to achieve a 40% reduction in greenhouse gas emission from 1990 levels; 50% renewable energy, and 23% energy reduction from 2012 levels (Reforming the Energy Vision, http://rev.ny.gov/).

In NYSEG's territory, Sustainable Westchester chose Constellation Energy to provide the power. Similar to the programs in MCE and Lowell, the program operates under an opt-out system (Westchester Power).

Proposed benefits for Westchester CCA include price stability for a fixed contract term and flexibility to accommodate different preferences, such as the proportion of clean energy provided. Mostly importantly, the fixed price will guard against fluctuating conventional fuel prices. Also, as the cost of renewables has been declining over the years, an increasing mix of renewables will become more affordable over time (Westchester Power).

Sustainable Westchester estimates that the program will collectively save $4–5 million/year throughout the life of the contract (De Avila, 2016). Statistics from the first few months have already shown the program's competitive price for both conventional and renewable sources. NYSEG territory started at a fixed standard rate of 6.95 cents/kWh, higher than the utility supply, but has caught up and the average through October is 0.16 cents/kWh better. The green option fixed rate 7.085 cents has just now moved to break even. Rates for the ConEdison territory started at 7.381 cents/kWh and have saved 0.66 cent/kWh on average. The green energy rate of 7.681 cent/kWh is one-third of a cent cheaper than traditional utility power (Westchester Power, Rates).

In regards to renewable energy options, 14 out of 20 participating municipalities have chosen to go for a 100% green energy supply. Individual consumers in the rest of the communities can still choose to receive 100% renewable energy supply by filling out an online form on Westchester Power's website.

Under the program, only solar, wind, and hydro qualify as green power. In addition, the procured renewable energy is backed by Green-e certificates. Green-e Energy is a voluntary certification program with the Center for Resource Solutions in California; a neutral third party to ensure the renewable energy credits are bundled with the sold energy, and there is no double counting (Center for Resource Solutions). Current rooftop solar panels users in the participating municipalities will be automatically enrolled and continue to receive net metering credits at the retail rate from ConEdison or NYSEG. (Westchester Power, FAQs). Moreover, Westchester Power has also partnered with Bedford 2020, a local nonprofit organization to assist in local education and outreach to achieve "100% opt up to renewable energy" for the program (Westchester Power, About). The program is trying to bring Westchester County to 100% renewable electricity by 2030 (Hales, 2016).

7.4 Services Beyond Energy Procurement: Community Solar, Demand Response, and Microgrids

The Westchester CCA contract includes some unique provisions that allow Sustainable Westchester to evolve beyond more traditional models. "We see community aggregation as a vehicle to get the benefits of using distributed resources, whether energy efficiency resources or supply resources, and thinking about the community as a whole," said Audrey Zibelman, chair of the New York Public Service Commission (Tweed, 2016). Unlike the much more mature CCA version in Marin County, Westchester's CCA hasn't established a feed-in tariff or other supply options to directly procure renewables for the long term. Due to the flexibility of the Westchester contract, this could change in the near future, as under the contracts, if Westchester decides to build community solar, it has reserved the rights to displace its power supply (Tweed, 2016).

A similar provision of their power supply agreement provides leeway to Sustainable Westchester to earn money on demand response. The bulk purchase is based on covering the power needs of the customers throughout the year, including a few peak days. Sustainable Westchester calculated that more than $30 million is spent on consumption during one peak hour per year. If Sustainable Westchester can reduce capacity during peak hours, the supplier will be allowed to sell that excess capacity into the wholesale market. The money from that demand–response transaction would then go back to Sustainable Westchester. The CCA program has a vested interest not only in offering efficiency tips, but also in selling hardware, such as smart thermostats (Tweed, 2016).

To reduce peak loads, researchers for Sustainable Westchester have thought about "collective demand response," to encourage lower usage electricity during peak demand hours and rebate the customers for any savings. For example, during hot summer days when power demand is high, the grid experiences greater demand and increased costs. According to a researcher with Sustainable Westchester, "If we can predict when we'll have peak demand, we can incentivize people to shut down their homes or businesses and go to the movie, an ice cream shop, or go to happy hour. And they might get a coupon to do so." To provide customer such incentives, researchers have even thought about joining forces with local businesses to provide coupons (Schiller, 2016).

The next step might be to move further toward establishing a local microgrid. A microgrid is "a group of interconnected loads and DERs within clearly defined electrical boundaries that acts as a single controllable entity with respect to the grid. A microgrid can connect and disconnect from the grid to enable it to operate in both grid-connected or island mode" (Smith and Ton, 2013). There are a number of reasons that customers would want to deploy a microgrid, including "improving system resilience and reliability, reducing operational costs, and improving the environmental characteristics of their energy supply." (Smith and Ton, 2013). Microgrids can operate independently from the larger system because they are composed of an energy supply source and electric infrastructure to distribute energy from its generation resources.

Part of the Sustainable Westchester vision is for "a web of microgrids" throughout the county. Following Hurricane Sandy, strong interest in microgrids in New York, led the New York State Energy Research and Development Authority (NYSERDA) to create the NY Prize competition to encourage community microgrids. NY Prize is a $40 million dollar multistage competition supporting the design and construction of community microgrids that improve local electrical distribution system performance and resiliency in normal operating configurations and during times of electrical grid outages.

Sustainable Westchester, assisted 10 of its member municipalities in applying, and all of them were successful in winning awards of about $100,000 each by the NY Prize to support feasibility studies. As one illustrative example, the Village of Croton-on-Hudson study "will follow the model of a nested microgrid in which there will be two main geographical areas in the system, each fed by their own portfolio of distributed generation (combined heat and power, solar, and energy storage) and each capable of staying powered in island mode during a grid outage. The proposed microgrid would provide power to a municipal building, library, three fire stations, three schools and district office, medical clinic, grocery/pharmacy, and gas station" (Mikulak, 2015).

Sustainable Westchester has big visions for how the county will combine microgrids, community aggregation, demand management, and community solar generation to create a dynamic local energy market. Looking to Westchester from a statewide point of view, CCA can become a driving vehicle to implement the values of the state's REV, which is aimed at integrating "distributed energy resources (DERs) into the planning and operation of the system" (Jones et al., 2016). Audrey Zibelman, who contributed this volume's introduction, has been the leader of the REV process in New York as Chair of the New York Public Service Commission.

Westchester's CCA program is too new to critic whether it is a successful program. Though there has not been any flag raised yet; however, moving forward, it is important for Westchester to be mindful about the issues that other CCA programs have already experienced. Hopefully the anticipated success of Westchester will attract more communities into joining the CCA program or start their own.

8   COMPARISON OF COMMUNITY CHOICE AGGREGATION CASES

While each of the CCA efforts in Marin County, the City of Lowell, and Westchester County employ a version of a green choice program through unbundled and bundled RECs, these case studies, particularly Marin and Westchester, demonstrate the opportunities to expand the community's energy resource potential of CCAs.

Although MCE has historically partly relied on unbundled RECs, it is taking steps through its net metering program and feed-in tariff program to increase local energy generation. Additionally, MCE is developing local community solar facilities. MCE customers' high renewable energy purchase rate and the program's commitment to move MCE to 100% renewable energy from renewables with bundled RECs are inspiring. Microgrid and grid resilience are not yet priorities for MCE, but preliminary steps to develop energy storage and increase resilience are underway. The next logical step would be for MCE to consider working with PG&E to develop microgrid pilot projects in the MCE area.

Sustainable Westchester, while still at an early stage of development, seems to be anticipating a more integrated approach that, both, explores strategies that help reduce the community's carbon footprint and increase local energy resilience.

Glenn Weinberg, Joule Assets manager of special projects, calls the Westchester model the "triple threat—a real new community energy paradigm." Community choice energy and demand response aggregations, microgrids, and community solar are being used separately and in various communities. However, bringing them together is a more holistic approach and is greater than the sum of the separate parts, offering green energy and increased reliability at a better economy of scale for the community. According to Weinberg, "The value of each is dynamically enhanced by the other" (Wood, 2015).

Local solar, without the more comprehensive control structure that comes with a microgrid, does not enhance reliability because, when the grid goes down, so does the local community solar. The presence of the microgrid allows the community to reduce its carbon footprint and increase its resilience. "Microgrids make ideal sites for shared renewables. A microgrid gives the host the flexibility to sell excess power during normal grid conditions, and distribute power to loads within the microgrid, when in islanding mode," said Sustainable Westchester in a filing before the Public Service Commission (Wood, 2015). From these case studies, CCA can solely be focused on community control of the electric supplier, or it can launch a community down a low-carbon path that may lead to increasing resilience.

9   CONCLUSIONS

Given the challenges of a changing climate, the opportunities from CCA are promising news for both the planet and the local community. Technological and market forces are transforming the centralized vertically integrated grid. Communities interested in achieving local sustainability cannot simply be observers of these changes. CCA is an opportunity for communities to engage with policymakers, the local utility, and third-party suppliers of DERs to take control of this transformation to achieve local energy goals, including sustainability, affordability, and resilience.

REFERENCES

Blansfield, J., Jones, K., 2014. Industry response to revenue erosion from solar PVs. In: Fereidoon, P., Sioshansi (Eds.), Distributed Generation and Its Implications for the Utility Industry. Academic Press, Oxford.

Bloom, M., 2016. Pilot energy storage project installed at College of Marin in California. Renewable Energy World. Available from: http://www.renewableenergyworld.com/articles/2016/07/pilot-energy-storage-project-installed-at-college-of-marin-in-california.html

Colonial Power Group, 2012. Lowell Community Choice Power Supply Program rates. Available from: http://www.colonialpowergroup.com/lowell/

Colonial Power Group, 2014. City of Lowell becomes first Massachusetts municipality to achieve carbon neutral electrical consumption for the entire community. Available from: http://colonial-powergroup.squarespace.com/lowell-carbon-neutral/?SSScrollPosition=342

Colonial Power Group. City of Lowell Community Choice Power Supply Program: aggregation plan. Available from: http://static1.1.sqspcdn.com/static/f/781687/26030517/1426006314233/Lowell+-+Aggregation+Plan+Approved+by+DPU+12-124.docx?token=9kMoQoNXzbDzRS JvjltTpbu3vCI%3D

Center for Resource Solutions. Programs: Green-e. Available from: http://resource-solutions.org/programs/green-e/

De Avila, J., 2016. For Westchester Towns, a new energy contract. Available from: http://www.wsj.com/articles/for-westchester-towns-a-new-type-of-energy-contract-1457487622

Department of Public Service, 2015. PSC OK's state's first Community Choice Aggregation Pilot Program. Available from: http://www3.dps.ny.gov/pscweb/WebFileRoom.nsf/Web/C9DCDFF 7232D6C4185257DF80063C456/$File/pr15020.pdf

Giamusso, D., 2015. Westchester group pushes local selection of energy sources. Available from: http://www.politico.com/states/new-york/albany/story/2015/01/westchester-group-pushes-local-selection-of-energy-sources-018553

Hales, R., 2016. Sustainable Westchester leads New York into CCA. Cleantechnica. Available from: https://cleantechnica.com/2016/03/18/sustainable-westchester-leads-new-york-cca/

Jones, K., James, M., 2017. Distributed renewables in the new economy: lessons from community solar development in Vermont. In: Scanlan, M. (Ed.), Law and Policy for a New Economy: Sustainable, Just, and Democratic. Edward Elgar, Northhampton.

Jones, K., Curtis, T., Thege, M., Sauer, D., Roche, M., 2016. Distributed utility: conflicts and opportunities between incumbent utilities, suppliers, and emerging new entrants. In: Sioshansi, F.P. (Ed.), Future of Utilities: Utilities of the Future: How Technological Innovations in Distributed Energy Resources will Reshape the Electric Power Sector. Academic Press, Amsterdam.

Marin Clean Energy, 2015. Annual Update: Integrated Resource Plan. Available from: https://www.mcecleanenergy.org/wp-content/uploads/2016/01/Marin-Clean-Energy-2015-Integrated-Re-source-Plan_FINAL-BOARD-APPROVED.pdf

Marin Clean Energy, 2016. Advice Letter MCE E-14: Re: Refiling Marin Clean Energy's Biannual Energy Storage Procurement Compliance Report. Available from: https://www.mcecleanen-ergy.org/wp-content/uploads/2016/01/Marin-Clean-Energy-2015-Integrated-Resource-Plan_FINAL-BOARD-APPROVED.pdf

Massachusetts General Law, 2009. c. 164 §134.

Mikulak, R., 2015. Mamaroneck wins NY Prize Community Grid Competition Awards. Mamaroneck Daily Voice. Available form: http://mamaroneck.dailyvoice.com/news/mamaroneck-wins-ny-prize-community-grid-competition-awards/474536/

O'Shaughnessy, E., 2016. Community Solar, status, trends, legal and financial issues. NREL. Available from: https://www.epa.gov/sites/production/files/2016-03/documents/webinar_20160309_oshaughnessy.pdf

Sato, H., 2016. Lowell Sun: contract will cut Lowell electricity rates. Available from: http://www.lowellsun.com/todaysheadlines/ci_25121480/contract-will-cut-lowell-electricity-rates

Schiller, B., 2016. How collective energy buying is helping communities to go 100% renewables. Available from: https://www.fastcoexist.com/3058324/how-collective-energy-buying-is-helping-communities-go-100-renewable

Smith, M., Ton, D., 2013. Key Connections. IEEE Power & Energy Magazine, July/Aug.

Stepanicich, G., 2013. Marin Energy Authority: A Community Choice Aggregation Program for electricity service. Trends 44 (5).

The Commonwealth of Massachusetts Department of Public Utilities, 2014. Petition for City of Lowell for Approval of an Amendment to its Municipal Aggregation Plan Pursuant to M.G.L. c. 164 §134. Available from: http://web1.env.state.ma.us/dpu/fileroomapi/api/attachments/get/?path=14-100%2Flowell_initial_filing_81814.pdf

Tweed, K., 2016. New York Towns aim for solar PPA and smart thermostats under Community Choice Aggregation. Greentech Media. Available from: https://www.greentechmedia.com/articles/read/New-York-Town-Aims-For-Solar-PPAs-and-Smart-Thermostats-Under-Community-Cho

United States Department of Energy, 2016. Green power markets: community choice aggregation. Available from: http://apps3.eere.energy.gov/greenpower/markets/community_choice.shtml

Westchester Power. About Westchester. Available from: http://www3.westchestergov.com/about-westchester

Westchester Power. Frequently asked questions. Available from: http://www.westchesterpower.org/faqs/

Wood, E., 2015. The triple threat: community choice, microgrids and community solar. Microgrid Knowledge. Available from: https://microgridknowledge.com/triple-threat-community-choice-microgrids-and-community-solar/

FURTHER READING

Environment, Economics, and Society, 2016. A Look at three Community Choice Aggregations Systems in the US. Available from: http://eesi.us/blog/2016/6/6/a-look-at-three-community-choice-aggregation-systems-in-the-us

Feldman, D., Brockway, A.M., Ulrich, E., Margolis, R., 2015. Shared solar: current landscape, market potential, and the impact of Federal Securities Regulation. Available from: http://www.nrel.gov/docs/fy15osti/63892.pdf

Hernandez, M., 2013. Solar Power to the people: the rise of rooftop solar among the middle class. Center for American Progress. Available from: https://www.americanprogress.org/issues/green/report/2013/10/21/76013/solar-power-to-the-people-the-rise-of-rooftop-solar-among-the-middle-class/

Marin Clean Energy, 2016. Marin Energy Authority—Joint Powers Agreement. Effective December 19, 2008; as further amended by Amendment No. 10 dated April 21, 2016. Available from: https://www.mcecleanenergy.org/wp-content/uploads/2015/11/16_MCE_Advice_Letter_14-E.pdf

The City of Lowell, 2016. Public notice: The City of Lowell's Community Choice Power Supply Program. Available from: http://www.lowellma.gov/dpd/housing/Pages/Lowell-Community-Choice-Power-Supply-Program.aspx

United States Department of Energy, 2012. A guide to community shared solar: utility, private and nonprofit project development 3. Available from: http://www.nrel.gov/docs/fy12osti/54570.pdf

Chapter 5

Grid Versus Distributed Solar: What Does Australia's Experience Say About the Competitiveness of Distributed Energy?

Bruce Mountain* and Russell Harris**

*Carbon and Energy Markets (CME), Melbourne, VIC, Australia; **Wollemi Consulting, Melbourne, VIC, Australia

1 INTRODUCTION

There is an active debate in Australia and in many other countries on the economics of distributed solar photovoltaics (PV) relative to grid-supplied electricity. The term "grid parity" in this context refers to the point at which distributed electricity is as expensive—to the user—as grid-supplied electricity.

The calculations are complex. They need to take account of capital and/or production subsidies, the amount of grid-supplied electricity displaced by PV, and the price of grid-supplied electricity (which often has at least two components to the charge and in many retail offers more than two).

The debate on the economics of PV relative to grid-only is an important discussion. It encompasses the wisdom of customers' investment in distributed resources; and the knock-on effect of their choices that is, the volume of electricity sold on the grid, future expenditure on network investment, and the residual demand for centrally dispatched electricity production.

This chapter explores the economics of electricity supplied by the grid, relative to the combination of grid and rooftop PV for households in Melbourne, Australia, in November 2016.

In comparison to other countries, Australia has experienced rapid expansion in household solar PV installations, with relatively little commercial or utility scale activity. Total solar PV capacity installed in Australia has recently

Innovation and Disruption at the Grid's Edge. http://dx.doi.org/10.1016/B978-0-12-811758-3.00005-X
83

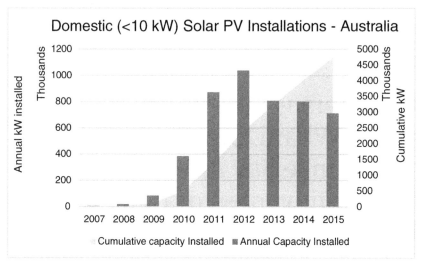

FIGURE 5.1   Rooftop photovoltaics (PV) installations in Australia.

exceeded 5000 kW as shown in Fig. 5.1. More than 80% of this capacity, equivalent to around 9% of Australia's total grid-connected generating capacity, is on domestic rooftops.[1] Rapid rooftop PV expansion was driven by the combination of declining PV costs, policy to promote renewables, and rapid increases in electricity prices (Mountain and Szuster, 2014).

This chapter examines the economics of installing rooftop PV in Melbourne, the capital of the state of Victoria. Victoria is one of five regions in the National Electricity Market covering the south and eastern states of Australia.[2] The Victorian residential electricity market has been opened to competition since the late 1990s and all retail price controls were withdrawn in 2009. There are currently 22 retailers that in October 2016 offered 2017 different retail offers to a little over 2.4 million households in the 5 distribution areas in Victoria. Of these 2.4 million homes, some 298,000 (12.5%) have installed 957 MW of solar PV generation capacity since the start of 2010.[3]

In Australia, as elsewhere, the contest between PV and grid-supplied electricity is actively debated by consumers, regulators, and the industry. The Victorian Government recently announced that it intends to increase the minimum retailer feed-in rates for electricity that is exported from rooftop PV to the grid. And, following advice from the Victorian Essential Services Commission, temporally and spatially differentiated regulated minimum feed-in rates will be introduced (Essential Services Commission, 2017).

1. Annual Report 2015, Clean Energy Council Australia.
2. The other four states are New South Wales, South Australia, Queensland, and Tasmania.
3. Australian Government data on Small-Scale Technology Certificate creation, Clean Energy Regulator (http://www.cleanenergyregulator.gov.au/).

In this book, Biggar and Dimasi point to the significance of tariff structures for efficient investment in distributed generation and express concern that the growth of distributed generation in Australia reflects distortions in tariffs and feed-in rates.

Analysts and scholars debate issues of tariff design and investment incentives but often in the absence of good quality information on the existing situation. This chapter seeks to address that and, using the retail market data, assess the extent to which rooftop PV investment incentives will be affected by changes in the feed-in rates for solar exports, or changes in tariff structures by increasing or reducing the fixed component of retail offers.

The chapter starts with a description of the retail and rooftop solar markets and then explains the methodology used. This is followed by the results of the analysis, a discussion of those results, and finally, conclusions.

2   VICTORIA'S ELECTRICITY MARKET

Victoria's 22 electricity retailers have an obligation to publish electricity price fact sheets setting out the terms of all generally available retail offers. The MarkIntell database (www.markintell.com.au) records and analyzes the data in all these fact sheets. It shows that in November 2016 the 22 electricity retailers together made 2017 different retails offers to households in 5 distribution zones that cover Melbourne. A count of the number of offers, by retailer, in each of the five distribution areas is shown in Table 5.1.

Retail offers are classified as either "Standing" or "Market." Standing Offers are regulated offers that retailers are required to make and they have prices that are almost always higher than Market Offers. Around 12% of all customers remain on Standing offers. Market Offers often include conditional discounts (such as for paying bills on time or using direct-debit to pay bills) that can result in bill reductions by as much as 20% relative to Standing offers.

A remarkable range of electricity tariffs are offered by retailers in Victoria including:

- two-part (cents per day plus single variable charge);
- two-part seasonal (summer and nonsummer differentiation of variable charges);
- time-of-use (peak and off-peak variable charges plus a daily charge) in which the peak period may be weekdays only or the full weak;
- flexible time-of-use (peak, shoulder, and off-peak variable charges plus a daily charge);
- block tariffs (block sizes vary and blocks may be measured in days, months, or quarters); and combinations of block and time-of-use rates or block and flexible tariffs; and
- combinations of time of use and demand charges (calculated as cents per kW or kVA per day where demand is measured during defined summer and nonsummer periods).

TABLE 5.1 Count of Retail Offers, by Distribution Zone, in Victoria

		Distributor				
		Ausnet	Citipower	Jemena	Powercor	United Energy
Retailer	AGL	16	17	17	24	18
	Alinta	6	6	6	8	6
	BlueNRG	9	9	9	7	9
	Click	39	31	38	32	39
	Commander	8	8	8	10	10
	Covau	8	8	8	8	8
	Diamond	12	12	12	16	12
	Dodo	8	8	8	10	10
	EA	15	15	15	15	15
	GloBird	18	18	18	18	24
	Lumo	48	48	48	48	48
	Momentum	15	12	15	15	18
	OPG	26	26	26	26	26
	Origin	15	15	15	20	15
	Pacific	3	4	4	3	2
	People Energy	16	16	16	16	16
	Powerdirect	6	6	5	6	6
	Powershop	9	9	9	12	9
	Qenergy	8	8	8	6	8
	Red Energy	12	12	12	12	12
	Simply	117	170	170	173	185
	Sumo	16	16	16	16	16

Source: MarkIntell (www.markintell.com.au).

Almost all Melbourne households have half-hourly interval kWh meters with remote read capability following a government-mandated installation of these meters by the five regulated network service providers. Despite this, 92% of all households remain on two-part or block tariffs whose rates do not vary by the time of use.

Following recent legislation, retailers are not allowed to discriminate between customers that have installed PV and those that have not. In other words, the same offers must be available to both cohorts.

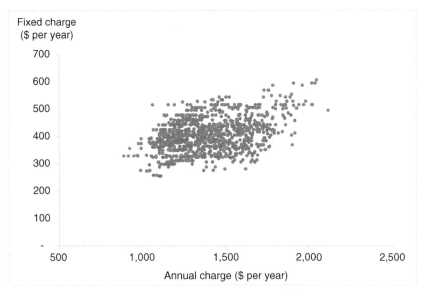

FIGURE 5.2   **Annual charges and fixed charges, all Market Offers, October 2016.** *(Source: MarkIntell[Insight], authors' analysis.)*

The various retail offers can result in significantly different bills. It is also notable that fixed charges make up a large proportion of customer bills. This is shown in Fig. 5.3 which plots the annual bill (on the X-axis) against the fixed charge (the fixed part of the bill) for all 2017 residential offers open to new customers in the market on October 25, 2016 assuming a customer with annual consumption of 4.8 MWh.[4]

Summary statistics on the data shown in Fig. 5.2 is shown in Table 5.2.

Residential electricity offers bundle all network and nonnetwork charges. The network charges vary significantly in the five distribution areas so it is necessary to strip out these costs when comparing offers. Fig. 5.3 is indicative of the relativity of the components of the residential electricity bill[5] based on the simple average of the relevant charges in all Market Offers[6] in each of the five distribution areas.

To promote the installation of domestic PV systems, various state governments including Victoria instituted feed-in tariffs at a range of rates. While

---

4. There are also 19 offers in 1 network service provider's area that have no fixed charges, but these offers are not available to new customers.

5. Assuming 4.8 MWh annual consumption, no solar, wholesale prices of $55 per MWh and Large Scale Generation Certificate and Small Scale Renewable Energy Scheme prices of $70 and $40, respectively). It is also assumed that conditional discounts in market offers are achieved. The retail proportion of the bill would be higher than shown in Fig. 5.2 if Standing Offers or Market Offers (with conditional offers not included) were used.

6. Assuming, as before that conditional discounts are received.

TABLE 5.2 Summary Statistics: Annual Charges and Fixed Charges, All Market Offers, October 2016

	Annual bill ($ per year)	Fixed charge ($ per year)	Average annual price (cents per kWh)
Highest	1987	583	41.4
Lowest	713	252	14.9
Average	1313	386	27.4
Median	1294	377	26.9

To be precise, only market offers open to new customers are considered. Hence Climate Saver tariffs in Powercor area have been excluded. All monetary values are in Australian Dollars (AUD). In November 2016, $1 AUD = $0.72 USD.
Source: MarkIntell[Insight], authors' analysis.

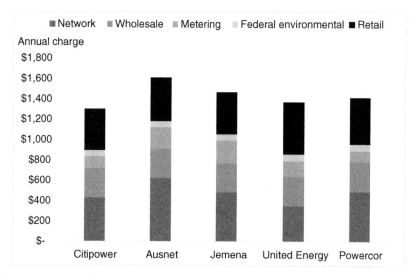

FIGURE 5.3 Residential electricity bill components, all Market Offers, November 2016. *(Source: MarkIntell[Insight], authors' analysis.)*

these are now closed to new participants, a significant number of consumers have benefited from the schemes.

In Victoria around 88,000 homes[7] receive a premium feed-in tariff of 60 cents per kWh plus retailer feed-in rate, approximately 20,000[8] receive

---

7. http://earthresources.vic.gov.au/energy/environment-and-community/victorian-feed-in-tariff/closed-schemes/premium-feed-in-tariff
8. http://earthresources.vic.gov.au/energy/environment-and-community/victorian-feed-in-tariff/closed-schemes/transitional-feed-in-tariff/transitional-feed-in-tariff-faq

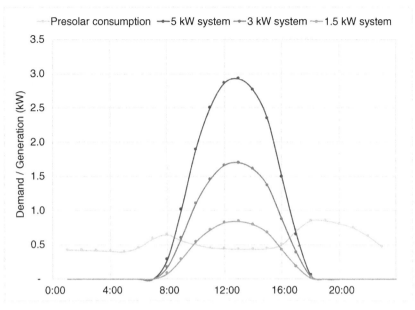

FIGURE 5.4   Typical daily load profile before and after PV. *(Source: SAM [System Advisor Model (2016:3:14), National Renewable Energy Laboratory], authors' analysis.)*

a rate of 25 cents per kWh which will expire at the end of 2016. Around 205,000[9] houses—two-thirds of the total number of households with rooftop PV—only receive a retailer feed-in rate for their exports. The retailer feed-in rate has a regulated floor price that is currently 5 cents per kWh. Although a few retailers offer more the profile of an average daily consumption and solar rooftop production for various sizes of solar installation is shown in Fig. 5.4.

Table 5.3 presents information on the capital cost of PV before and after federal capital subsidies. The installed cost per watt in Australia seems to lower than we have observed in many other countries. This seems to reflect the very competitive market for PV supply and installation. In addition, as we discuss later, the subsidies in Australia, per watt, are now much lower than we observed in other countries.

Table 5.4 translates these capital costs into annual average prices per kWh produced over the life of the PV system.[10] Table 5.4 shows quite significant scale economy so that the average cost of electricity from a 5 kW system being about half the average cost of a 1.5 kW PV system.

---

9. Small Technology Certificate creation data, Australian Clean Energy Regulator.
10. The assumptions used in this calculation are set out in footnote 11.

TABLE 5.3 Capital Costs for PV Systems of Various Sizes, Victoria

System size (kW)	1.5	2	3	4	5	10
Installed cost presubsidy ($)	4,394	5,309	6,904	8,397	9,933	20,539
Capital subsidy ($)	1,014	1,365	2,067	2,769	3,432	6,903
Installed cost after capital subsidy ($)	3,380	3,944	4,837	5,628	6,501	13,636
Installed cost per Watt presubsidy ($ per watt)	2.93	2.65	2.30	2.10	1.99	2.05
Installed cost per Watt postsubsidy ($ per watt)	2.25	1.97	1.61	1.41	1.30	1.36

Source: Solar Choice, Renewable Energy Regulator, authors' analysis.

TABLE 5.4 Capital Costs and Average Prices, PV Systems of Various Systems, Victoria

System size (kW)	1.5	2	3	4	5	10
Installed cost per Watt postsubsidy ($ per watt)	2.25	1.97	1.61	1.41	1.30	1.36
Average cost (cents per kWh)	13.8	12.2	9.9	8.6	8	8.3

Source: Solar Choice (www.solarchoice.com.au), Renewable Energy Regulator, authors' analysis.

The installed costs of PV systems vary widely. For example, the Solar Choice website PV price data shows that installed price of a 3 kW system ranges from $1.18 per watt to $2.14 per watt with a median price of $1.56 and an average of $1.61 per watt. This variance reflects differences in the cost of competing systems and differences in the cost of installing systems on houses of different types.

Although installed system prices continue to fall across all segments, it is difficult to understand the reasons for the variance between highest and lowest prices.

The higher prices may reflect customer preferences for local, trusted installers using higher quality components. Lack of consumer knowledge is also a factor.

3   ANALYTICAL METHODOLOGY

The analysis in this chapter compares annual bills before and after the installation of PV assuming that customers remain on the same retail offer before and after installing PV. By installing rooftop PV households can reduce the amount of electricity they buy from the grid, and earn revenue from the excess production that is exported to the grid. Whether they benefit, financially, from installing PV will depend on the price they pay for PV, the amount they pay for electricity from the grid before and after installing the PV, and the income they obtain from the surplus electricity they sell back to the grid.

The retail offer dataset used in this analysis is a subset of all retail offers. Specifically, since 88% of households in Victoria are on Market Offers, only Market Offers (1602 offers out of a total of 2400 offers) are included. In addition, it is assumed that in those Market Offers, consumers always meet the conditions of the conditional discounts in those offers. This reduces bills, typically, by 10%–20% per year.

For the purpose of this comparison, the cost of rooftop PV, typically paid by households on installation, is converted into an annual charge (annuity).[11]

The algebra of the calculation is expressed mathematically in Eq. (5.1) for retail offers, i, and solar systems, s.

$$\text{Bill reduction}_{i,s} = \text{Bill before solar}_i - \text{Bill after solar}_{i,s} - \text{Solar annuity}_s \quad (5.1)$$

where Bill before solar$_i$ = Consumption before solar * tariff$_i$ _, Bill after solar$_{i,s}$ = Consumption after solar$_s$ * tariff$_i$ − Solar export$_s$ * feed-in-tariff$_i$, Solar annuity$_s$ is the annual charge for solar systems of 1.5, 3, and 5 kW.

The analysis consists of a Base Case which uses the actual feed-in tariffs paid by retailers, and two sensitivities. The sensitivities focus on the effect, in terms of payback periods, to households that install PV. The sensitivities examine the impact of fixed charges in retail offers, and higher feed-in rates on payback periods. Views in Victoria differ on these issues (the size of fixed charges) and feed-in rates. The chapter does not argue the level of these charges and rates, but rather quantifies the impact of changes in fixed charges and feed-in rates on the viability of rooftop PV to households that install them.

---

11. Assuming a 2% real cost of capital, 20-year life, zero residual value, and maintenance charges equal to 20% of the initial outlay. The outlay is based on the installed cost after subsidy shown in Table 5.2. Maintenance charges include replacement of the inverter within the life of the system, at current (2016) prices. Although the actual life of the system is nominally 25 years (and likely longer), the 20-year value conservatively reflects allowance for system faults and performance degradation.

3.1 Sensitivity One: Increase Feed-In Tariff by 5 Cents Per kWh

This sensitivity raises the feed-in tariff by 5 cents per kWh. This makes solar more valuable since it increases the value of the electricity exported to the grid. This is expressed mathematically in Eq. (5.2) below:

$$\text{Bill reduction}_{i,s} = \text{Eq. (5.1)} - \text{Solar export}_s * \text{additional feed-in-tariff} \qquad (5.2)$$

3.2 Sensitivity Two: Convert Fixed Charges Into Variable Charges and Increase Feed-In Tariffs by 5 Cents Per kWh

This sensitivity assumes that the daily charges in retail tariffs are substituted by variable charges that provide the same annual revenue to the retailer before solar, assuming 4973 kWh annual consumption. This results in the same annual bills (before solar) but the tariffs have higher variable charges to offset the elimination of fixed charges. Eliminating fixed charges makes solar more valuable since it will be displacing higher priced consumption from the grid. This sensitivity is expressed mathematically in Eq. (5.3) below:

$$\begin{aligned}
\text{Bill reduction}_{i,s} = \text{Eq. (5.1)} &- \text{Additional variable charge}_i \\
&* \text{consumption displaced by solar}_s \qquad (5.3) \\
&- \text{Solar export}_s * \text{additional feed-in-tariff}
\end{aligned}$$

where Additional variable charge is the annual fixed charge divided by 4973 kWh; Consumption displaced by solar is the annual residential consumption that has been displaced by solar production.

In this analysis, the typical household consumption before solar is 4973 kWh per year, based on information provided by the Australian Energy Regulator's "Energy Made Easy" online benchmark tool.[12] An hourly annual consumption profile based on the annual and seasonal benchmarks for a four-person household in Melbourne was synthesized. Production by PV for the 1.5, 3, and 5 kW cases is based on The ASHRAE International Weather for Energy Calculations Version 1.1 (IWEC) TMY (20 year) dataset for Melbourne with output calculated using SAM.[13] Fig. 5.5 shows the profile of demand and PV production used in this analysis.

The chart shows that for the typical daily profile, a 5 kW PV system (on a north-facing roof) produces more electricity than a household consumes for most daylight hours. The consumption and load profile data before and after PV used to calculate all retail offers in this analysis is shown in Table 5.4. The most significant impact from the installation of a 5 kW system, compared to a 1.5 kW system, is that excess solar production increases from 0.69 to 5.4 MWh per year.

---

12. https://www.energymadeeasy.gov.au/benchmark
13. System Advisor Model (2016:3:14), National Renewable Energy Laboratory.

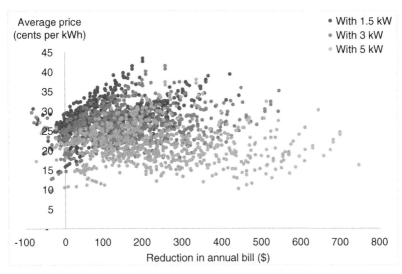

FIGURE 5.5 Reduction in annual bill versus average price before and after 1.5, 3, and 5 kW PV.

By contrast, the grid consumption with a 5 kW rather than 1.5 kW system shows a much smaller change (2.98–3.48 MWh per year) (Table 5.5).

4 RESULTS

The purpose of the analysis is to assess the viability of rooftop solar to households. This is quantified as the reduction in annual bills and consequential payback periods.

Fig. 5.5 plots the reduction in the annual bill (on the *x*-axis) against the annual average price of electricity, after accounting for the annual amortization of the investment in the PV. The scatter shows the effect of different retail offers. Consistent with the diversity of retail offers, it shows that installing PV can lead to a wide range of bill reductions (actually increases in some cases) depending on the households' offer. This is visible in the breadth of the scatter plot.

Table 5.6 identifies the highest, lowest, and median bill reductions. The size of the bill reduction rises with larger systems, indicating that the saving from displaced consumption and higher grid export more than off-sets the higher cost of the larger systems. This is consistent with the information in Table 5.4 which shows large economies of scale between 1.5 and 5 kW systems. This explains the greater popularity of larger PV systems in contemporary installations in Victoria.

TABLE 5.5 Consumption, Export, and Load Profile Data Pre- and Post-PV Installation

	Presolar	Postsolar (1.5 kW)	Postsolar (3 kW)	Postsolar (5 kW)
Grid consumption (MWh)	4.974	3.482	3.233	2.984
Solar export (MWh)		0.692	2.524	5.426
Flex. Peak (%)	26.1	22	21	21
Flex. Shoulder (%)	10.4	9	8	8
Flex. Off-peak (%)	13.5	19	20	21
Peak (7-day) (%)	73	53	59	58
Off-peak (7-day) (%)	27	47	41	42
Peak (5-day) (%)	52	38	43	41
Off-peak (5-day) (%)	48	62	57	59

Source: SAM [All monetary values are in Australian Dollars (AUD). In November 2016, $1 AUD = $0.72 USD.], authors' analysis.

TABLE 5.6 Summary Statistics on Difference in Bills After PV Installation

Statistic	1.5 kW	3 kW	5 kW
Highest ($)	441	544	744
Lowest ($)	−78	−83	−71
Median ($)	92	118	171

Viability of PV installation can be quantified in terms of the payback period—how long it takes for the cumulative savings to exceed the capital outlays needed to deliver those savings.

Table 5.7 examines the payback period, that is, the relationship between bill reductions and the outlays needed to deliver those reductions.[14] It shows that the fastest and median payback period is similar regardless of system size. In other words, the higher cost of the larger PV size is almost exactly offset by the combination of lower grid purchases (when PV production is less than household demand) and higher exports to the grid (when PV production exceeds household demand).

---

14. A "simple" payback is used that is, the undiscounted ratio of outlay to annual bill reduction. Also, as noted earlier, for some retail offers, customers are worse off after installing PV. In this case the installation of PV will never pay off. These are not counted in the results shown in Table 5.6.

TABLE 5.7 Summary Statistics on Payback Period (Years)

Statistic	1.5 kW	3 kW	5 kW
Slowest	33	28	19
Fastest	6	7	6
Median	14	14	12

TABLE 5.8 Summary Statistics on Payback Period (Years)

	Base case	Fixed converted to variable	FiT 5 cents higher	Fixed absorbed in variable and FiT 5 cents higher
Slowest	19	18	12	11
Fastest	6	6	5	5
Median	12	10	8	7

Table 5.8 sets out the results of the sensitivity analysis. It shows the slowest, fastest, and median payback periods in the base case. This is then compared to the first case that the fixed charges in network tariffs is converted into a variable charge; the second case that the feed-in tariff is 5 cents per kWh higher; and the third case that combines higher feed-in rates and the conversion of fixed to variable charges.

5   IS ROOFTOP PV A GOOD INVESTMENT IN VICTORIA?

The structure of electricity tariffs and the level of feed-in rates for grid exports from distributed generation has received considerable attention in Australia (and elsewhere) corresponding to the rise of PV. In Victoria, the Essential Services Commission of Victoria (Essential Services Commission, 2017) suggests that distributed generators do not provide benefits to the distribution networks since there is already excess capacity on those networks. However, in the time-of-use tariffs used by Victoria's network service providers, the variable charge for network services during peak (day time) periods is around 4 times the rate that applies in off-peak periods. In other words, the network tariffs provide powerful incentives for households (and small businesses) to shift demand to off-peak periods.

Presumably this reflects the network providers' view that consumption during peak periods when networks are more likely to be congested should attract higher charges. Yet PV that exports energy to the grid during those peak periods and, by superposition, effectively expands network capacity during those peak periods, is not estimated, by the Commission, to provide a valuable service to the distributors and their other captive customers. This view is not universally shared as reflected in some of the accompanying chapters in this volume.

Economic theory on these issues often provides inconclusive advice. Specifically, normative and positive theories of regulation offer quite different perspectives on the appropriate structure of electricity tariffs. For example, contrast the arguments in set out in Hotelling (1938), Littlechild (1975), and Coase (1970). Even normative approaches are inconclusive, for example, contrast Boiteux (1960), Houthakker (1951), and Bonbright (1961) with Vickrey (1985), Laffont and Tirole (1993), and Joskow and Noll (1981). Similarly, in relation to feed-in rates, plausible theoretical approaches can justify a wide range of feed-in rates.

In the context of contentious theoretical approaches to electricity pricing, this analysis has sought to draw out relevant empirical evidence on how profitable PV is to households that invest in it, in the context of the current capacity subsidy, retail electricity prices (and the structure of those prices), and feed-in rates for PV electricity exported to the grid.

The analysis shows that the payback periods vary quite widely depending on the retail electricity offer that the customer has selected. The median payback period is 12 years in the Base Case. The payback period reflects several factors:

- Existing feed-in rates for exported electricity are typically inferior to the average cost of PV. This means exported PV electricity receives a price that is typically below its average PV costs. The value that PV delivers is therefore dominated by its ability to replace much more expensive grid-delivered electricity.[15]

- Although the marginal price of grid-supplied electricity is much higher than average PV costs, PV only displaces around a third of the typical household's annual electricity consumption. Therefore though the price difference between average PV costs and variable grid charges is large, the limited volume of displaced consumption undermines the financial benefits that households gain from installing PV. Note also that time-of-use weekend (off-peak) grid prices are much cheaper than prices in peak periods, so that the benefit of displaced electricity on 2 days of the week is lower.

- High fixed charges in retail plans undermine the savings that households accrue from installing solar. Fixed charges are a larger proportion of customer bills in Victoria than in the other retail markets in Australia. They are also higher than appears to be the case in other countries and most states of the United States, which often have small or no fixed charges.

---

15. Although we consider the customer's use profile as unchanging in this analysis, there are many reports of customers moving consumption into the PV generation period to reduce exports and maximize their returns. This is done behaviorally (e.g., running the dishwasher during the day) or by installing devices that turn on domestic hot water storage heating if the household approaches export status.

The combination of these factors means that though the average price of solar is significantly lower than the marginal or average price of grid supply, on the median offer, the expected payback period in Victoria is around 12 years.

The sensitivity analysis (on 5 kW solar) quantified the effect of fixed charges and different feed-in rates on payback periods. The analysis found that both of these factors had a reasonably significant impact on payback periods. Raising feed-in rates by 5 cents per kWh reduced payback periods from 12 year on the median offer to 8 years. Converting fixed charges into variable charges in addition to raising feed-in rates by 5 cents per kWh reduces the payback period on the median offer to 7 years.

There are several caveats to this analysis:

- First, the dataset of retail offers included only Market Offers and in addition assumed that in these offers all conditional discounts are received. Market Offers in Victoria often promise large conditional discounts but customers often do not satisfy these conditions. Furthermore the term of those discounts are limited and so after having attracted customers with discounted offers, those discounts are often progressively reduced or eliminated. If we had selected a dataset that included the higher priced Standing Offers or Market Offers in which conditional discounts had not been received, payback periods would be shorter than those reported.

- Second, the dataset of retail offers covers those offers available to new customers or existing customers that specifically request them (and are not prevented from doing so under the terms of their current offers). The actual prices that customers pay will reflect their historic offers. We surmise that in many and probably most cases these prices will likely be higher than the offers the retailers make to attract new customers. In such cases, payback periods will be shorter than reported.

- Third, tracking of retail offers over time shows that many retailers frequently change their offers, sometimes by large amounts.

- Fourth, the approach here assumes that the customer remains on the same offer before and after the installation of PV. Customers that are able to survey the market and select the best offer for them after the installation of PV may be able to reduce payback periods further.

- Finally, as noted, some PV systems are more expensive than others. Had the cheapest rather than the median offer been selected, payback periods would be about 30% quicker, or had he most expensive been selected, payback periods would be about 40% slower.

For these reasons, analysis in this chapter should be updated frequently to track the variation in payback periods as a function not just of the rapid changes in PV costs but also the changing retail offers. We suggest that in all retail markets in which there is a diversity of offers, analysis, such as here will be valuable in properly assessing the relative economics of PV to grid.

Finally, in understanding the situation in Victoria relative to that in other countries, it is interesting to contrast the subsidies paid to households that install rooftop PV in Victoria, with those paid to households for the installation of rooftop PV in the United States. In a recent report prepared for the Consumer Energy Alliance, Borlick Associates (2016) calculates the total subsidy paid to rooftop solar for installations between 3.9 and 6 kW in 15 states in the United States. For 3.9 kW systems, Borlik Associates calculates that these subsidies range between a low of US\$ 2.06 per watt (in Georgia) to US\$ 9.39 per watt (in California). In Victoria, the total subsidy paid to rooftop PV systems is US\$ 0.50 per watt—in other words about one-quarter of the amount per watt in Georgia and about 1/20th of the amount in California.

6 CONCLUSIONS

This chapter has examined tariff structures and feed-in rates and their impact on PV payback periods based on the situation in the Victorian electricity market today. It finds that there is a large variance in the payback period for the installation of rooftop PV depending on the retail electricity offer that the customer has selected. The payback period on the median offer is 12 years.

Households can significantly reduce the price they pay for electricity—and affect the viability of their investment in PV—by selecting electricity tariff offers that are most suitable to them. The analysis also found that payback periods are sensitive to retail tariff structures and feed-in rates:

- On the level of the feed-in tariff, the study found that raising these by 5 cents per kWh reduces payback periods on the median offer from 12 to 8 years.
- Converting fixed charges in retail offers into variable charges reduces payback periods from 12 to 7 years.

But a key point that can be taken from this study is that to ensure a comprehensive and meaningful analysis of the economics of PV, it is essential that the range of retail offers is examined. Victoria shows that these vary widely. Similarly wide retail offer variation is seen in other parts of the Australia.

The apparently long payback period in Victoria (12 years on the median offer) and the evidence of continued robust demand for PV installation in Victorian households challenges the conception that short payback periods are needed to ensure demand for rooftop PV in residential applications.

The apparently long payback period also challenges the notion, expressed by some commentators in Australia, that residential solar uptake is providing windfall gains for households that install solar. Mountain and Szuster (2014) dispelled this notion during the period in which rooftop solar received generous policy support. This analysis suggests that the conclusion remains valid in the period since policy support has substantially diminished.

Since the average price (measured as the Levelized Cost of Electricity) of electricity produced by rooftop PV is approximately comparable to the typical feed-in rate, but significantly below the variable price of grid-supply, there are powerful incentives for households with PV to maximize consumption when the sun shines, or to invest in storage capacity that will allow them to shift production and thereby extend their gains from PV production. This explains the high levels of interest in battery storage in Australia.

This study has not attempted to examine the economics of distributed battery storage for new customers (with or without PV) or for the addition of storage for those households that have installed PV. An initial examination of this in South Australia (Mountain, 2016) suggests that the combination of PV plus storage will reduce electricity bills even for households that are able to access the cheapest grid-supplied electricity. It will be valuable to extend this analysis to Victoria.

REFERENCES

Boiteux, M., 1960. Peak-load pricing. J. Bus. 33 (2), 157–179.

Bonbright, J.C., 1961. Principles of Public Utility Rates. Colombia University Press, New York, NY.

Borlick Associates, 2016. Incentivising Solar Energy: An In-depth Analysis of U.S. Solar Incentives. A report prepared for the Consumer Energy Alliance. Consumer Energy Alliance, Houston, TX.

Coase, R., 1970. The theory of public utility pricing and its application. Bell J. Econ. 1, 113–129.

Essential Services Commission, 2017. The Network Value of Distributed Generation, Distributed Generation Inquiry Stage 2 Draft Report. Melbourne.

Hotelling, H., 1938. The general welfare in relation to problems of taxation and of railway and utility rates. Econometrica 6 (3), 242–269.

Houthakker, H.S., 1951. Electricity tariffs in theory and practice. Econ. J. 61 (241), 1–25.

Joskow, P.L., Noll, R.G., 1981. Regulation in theory and practice: an overview. In: Fromm, G. (Ed.), Studies in Public Regulation. MIT Press, Cambridge, MA.

Laffont, J.-J., Tirole, J., 1993. A Theory of Incentives in Procurement and Regulation. MIT Press, Cambridge, MA.

Littlechild, S.C., 1975. Two-part tariffs and consumption externalities. Bell J. Econ. 6 (2), 661–670.

Mountain, B.R., 2016. Will they go their own way: rooftop solar and batteries in South Australia. Energy SpectrumCornwall Energy Associates, England.

Mountain, B.R., Szuster, P., 2014. Australia's solar roofs: disruption on the fringes or the beginning of a new order. In: Sionshansi, F.P. (Ed.), Distributed Generation and its Implications for the Utility Industry. Academic Press, San Francisco, CA.

Vickrey, W., 1985. The fallacy of using long run cost for peak load pricing. Q. J. Econ. 100 (4), 1331–1334.

Chapter 6

Powering the Driverless Electric Car of the Future

Jeremy Webb and Clevo Wilson
Queensland University of Technology, Brisbane, QLD, Australia

1 INTRODUCTION

This book's underlying theme is innovation and disruption in the power sector and their impact on the grid, particularly the distribution network. It is fair to say that of all technological innovations likely to impact the power sector, the rise of electric vehicles (EVs) and the evolution of the autonomous versions thereof are among the most noteworthy.

The automotive industry is entering an era of unprecedented transformative change, a key element of which, is the increasingly rapid uptake of electric cars (EVs).[1] The annual doubling of EV sales (Fig. 6.1) has pushed the global population over the 1 million mark for the first time in 2016 (1.25 million). While this represents only 0.1% of the current global population of around 1 billion passenger cars, an ongoing exponential rise in sales and ultimately wholesale replacement of the internal combustion engine automobile (ICV) is being seen as increasingly inevitable.

Given how radically technological and socioeconomic factors are changing, there can be no great precision in forecasts about EVs' penetration of the ICV's century-old overly long dominance—as is evidenced by widely ranging estimates. A Goldman Sachs (2016) report puts EV sales in 2025 at 25% of the light car market, while Bloomberg (2016) estimates the EV total at 41 million by 2040, representing 35% of new light vehicle sales and 90 times that sold in 2015. Orton et al. quote a projection of 22% EV penetration in Australia by 2030. The International Energy Agency (2016a)

---

1. For the purposes of this chapter, EVs include both plug-in hybrids (PHEVs) and battery only vehicles (BEVs).

Innovation and Disruption at the Grid's Edge. http://dx.doi.org/10.1016/B978-0-12-811758-3.00006-1

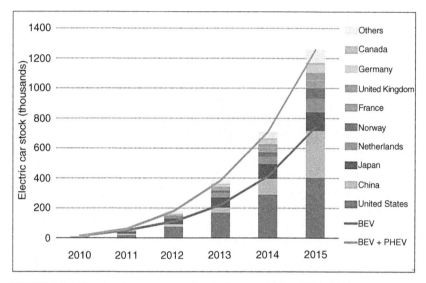

FIGURE 6.1   Global and country uptake of electric vehicles (EVs), 2010–15. Note: The figure's legend for different countries (bottom to top) is replicated in the same order on chart's bars from bottom to top. The top graph line represents BEV and the bottom graph line BEV + PHEV. The EV stock shown here is primarily estimated on the basis of cumulative sales since 2005. *(Source: International Energy Agency (2016a).)*

projects a global population of 60 million vehicles in 2025 and a more than doubling to 140 million by 2040. We consider this is the more likely scenario. These transformative changes are now becoming more visible with a rising uptake of EVs as their price falls—a product in turn of growing competition in the EV market. Accelerating EV ownership is equally a product of consumers becoming more aware of how much cheaper EVs are to run and maintain.

In concert with the uptake of EVs, the prospect is being created of battery-powered vehicles (BEVs) becoming an integral part of electricity grids in terms of both representing sources of demand and supply and thereby creating a new class of prosumers.

Of critical importance then for utilities is an understanding of the powerful technological, social, commercial, environmental, and governmental drivers of this uptake and of the certainty with which the uptake can be predicted.

Section 2 of this chapter describes the phenomenon of peak car and the likely progression toward shared electric autonomous cars. Section 3 investigates the scale of current and likely future EV cost reductions while Section 4 reviews developments in the deployment of EV fueling infrastructure. Section 5 examines the commercial drivers of rising EV sales and the role of multinationals in EV uptake. Section 6 looks at the current developments in battery technology and prices while Section 7 looks at the effect of the rollout of EV refueling infrastructure on power grids and whether a constraint on EV adoption will be created. Section 8 looks at new developments in battery technology and Section 9 describes the current and likely government role in incentivizing the adoption of EVs followed by the chapter's conclusions.

2 PEAK CAR AND EVS

In a remarkably broad cross-section of cities in the United States, Western Europe, and Australia, the upward trend in automotive usage in urban areas now appears to be at an end and is now in decline. This phenomenon is being described in a number of studies as "peak car" (Metz, 2014; Webb et al., 2016; Briggs et al., 2015). It is generally defined as a leveling off and in many cases a decline in annual per capita kilometers traveled. In countries, such as the United States and Australia, the arrestation of the increase in kilometers traveled occurred in the late 1990s early 2000s. Its causes involve a complex of issues including:

- Urban reconfiguration: the recent trend to increasing densities in inner and CBD precincts
- Sociodemographic changes: a decreasing attachment to the automotive mode of transport particularly due to the rise of "digital natives"
- Cost and congestion: increasing aversion to the automotive mode of transport in dense urban environments and the rising cost of social and environmental negative externalities
- Internet-enabled car sharing and integrated public transport

What is clear from these trends is that the socioeconomic drivers of peak car are underpinning the integration and progression toward shared electric and autonomous cars. Thus a number of studies including Seba (2014); Tillemann (2014) are pointing to a process of transformative change with the simultaneous advent of shared cars (SVs), all EVs, autonomous vehicles (AVs), and their impending emergence as shared electric autonomous vehicles (SEAVs). Their collective arrival, is described by Jim McBride, technical leader in Ford's autonomous vehicles team as "… a paradigm shift that's not terribly dissimilar from [the shift from] horses and carriages going to cars. We're going to have cars driving without you …. That's a huge paradigm shift and it opens up a whole variety of new business models that weren't previously available" (ZDnet, 2016).

Impression of an autonomous vehicle. *(Source: Autonomousvehiclesymposium.com.)*

Key elements of the new emerging business model are, first, the way this new form of individualized transport is to be powered. There are convincing reasons, explored in this chapter, why wholesale electrification of the car's propulsion system is rapidly moving from a projection to a certainty. Such a development clearly has major flow-on consequences in terms of how this power is sourced, how it is stored, and how it is utilized.

The second major element is the prospect that EAVs as on-demand shared vehicles would become substantially cheaper than privately owned vehicles per km traveled and substantially cheaper than taxis. If the price differential is sufficient there will clearly be substitution of private ICVs for SEAVs with profound consequences for urban habitats. The current rule of thumb is that one SEAV can replace between 7 and 10 privately owned vehicles depending on the urban environment.

Autonomous substitution is also likely in public transport at first on bus routes where low patronage makes them currently uneconomic and therefore suitable for small autonomous public vehicles. Ultimately there is no reason why all buses should not be autonomous. Equally, AVs will enable hybrid systems, such as that of the Italian start-up Next Future Transport Inc. Being developed are autonomous modular public transport pods which can carry up to 10 people and which are interlinked as a single vehicle in high density routes in a city's CBD. However, once it reaches lower density suburban regions they separate out on individual routes.

Thus while SVs and AVs are not a *necessary* element in the evolution of EVs there are strong synergies between all three technologies which indicate BEVs will be a natural complement to SVs and AVs and their increasing dominance of individualized automotive transport.

Hybrid SEAV modular transport system. *(Source: NEXT Future Transportation Inc.)*

Without a human driver, SEAVs' design is set for a total reconfiguration. The far smaller and the far simpler EV power train (most importantly the absence of a gearbox) provides exceptional flexibility for such a redesign in which the motor can be positioned in a variety of positions given its size and the simplicity of connection to powering the car's wheels.

3 EV COST ADVANTAGES

A second major area of EV advantage lies in its running cost advantages. The US Department of Energy (2016) computes that even at the current historically low gas prices, EV power is 50% cheaper. That advantage does not change dramatically when, in the future, EVs become the dominant mode of transport given the current capacity of electricity grids (if charging is carefully managed) is estimated to be generally adequate to meet most of their power demand. Such a scenario is discussed later in this chapter.

It is also clear that nonrunning costs for electric cars are substantially less. The modern electric car does not have costly service intervals as is now necessary with IC engine cars. That is because the engine is far simpler mechanically, has no gearbox, drive shaft, and far fewer moving parts (an EV has around one-third of the moving parts of an ICV). The Rocky Mountain Institute (2016) estimates that within the pipeline are longer range lower cost EVs which will deliver in the next few years annual cost saving (EV over an ICV) in the United States of around $4000. For the country as a whole such savings could reach $200 billion annually once the EV population reached 50 million vehicles (Fig. 6.2).

Estimations by Tesla S model owners show that if their substantially lower fuel and maintenance costs are factored in together with government purchase subsidies, over a 5 year 100,000 km car life the cost is roughly the same as

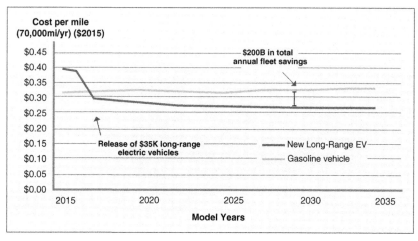

FIGURE 6.2 Comparative cost of electric and gasoline-powered vehicles. *(Source: Rocky Mountain Institute (2016).)*

owning a similarly priced luxury model ICV (Transport Evolved, 2016). Such cost savings will become greatly magnified as EVs become the standard propulsion system for SVs, which have far higher utilization levels (current individually owned vehicles are idle around 95% of the time). Here the capital cost of an EV becomes less of a factor in per km costs with lower fuel and other running cost becoming an overwhelming advantage. Moreover, given the accepted convergence of EVs/SVs and AVs in their SEAV form, EVs will be considerably lighter structurally given such a design is compatible with its radically lower accident rate. Such weight reductions—possibly around 20%—translate into an equivalent increase in fuel efficiency and a lower vehicle cost. Of course according to the extent to which batteries increase a vehicle's weight compared to that of an ICV the fuel saving would be commensurately reduced.

4 EV FUELING INFRASTRUCTURE

While, as outlined, there are compelling reasons why SAVs will adopt all electric power trains, there are, nevertheless, major technological hurdles which will have to be overcome before there is a wholesale uptake of EVs.

In meeting these targets there are, as the International Energy Agency (2016b) admits, particular barriers to overcome where BEVs are concerned.[2] This is important given there is an assumption that a fully fledged EV take-off will come only when BEVs are competitive in costs and have adequate range and a

---

2. Reviewed here are costs and capacity of batteries used in battery electric vehicles which we denote as EVs. Generally batteries used in hybrid HEVs are 30%–50% more expensive and not used in BEVs.

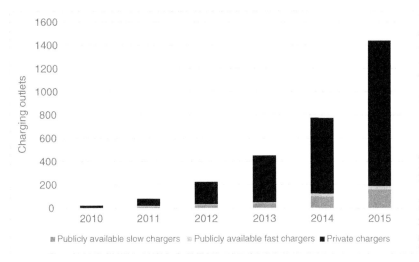

FIGURE 6.3   Global EV power outlets, 2010–15. Note: The left to right order of the legend for types of power chargers is replicated in the figure's bar charts from bottom to top. Private chargers are estimated assuming that each EV is coupled with a private charger. *(Source: International Energy Agency (2016b).)*

reasonably extensive refueling infrastructure. Hybrids (PEVs) and plug-in hybrids (PHEVs) are, for the major proportion of the automotive industry, recognized as a transitional technology, which will slide into obsolescence once the above hurdles are overcome. Prices of hybrids are already falling sharply, an indication that alternative technologies are "knocking on the door."

It is generally accepted that EV power outlets have and will continue to grow roughly in line with the growth of the EV population (Fig. 6.3). Thus the average growth in the number of publicly available charging outlets has more than doubled on an annual basis in the past 5 years. In 2015, EV power outlet growth was 71% compared to EV stock growth of 78% with roughly equal numbers of public slow and fast chargers. Moreover, many governments are providing substantial subsidies for the rollout of charging outlets including home charging outlets (e.g., in Denmark and the United Kingdom). In the United States some 36,000 publicly accessible outlets installed by 2015 had received government subsidies. A number of governments have also mandated the inclusion of EV chargers in new residential and commercial buildings (e.g., France).

In line with commitments to increase EV fleets, governments are equally making forward commitments to EV power outlet rollouts. In July 2016, the Obama administration announced a $4.5 billion loan guarantee scheme to back its plan for a national rollout of EV changing stations. China plans to have 5 million more outlets by 2030 while the EU has recommended at least one outlet per 10 vehicles and 8 million charging points by 2020. The International Energy Agency (2016b) estimates that, globally, the number of slow chargers would have to reach between 1 and 2.5 million and fast chargers between 0.1 and 0.5 million in 2020. By 2030 these figures would need to rise to between 8.4–19 million and 0.8–3.7 million, respectively—an overall 10-fold rise between 2015 and 2020 and 80–120-fold by 2030.

5 COMMERCIAL DRIVERS OF EV UPTAKE

Collectively then, these represent strong drivers of EV uptake notwithstanding the key remaining hurdles of EV range and battery cost. There is nevertheless evidence that the rate of technological progress in extending battery range and lowering costs will continue and create a fundamentally superior alternative to ICVs. Tesla has a market capitalization of over $US 30 billion (as of October 2016) not a great deal less than that of Ford's $US 47 billion and GM's $US 50 billion. Comparatively these are exceptional figures given the diminutive size and production volume of Tesla. As analysts are pointing out, the share price in large part reflects investors' view that Tesla is the global leader in a highly disruptive form of technology and which has the potential to profitably transform the automotive industry.

As Max Warburton, a Sanford C. Bernstein & Co. analyst in Singapore says, Tesla's shares are priced less on the company's own financials and more on how it's forcing competitors to boost spending on electric cars. "Tesla is massively disruptive…. It's valuation reflects the $30 billion problem it's created for the rest of the car industry" (Bloomberg, 2015)—a reference to the amount he says other automakers will spend chasing Tesla.

Uber as the leader in the booming sharing economy is equally a driver of the switch to EVs. With a market capitalization of $70 billion (making it by far the largest, globally of any start-up) its vision is of SEAVs taking over globally as the major form of individualized shared transport. The presence of some of the largest global companies, Google and Apple, in the domain of autonomous car also reinforces the conviction of the top end of the business community that SVs, EVs, and AVs will converge into a massively disruptive and profitable transport form—notwithstanding the evident need for further substantive technological progress to achieve an SEAV "killer app."

It has not been lost on stock market analysts that just as with the iPhone and other disruptive technologies, EVs' penetration of the automotive market is now very much a product of increasingly strong market forces. As Álvarez Pelegry notes in his chapter, Spanish EV numbers increased 20% in the year ending June 2016. That is, as expectations and competitive forces ramp up, so are resources directed into research which becomes more focused. Tesla's extraordinary success and the prospect of new entrants into the automotive industry with extremely deep pockets has been a key driver of these competitive forces. This threat is clearly increasing the speed with which existing car companies are scheduling electrification of their car fleets. Nissan and Renault have revealed a $US 5.4 billion investment in electric cars. Ford has said it will pour $4.5 billion into developing electric cars through to 2020 when its aim is to have a fleet which is at least 40% electric. GM has been no less focused: reportedly some 8600 designers and engineers are working on electric and other alternative propulsion systems. Volkswagen's plans to introduce 30 new electric models by 2025 carries with it the goal of producing at least 25% of its output with electric propulsion.

6 DEVELOPMENTS IN AUTOMOTIVE BATTERY TECHNOLOGY

The range delivered by and cost of batteries are clearly the key factors in marking the point at which EVs become competitive with ICVs. A study by Nykvist and Nilsson (2015) shows the cost of automotive batteries by leading producers falling from around $1000 in 2007 to around $300—an annual reduction of around 6%–9% a year. Since 2011 they have further calculated that battery capacity has risen by an average of 100%. This exponential fall is, they point out, well in excess of most projections which had the current costs as only being realized by 2020.

Such reductions have been a product not only of R and D improvements, but also from economies of scale (given sales in this period have been doubling each year) and the sidelining of less efficient battery models. Moreover, there are good prospects for further R and D improvements for the industry standard lithium ion (Li-ion) battery in terms of anode and cathode materials, separator thickness and stability, and electrolyte composition. For these reasons, Nykvist and Nilsson (2015) project a continuing cost decline in automotive batteries being produced by leading manufacturers of around 9% annually which would bring costs down to around $230 kWh by 2017/2018 (Fig. 6.4). This represents a level

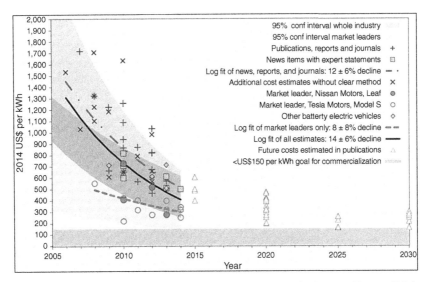

FIGURE 6.4 **Projections for reduction in kWh costs of automotive battery.** *(Source: Nykvist and Nilsson (2015).)*

significantly below projections in most peer reviewed studies and is on a par with more optimistic projections by nonacademic analysts, such as McKinsey.

There has been a general consensus that EV/ICV parity at the entry level/ mass market auto sector will be reached when car batteries reach a cost level of $US 100–150 kWh. The International Energy Agency (2016a) and analysts such as Morgan Stanley (2015) have projected parity for Tesla level cars with comparable ICVs at $US 300 kWh (assuming a gasoline price of around $3.50 a gallon) over the lifespans of 10 years. At these levels the battery represents around 25% of the cost of an EV—a figure which applies to both the larger Tesla S and the compact Nissan Leaf.

Even such bullish projections were again shown to be conservative in mid-2016 when GM (Chevy Bolt), Nissan (Leaf), Renault (Zoe) announced compacts with almost double the previous range of over 200 miles (300 km). These Li-ion batteries boast only a very modest increase in weight and size for an almost doubling of output. GM has claimed a kWh cell price of around $150 although industry experts puts the price at closer to $US 200 kWh once packaged as a car battery while Tesla claims a packaged battery cost of around $190 kWh. Some industry experts suggest these figures are optimistic and put the cost at around $215 and $250 for GM and Tesla, respectively.

Even if these higher estimates are closer to reality, the further falls of 40%–60% needed to reach the gold standard of $100–150 kWh will clearly be reached if the current annual cost reductions of around 9% continue as projected. This assumes that the Li-ion battery remains the battery of choice for major manufacturers. Not everyone agrees, however. The Li-ion battery has been around since the 1990s and some experts feel that there is not a great deal of scope for further R and D based major increases in its density and cost reduction. These doubts were, however, expressed before LG Chem and before them Panasonic were able to dramatically increase the Li-ion's power weight ratio.

Nevertheless, with the exponential growth increases of the global EV population (a stock of around 20 million BEVS and PHEVs is projected by 2020) (Fig. 6.5), battery megafactories are being created, such as those of Tesla, and of

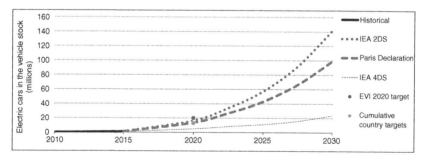

FIGURE 6.5 Projected uptake of HEV/PHEV and battery-powered electric vehicle (BEV). *(Source: International Energy Agency (2016c).)*

Korea's LG Chem and Samsung. They promise substantial economies of scale which are projected to pull battery kWh costs down by as much as 30%. Some of the more bullish predictions have Tesla's gigafactory, driving down auto battery costs by as much as 50% via economies of scale, supply chain optimization, increased automation, and production domestication (Cleantechnica, 2015).

7   EV REFUELING INFRASTRUCTURE: INTERFACE WITH THE GRID

The closing in on the parity target is, of course, putting progressively greater focus on another crucial barrier to EV adoption, ease of refueling. In part this is being solved by cost parity in that the lowering of EV battery cost is being achieved by developing batteries with higher power density and range, but with only marginal increases in size, weight, and cost. Thus the new range extended ZOE/Leaf and GM Bolt means consumer concerns about both the time taken to refuel and the size of the refueling network will be commensurately diminished. This is particularly so in urban environments where trip lengths are modest and charging points more numerous.

US census data (US Department of Transportation, 2009) show that 95% of all trips by car are less than 30 miles (48 km) and 95% less than 70 miles (112 km) with a weighted average of 8.5 miles (13 km) in urban areas and 12.1 miles (19 km) in rural environments. Moreover, with considerably lower per trip distances traveled in Europe and Asia/Pacific cities, range anxiety would seem to be even less of an issue.

On the other hand, with price parity no longer the key hurdle for large-scale uptake of EVs, the real and perceived lack of charging infrastructure and the real and perceived issues surrounding charging time for Li-ion batteries is destined to become a central issue in consumer's powertrain preference. This is being borne out by surveys of the public's attitudes to purchasing EVs. The greatest concern shown by potential buyers (1/3) is range, while adequacy of charging networks ranked third after cost (27%) (Egbue and Long, 2012).

Indicated, therefore, is that over the next 3 years as the issue of EV price recedes as a disincentive to EV uptake, the interface with charging networks and the linked-in issue of charging time and duration will be at the forefront. As already noted, the current rollout is uneven although on average, keeping pace with EV sales. It is no secret, however, that as EV numbers gain critical mass, the charging infrastructure rollout will require substantial investment and regulation to ensure that the new EV interface does not disrupt conventional supply channels.

Depending on the structure of national grids, capacity and/or feedback issues will need to be addressed in a comprehensive national rollout of EV charging networks. Where constraints on grid capacity are created by large-scale EV uptake, the opportunity to use EVs as an ancillary grid storage instrument can present its own set of issues.

In a Spanish study (Fernandez et al., 2011) an assessment is made of the impact of plug-in EVs on distribution networks where EVs are used for grid backup. Their model takes three scenarios of EV penetration which correspond to 35% in 2035, 51% in 2030, and 62% in 2050. Also assumed is that 85% of cars charge in off-peak hours (midnight to 6:00 a.m.) and 45% during (Spanish) peak hours (1600–2100). At peak hours, it is assumed that 40% of EVs are connected, 90% in a fast charging mode, and 10% are injecting power into the grid.

Under the assumption 62% of cars are EVs, the required network reinforcement reaches 19% of total actual network costs for the high population density region surveyed (1800 people per sq. km and car ownership rate of 10%). However only around a 3% reinforcement is required for the rural region surveyed (15 per sq. km and a 50% car ownership rate). The required investment is higher in the urban area with high load density and high installation costs. But by the use of smart charging strategies it is shown that up to 60%–70% of the increased investment can be avoided. In addition, if strategies are deployed so that some of the PEVs that were charged at peak hours are charged at off-peak hours, up to 35% of the required investment is also avoided.

As battery storage capacity increases in concert with a rapid uptake of EVs, utilities, other enterprises and governments involved in the rollout of charging infrastructure are likely to find a commensurate increase in off-peak home charging from low-cost single-phase terminals. A commensurate decrease in the need to populate urban areas with higher cost fast charging points can equally be expected. Woodhouse and Bradbury (Chapter 2) provide estimates that in the United Kingdom, charging a 30 kWh EV battery from residential solar, can produce a saving—based on typical day-night tariffs—of €5 or more per charge. A further outcome to note is that as home charging uptake rises, public charging infrastructure resources can be refocused on decreasing the gap between long-distance charging points.

Rising use of home charging for EVs is firmly in the sights of Tesla's Elon Musk in his meshing of rooftop solar (through the acquisition of PV installer Solarcity), the sale of home storage batteries and use of the storage capacity of electric cars. Such a package, if commercially successful, would greatly increase the capacity of EV owners to stay off the grid.

However, as Orton et al. observe in Chapter 16, charging an EV from a residence increases power uptake between 30% and 50% which means solar arrays will need to be substantial. One calculation (Energysage.com, 2016) is that an extra $2000 investment in solar panels (over and above that installed for household consumption) could provide sufficient power to meet a compact EV's annual power needs. That is based on an average daily travel distance of 25 miles and a Tesla Model S fuel economy rating of 34 kWh/100 miles. On this basis it is calculated that homeowners could break even on their rooftop solar-for-EV investment in 7–10 years. There is however considerable debate about whether Tesla's rooftop/EV package is indeed currently an economic proposition – an

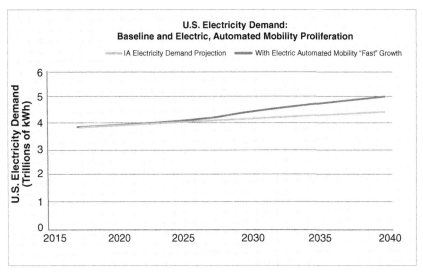

U.S. Electricity Demand:
Baseline and Electric, Automated Mobility Proliferation

IA Electricity Demand Projection   With Electric Automated Mobility "Fast" Growth

FIGURE 6.6 **Projections for EV electricity demand.** *(Source: Rocky Mountain Institute (2016).)*

issue which is difficult to resolve without proper specification of the systems being used by Tesla.

To the extent that greater battery power densities increase feedback into the grid (V2G), a great deal more attention will need to be paid to technical and regulatory infrastructures. US estimates of the effect of high levels of EV ownership indicate that the existing infrastructure can largely accommodate such an eventuality. The RMI estimates that by 2035, EV uptake could produce a 10% increase in electricity demand (Fig. 6.6). Orton et al. quote an Australian study which indicates even if EV penetration was substantial (over 20%) the effect on the national electricity market would be modest—about a 4% increase.

However, with smart charging, including the use of V2G, researchers at the US Department of Energy's Pacific Northwest National Laboratory estimated that, overall, US power grids have sufficient excess capacity to support over 150 million battery-powered cars, or about 75% of all cars on the road. However, it is noted that, to the extent that homes will begin to install high voltage quick chargers, the capacity of residential substations to handle such an increase may become an area of concern.

As noted in a number of studies (Putrus et al., 2009; Daim et al., 2016), the significant increase in distributed supply and reverse power flow created by EVs can have deleterious effects on a grid's voltage levels and fault currents, and thereby bring about malfunctions in the grid's protection system, and generate phase imbalance. That issue has been showing up in Australian cities which have some of the highest residential solar panel ownership rates in the world. A number of utilities are citing problems of regulating grid voltage and associated power trip offs—where solar feedback to the grid is at a high level—as a reason for limiting power to grid feedback.

While smart grids have already been developed to handle these eventualities, in most cases they have not been implanted, given at the current level of distributed feedback, they do not warrant the investment. Moreover, as the recent International Energy Agency report (2016b) points out, as the demand for fast charging three-phase outlets rises, without adequate location and time signals in electricity pricing, this could become a real concern for distribution networks.

8 THE FUTURE OF EV BATTERY TECHNOLOGY

While kWh price parity between BEVs and ICVs is in sight, there remains the fact that in terms of power density, the Li-ion—the standard bearer for the EV world—is no match for gasoline (Table 6.1). But, given gasoline's actual in-car efficiency is only around 15% against electric propulsion's 70%, the Li-ion

TABLE 6.1 International Comparison of Policy Support for EV Infrastructure

	EV purchase incentives				EV use and circulation incentives				Waivers on access restrictions			Tailpipe emissions standards		Market share of electric cars in 2015
	Rebates at registration/sale	Sales tax exemptions (excl. VAT)	VAT exemptions	Tax credits	Circulation tax exemptions	Waivers on fees (e.g. tolls, parking, ferries)	Electricity supply reductions/exemptions	Tax credits (company cars)	Access to bus lanes	Access to HOV lanes	Access to restricted traffic zones*	Fuel economy standards/regulations including elements	Road vehicles tailpipe pollutant emissions standards	
Canada													Tier 2	0.4%
China													China 5	1.0%
Denmark													Euro 6	2.2%
France													Euro 6	1.2%
Germany													Euro 6	0.7%
India													Bharat 3	0.1%
Italy													Euro 6	0.1%
Japan													JPN 2009	0.6%
Netherlands													Euro 6	9.7%
Norway													Euro 6	23.3%
Portugal													Euro 6	0.7%
South Korea													Kor 3	0.2%
Spain													Euro 6	0.2%
Sweden													Euro 6	2.4%
United Kingdom													Euro 6	1.0%
United States													Tier 2	0.7%

Legend:
- No policy
- Targeted policy **
- Widespread policy***
- Nationwide policy
- General fuel economy standard, indirectly favouring EV deployment
- Euro 6 Pollutant emissions standard in place in 2015

Notes: * Such as environmental/low emission zones; ** policy implemented in certain geographical areas (e.g., specific states/regions/municipalities), affecting less than 50% of the country's inhabitants; *** policy implemented in certain geographical areas (e.g., specific states/regions/municipalities), affecting more than 50% of the country's inhabitants.
Source: International Energy Agency (2016d).

battery's 450 Wh/g energy density is in effect competing against gasoline's effective density of around 2500 Wh/kg.

On the other hand there are a number of other battery types which could well exceed gasoline's effective energy density. These are largely populated with metal air and iron air. Thus Li-air and Al-air have theoretical energy densities of 11,140 and 800 Wh/kg, respectively although achieved densities are currently only 700 and 1,300 Wh/kg. Al-ion and Air-ion have theoretical densities of 1060 and 406 Wh/kg, respectively but which have so far achieved 40 and 110–175 Wh/kg (Energy without Carbon, 2016). Thus despite their exceptional theoretical potential, they pose still challenging problems for commercialization. As one commentator has observed:

> Creating higher energy densities is easy. But mass producing batteries that won't explode, suffer thermal runaway, are durable, don't produce toxic byproducts during the charge/discharge cycle, and are inexpensive is extremely difficult. Many of these problems have held lithium-air batteries back from commercialization—such designs degrade rapidly when recharged, are damaged by the presence of water (a problem on planet Earth), or don't hold up under multiple charge cycles. A battery that only retains 90% of its charge after 5 cycles and degrades to 50% charge after 10 cycles, again, isn't very useful. (Putrus et al., 2009).

Some high technology solutions are, nevertheless, showing considerably promise—perhaps the most notable and exciting are the progressive improvement to supercapacitors. This form of storing energy depends not on a chemical reaction but on an electrostatic means of storing electrons. In this field graphene supercapacitors hold particular promise (Notarianni et al., 2014). Their notable characteristic is the capability to provide power and be recharged extremely rapidly. Their drawback is that, until recently at least, they have been poor in power density and considerably less so than the Li-ion battery. However, given graphene supercapacitors involve the use of layers of ultrathin (atom thick) graphene plates which have no electrolyte, they can be placed in a variety of configurations within a car.

Research carried out by a team based at the Queensland University of Technology (QUT) (Zhao et al., 2015; Liu et al., 2015) provides an illustration of the potential for cars to be partly or eventually wholly powered by such supercapacitors. Research team member Dr. Jinzhang Liu says, that while graphene based storage can be used as a power booster for cars, ultimately, he speculates:

> .... the supercapacitor will be developed to store more energy than a Li-Ion battery while retaining the ability to release its energy up to 10 times faster—meaning the car could be entirely powered by the supercapacitors in its body panels. After one full charge this car should be able to run up to 500 km—similar to a petrol-powered car and more than double the current limit of an electric car.

That vision may be already backed up by some recent research reports from the Gwangju Institute of Science and Technology in South Korea (Zhong et al., 2016) which claims an energy density for graphene supercapacitors of a remarkable 131 watt-hours per kilogram (Wh/kg)—nearly 4 times that previously achieved and on par with Li-ion batteries. If achieved, this would open the way for cars to be charged within several minutes.

Another particular advantage of graphene supercapacitors is cost. As QUT's Professor Motta points out:

> *We are using cheap carbon materials to make supercapacitors and the price of industry scale production will be low. The price of Li-ionon batteries cannot decrease a lot because the price of Lithium remains high. This technique does not rely on metals and other toxic materials either, so it is environmentally friendly if it needs to be disposed of. (Motta, 2014).*

The critical issue remains, however, the expected length of the research to commercialization pathway. ULCA Professor Richard Kaner who is affiliated with Nanotech Energy—a startup working to commercialize graphene-based batteries—puts the time frame at 3–5 years (EEnergy Informer, October 2016). As a rule of thumb, development paths are invariably longer than predicted, particularly for radically new technology. This view would seem to be that of the major global EV battery producers—Tesla/Panasonic LG Chem and Samsung—who collectively have initiated a multibillion dollar investment plunge in large-scale Li-ion manufacturing facilities.

There is an added significance to this research in that it appears to signal that other forms of alternative power for low-carbon vehicles are not being viewed as commercially viable. In particular hydrogen-powered cars—the only type of car being currently produced albeit in very small numbers—presents a range of problems in terms of generation, transporting, distribution, and ultimately energy efficiency. For some time, studies of the relative energy efficiency of fuel cell and battery EVs (Eaves and Eaves, 2004), have shown the former to be far less efficient. They indicate that if the relatively power inefficient process of electrolysis is used to produce hydrogen and the not insubstantial losses involved in compression, storage, and distribution are taken into account (given it is a power carrier not a power source as such) fuel cells are no competitor with electricity. Moreover while the electricity grid is at least a largely readymade distribution system for EVs, hydrogen needs a whole new and very costly infrastructure.

Seba (2014) is quoted as saying: "Assuming that at some point fuel-cells will be cheap and hydrogen production will reach critical mass, it will still be at least three times more expensive to power a hydrogen-powered car than an EV." (The Green Optimist, 2015)

9 GOVERNMENT AND REGULATORY DRIVES OF EV UPTAKE

It needs to be accepted that the uptake of EVs will not be driven simply by the extraordinary commercial opportunities they present. There are good reasons for governments and regulatory authorities to continue their already widespread incentivization of the use of electric cars (Table 6.1), although, it is argued, in an evolving form. Thus as the expected running and outlay costs of EVs fall, direct subsidization of EV car purchases are likely to be correspondingly reduced. However, internationally, sustaining a rapid uptake of EVs is still very much a government priority if the tough GHG emission targets flowing from the Paris agreement on climate are to be met (Table 6.1). For example, they require the United States and Australia to reduce GHG emissions between 26% and 28% by 2025, while the EU is committed to a more ambitious target of 80%–85% by 2050 (over 1990 levels). This will require the current annual rate of emission reduction in the United States and Australia to be more than double.

With approximately 23% of GHG emissions coming from transport in developed countries, a switch from gasoline to electricity has the potential to make a substantial contribution to achieving such reductions for governments seeking cost effective socially acceptable means.

With low hanging fruit already being harvested, it is acknowledged that these targets will largely have to come from the power generation and transport sectors. For governments, transport is particularly appealing given electrification of the automotive fleet carries with it the longer term prospect of substantially reducing the costs of this mode.

The Paris agreement has set out targets for GHG reductions within the transport sector in the form of a global EV fleet of 100 million in 2030. A European Environmental Agency (2016) report estimates that if 80% of EU cars were EVs in 2050, a 10% reduction in total emissions would occur taking into account emissions produced by the increased generation of electricity to meet EV demand.

For its part, the IEA says transport should account for approximately one fifth of the overall targeted global reductions which could be achieved if the global EV population reaches 140 million by 2030 and 900 million by 2050 (Fig. 6.7) This would be equivalent to 10 and 40% of light-duty passenger vehicles' stock in 2030 and 2050 respectively and 20 and 40% of sales, respectively.

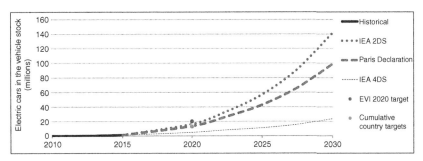

FIGURE 6.7   COP21 and IEA EV deployment targets. Note: 2DS = 2°C Scenario; 4DS = 4°C Scenario. (*Source: International Energy Agency (2016c).*)

This means an annual increase in sales of over 25% by 2025 and a 7%–10% increase between 2003 and 2005.

The IEA argues that such a growth is achievable notwithstanding cost, range, and infrastructure issues associated with EVs. Indeed a group of 16 governments the countries of which account for 80% of all EVs currently sold and which includes the United States, China, Norway, and major EU members, have committed to a collective fleet of 20 million EVs by 2020. Key members[3] have committed to specific growth rates which represent an average annual EV growth rate of 60% between 2015 and 2025 and a total EV population of 13 million vehicles.

Government regulatory regimes particularly where they tighten fuel economy and pollution standards are also powerful drivers of electrification of the private transport mode. Thus in all major EV markets, governments are legislating increasingly stringent standards. By 2021, the fleet average to be achieved by all new cars is 95g of CO_2 per kilometer. This translates to a fuel consumption of around 4.1 L/100 km. Thus the 2021 targets represent a reduction of 40% over the 2007 fleet average of 158.7 g/km. In the United States, the 34.1 miles per gallon (mpg) for 2016 model year passenger cars rises to 54.5 mpg in 2025. In China car fleet average economy has to improve from 6.9 to 5.1/100 km between 2016 and 2020, a 26% improvement.

The push for a rapid replacement of ICVs with EVs has become a particular priority of the Chinese Government where, far more than any other major automotive producer/consumer nation, it has been legislating and incentivizing industry to substantially and rapidly electrify both private and public transport. As a result China is now the world's largest and fastest growing market for EVs (Fig. 6.8). The Government's strategy—as ably outlined in Levi Tillemann's study (2014)—is, not immodestly, to leapfrog into the EV era, and thereby become a world leader in EV manufacturing while simultaneously effecting a radical reduction in GHG and associated automotive pollution levels. While China has so far demonstrably failed to develop a sufficient home-grown technological base to effect such a transformation with the degree of autonomy it badly wants, the linking of the powerful forces of government and industry indicates the growth of EV sales are likely to continue to rise at an exponential rate.

10 CONCLUSIONS

Given the highly unpredictable rate of change in technology and equally problematic estimations of commercialization times, any projections of EV uptake is subject to a high probability of error. Nevertheless, this review of developments does indicate a high degree of certainty that there will be a very rapid uptake of EVs in the next few decades. This assumption is based first, on the recent and

---

3. Austria, China, Denmark, France, Germany, India, Ireland, Japan, the Netherlands, Portugal, South Korea, Spain, United Kingdom, United States.

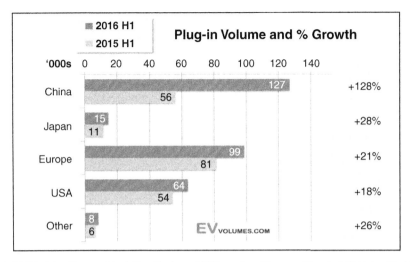

FIGURE 6.8 **Geographical distribution of EV uptake.** *(Source: Electrical Vehicles World Sales Data Base (2016).)*

radical improvement to EVs' range for entry-level vehicles (such as the LEAF, Zoe, and Bolt), and secondly a concurrent substantial fall in kWh costs—both key issues for prospective EV buyers. Equally, there is, among industry and academic analysts, a reasonable degree of consensus that the past rate of improvement to the industry standard Li-ion battery will continue to the point where, in the early 2020s, energy parity with the ICV will be achieved if not earlier. In the medium to long term, environmental issues are also likely to drive both consumers and governments toward EVs. Thus governments can be expected to play an important role in legislating and providing the necessary incentives for a rapid replacement of ICVs with EVs.

That is particularly the case for the Chinese government which has the unique capacity to partner the private sector into an ongoing and accelerated EV uptake in what is now the world's largest market. In this sense projections and targets may merge.

Now that range and, to some extent, price are lesser issues, charging facilities, charging time and power supply issues will inevitably become the focus of attention. While there is a high level of confidence in industry and government that smart charging can go a long way to facilitate a smooth transition to an EV-dominated transport system—careful regulatory and substantial financial investment will be needed to realize this potential. Thus not only will the balance of peak and off-peak charging need to be closely managed, but the investment needed to roll out adequate high voltage fast charging stations on a national scale will need to be carefully scheduled and, if needed, incentivized.

The extent to which there will be V2G integration of EVs into power grids is dependent on some complex cost benefit calculations. While there may be scope for V2G to become an alternative to increasing grid peak capacity, the accelerated depreciation of car batteries when used in this way needs to be taken into account.

In the longer term it must be assumed that the markets being opened up for batteries as power for cars and as a means for storage of residential PV solar power, means investment in battery R and D will undergo a further radical expansion. The sheer range of new battery developments and the evident potential to effect major increases in storage and decreases in cost, indicates the potential for EV uptake to be part of a radical spread of distributed power generation and storage. For utilities, there is in these projections, an encouraging scenario of increased demand for grid power from EVs, which if properly managed, can be sourced during nonpeak hours.

On the other hand if EV battery prices continue to fall as predicted so will (as is already the case) the cost of home storage batteries. In the longer term then there is a not so encouraging (Tesla promoted) scenario for utilities of households not only generating for their own use but partly or wholly for their automotive use. For the authors that seems a natural progression as costs of batteries and solar PV continue to decline and their efficiency rises. As Sioshansi (2015) notes such a PV/battery driven move off grid could reduce utility power from a necessity to a "nice to have" category.

But as Orton et al. note, it is still unclear whether car batteries will play a dual role with household batteries in storing rooftop solar. Elon Musk's vision (currently at least) is for householders to store and feed power from home batteries perhaps in part to ensure greater longevity of car batteries.

Clearly in store then is a complex and if not well managed, messy transition to such an off-grid environment for both cars and residences. Both utilities and network companies will need to develop quite new business models to survive in such a highly distributed environment. As noted in other chapters of this book they will need to become increasingly a service-oriented enabler of such a complex intermesh of on and off-grid power generation with a great deal of the capital investment in power generation shifting to battery manufacturers—and as projected in Chapter 19—PV solar manufacturers. For utilities one encouraging technology-driven scenario could be realized through eventual adoption of rapid charge graphene supercapacitors in EVs. That could open up the opportunity for competitive grid supplied power to a network of rapid charge stations.

REFERENCES

Autonomousvehiclesymposium.com. Available from: https://www.google.com.au/url?sa=i&rct=j &q=&esrc=s&source=imgres&cd=&cad=rja&uact=8&ved=0ahUKEwjW56S3tKDRAhWBj ZQKHYc9AXUQjhwIBQ&url=http%3A%2F%2Fwww.autonomousvehiclesymposium.com %2F&psig=AFQjCNEV7nstUHG6jAOAbqT69_gotLYpkA&ust=1483341658551175.
Bloomberg, 2015. Will Tesla Every Make Money?. Available from: http://www.bloomberg.com/ news/articles/2015-03-04/as-tesla-gears-up-for-suv-investors-ask-where-the-profits-are

Bloomberg, 2016. Here's How Electric Cars Will Cause the Next Oil Crisis, March, 2015. Available from: http://www.bloomberg.com/features/2016-ev-oil-crisis/

Briggs, M., Webb, J., Wilson, C., 2015. Automotive modal lock-in: the role of path dependence and large socio-economic regimes in market failure. Econ. Anal. Policy 45, 58–68.

Cleantechnica, 2015. Tesla Gigafactory & Battery Improvements Could Cut Battery Costs 50%, September, 2015. Available from: https://cleantechnica.com/2015/09/21/tesla-gigafactory-battery-improvements-could-cut-battery-costs-70/

Daim, T.U., Wang, X., Cowan, K., Shott, T., 2016. Technology roadmap for smart electric vehicle-to-grid (V2G) of residential chargers. J. Innov. Entrep. 5 (1), 1–13.

Eaves, S., Eaves, J., 2004. A cost comparison of fuel-cell and battery electric vehicles. J. Power Sources 130 (1), 208–212.

Egbue, O., Long, S., 2012. Barriers to widespread adoption of electric vehicles: an analysis of consumer attitudes and perceptions. Energy Policy 48, 727–729.

Electrical Vehicles World Sales Data Base. Available from: http://www.ev-volumes.com/country/total-world-plug-in-vehicle-volumes/

Energy without Carbon, 2016. Available from: http://www.energy-without-carbon.org/Batteries

Energysage.com, 2016.What-Does-It-Cost-To-Charge-A-Tesla-With-Solar-Energy?. Available from: http://news.energysage.com/what-does-it-cost-to-charge-a-tesla-with-solar-energy/

European Environmental Agency, 2016. Electric Vehicles and the Energy Sector—Impacts on Europe's Future Emissions, 26 September. Available from: http://www.eea.europa.eu/themes/transport/electric-vehicles/electric-vehicles-and-energy

Fernandez, L.P., San Roman, T.G., Cossent, R., Domingo, C.M., Frías, P., 2011. Assessment of the impact of plug-in electric vehicles on distribution networks. IEEE Trans. Power Syst. 26 (1), 206–213.

Goldman Sachs, 2016. The Changing Car. Available from: http://www.goldmansachs.com/our-thinking/technology-driving-innovation/cars-2025/

International Energy Agency, 2016a. Global EV Outlook 2016. International Energy Agency, Paris, p. 4. Available from: www.iea.org. Licence www.iea.org/t&c ©OEC/IEA.

International Energy Agency, 2016b. Global EV Outlook 2016. International Energy Agency, Paris, p. 26. Available from: www.iea.org. Licence www.iea.org/t&c ©OEC/IEA.

International Energy Agency, 2016c. Global EV Outlook 2016. International Energy Agency, Paris, p. 21. Available from: www.iea.org. Licence www.iea.org/t&c ©OEC/IEA.

International Energy Agency, 2016d. Global EV Outlook 2016. International Energy Agency, Paris, p. 29. Available from: www.iea.org. Licence www.iea.org/t&c ©OEC/IEA.

Liu, J., Mirri, F., Notarianni, M., Pasquali, M., Motta, N., 2015. High performance all-carbon thin film supercapacitors. J. Power Sources 274, 823–830.

Metz, D., 2014. Peak Car—The Future of Travel. Landor LINKS, London.

Morgan Stanley, 2015. Shared Mobility on the Road of the Future. Available from: http://www.morganstanley.com/ideas/c

Motta, N., 2014. QUT Science and Engineering News, November 2014. Available from: https://www.qut.edu.au/science-engineering/about/news-and-events/news?news-id=81661

NEXT Future Transportation Inc., 2016. Available from: https://www.next-future-mobility.com

Notarianni, M., Liu, J., Mirri, F., Pasquali, M., Motta, N., 2014. Graphene-based supercapacitor with carbon nanotube film as highly efficient current collector. Nanotechnology 25 (43), 435405.

Nykvist, B., Nilsson, M., 2015. Rapidly falling costs of battery packs for electric vehicles. Nat. Clim. Change 5 (4), 329–332.

Putrus, G.A., Suwanapingkarl, P., Johnston, D., Bentley, E.C., Narayana, M., 2009. Impact of electric vehicles on power distribution networks. In: 2009 IEEE Vehicle Power and Propulsion Conference, pp. 827–831.

Rocky Mountain Institute, 2016. Peak Car Ownership—The Market Opportunity of Electric Automated Mobility Services. Available from: http://www.rmi.org/peak_car_ownership

Seba, T., 2014. Clear Disruption of Energy and Transportation: How Silicon Valley Will Make Oil, Nuclear and Natural Gas Coal Electric Utilities and Conventional Cars Obsolete by 2030. Clean Planet Ventures Silicon Valley, California.

Sioshansi, F.P., 2015. Electricity utility business not as usual. Econ. Anal. Policy 48, 1–11.

The Green Optimist, 2015. Why Hydrogen Cars are More Efficient That Electric. Available from: http://www.greenoptimistic.com/hydrogen-cars-efficiency/#.WBwzzdi7qUm

Tillemann, L., 2014. The Great Race: The Global Quest for the Car of the Future. Simon and Schuster, New York, NY.

Transport Evolved, 2016. Teslanomics Returns as Owner Claims 18-Months of Cost-Free Tesla Model S Ownership. Available from: https://transportevolved.com/2015/03/24/teslanomics-returns-as-owner-claims-18-months-of-cost-free-tesla-model-s-ownership/

US Department of Energy, 2016. The eGallon: How Much Cheaper Is It to Drive on Electricity?. Available from: http://energy.gov/articles/egallon-how-much-cheaper-it-drive-electricity

US Department of Transportation, 2009. Summary of Travel Trends: 2009 National Household Travel Survey. Federal Highway Administration. Available from: http://nhts.ornl.gov/2009/pub/stt.pdf

Webb, J., Wilson, C., Briggs, M., 2016. Automotive modal lock-in: a theoretical framework for the analysis of peak car and beyond. QUT. Unpublished journal article submitted to Australasian Journal of Environmental Management.

ZDnet, 2016. Ford: Self-Driving Cars are Five Years Away From Changing the World. Available from: http://www.zdnet.com/article/ford-self-driving-cars-are-five-years-away-from-changing-the-world/

Zhao, Y., Ran, W., He, J., Huang, Y., Liu, Z., Liu, W., Tang, Y., Zhang, L., Gao, D., Gao, F., 2015. High-performance asymmetric supercapacitors based on multilayer MnO_2/graphene oxide nanoflakes and hierarchical porous carbon with enhanced cycling stability. Small 11 (11), 1310–1319.

Zhong, H., He, A., Lu, J., Sun, M., He, J., Zhang, J., 2016. Carboxymethyl chitosan/conducting polymer as water-soluble composite binder for $LiFePO_4$ cathode in lithium ion batteries. J. Power Sources 336, 107–114.

FURTHER READING

GTM Research, 2016. The Impact of Electric Vehicles on the Grid: Customer Adoption, Grid Load and Outlook. Available from: https://www.greentechmedia.com/research/report/the-impact-of-electric-vehicles-on-the-grid

Techradar.com. Available from: http://www.techradar.com/news/car-tech/google-self-driving-car-photos-1321820

Chapter 7

Regulations, Barriers, and Opportunities to the Growth of DERs in the Spanish Power Sector

Eloy Álvarez Pelegry
Orkestra, Bilbao, Spain

1 INTRODUCTION

Regulation and economics are key issues for technology development, and an appropriate combination of the two drives innovation. The recent history of Spain clearly shows how a combination of these factors has led to a major increase in the installed capacity of renewables,[1] while also engendering certain economic difficulties. The promotion of small and distributed photovoltaic (PV)-based generation, backed by favorable costs and concepts, such as consumer empowerment, is currently under debate, with particular emphasis on specific rules and economics.

These developments have all taken place within the context of a set of EU energy policies setting clear targets for the penetration of renewables in final energy consumption and a dramatic change in the economics of PV power, key elements in the development of distributed energy resources (DERs).

Section 2 of this chapter summarizes Spanish legislation on renewables since the 1980s; Section 3 discusses renewable energy plans. Section 4 describes the results of these measures and policies in terms of new installed capacity, renewable electricity production, and renewables in final use, as well as the economic implications of aggregated costs for the electricity system. Self-consumption has become a key issue in the development of DERs and their possible consequences for grids. Section 5 discusses the regulatory aspects of this technological arrangement.

---

1. "Renewables" here refers to wind, solar (photovoltaic and thermoelectric), minihydro, and biomass.

Innovation and Disruption at the Grid's Edge. http://dx.doi.org/10.1016/B978-0-12-811758-3.00007-3
123

Changes to the regulations now enable households to cover part of their electricity consumption at a reasonable price with small PV facilities, though their viability depends on location, capacity, and whether or not they are connected to the grid. Section 5 also provides an economic analysis, describing and analyzing different cases. A key issue is also considered in this section: the economics of PVs at a "utility scale."

Another key topic that may have a disruptive implication on electricity grids is the growth in electric vehicles (EVs). This issue is dealt with in Section 6. In view of the importance of batteries to the penetration and development of EVs, one part of this section analyzes the implications of developments in this area. The potential development of EVs is also analyzed, examining the investment needs followed by the chapter's conclusions.

2 REGULATION/LEGISLATION ON RENEWABLES IN SPAIN

From the outset, the development of renewable energy in Spain has proceeded on a trial-and-error basis, given that it was operating in largely uncharted waters. By the mid-1990s, when targets were far from being reached, ill-advised regulatory design had created a complex situation. The result was a number of very significant changes in the electricity system, which in turn affected numerous activities, including generation—where new taxes were introduced—and power transmission and distribution—for which remuneration was reduced.[2]

Regulatory promotion of renewables can be traced back to 1980 and the enactment of the Law on Energy Conservation (Ley 82/1980).[3] Under this legislation, power utilities were required to buy energy from small producers with domestic installations for which they paid a regulated domestic electricity tariff.

Royal Decree (RD) 2366/1994 set prices for wind, solar, and minihydro facilities with under 100 MVA capacity[4] and also established priority access to the grid for the electricity produced by this type of facility.

Subsequently, Law 54/1997 liberalized electricity generation in Spain and created the *Regimen Especial* (a special regime or framework to include power from renewables and cogeneration). The law also set a target of a 12% share of renewables in electricity by 2010.

The special regime was further developed by RD 2818/1998, which established payment rates for the various different forms of renewable technology. A PV plant with capacity of under 5 kW, for instance, was entitled to remuneration of €396 per MWh; a wind plant to €66 per MWh. The producer could choose between this regulated price and a *prima* (premium on pool price); this

---

2. This section is partly based on a summary of a more detailed description, taken from a study by Corrales et al. (2016).
3. Throughout the text, *Ley* is translated as "Law", rather than "Act" or "Bill," and *Real Decreto* and *Real Decreto Ley* as "Royal Decree" (RD) and "Royal Decree Law" (RDL), respectively.
4. At €2 per kW-month plus €696 per MWh.

took the form of an incentive added to the pool price, which came to €360 per MWh for PV and €32 per MWh for wind energy.

In 2004, RD 436/2004—which remained in force until 2007—incorporated economic incentives for solar thermoelectric plants. These were linked to an average reference tariff (*Tarifa Media de Referencia* or TMR). For example, PV plants of less than 100 kW enjoyed an incentive of 575% for the first 25 years and 460% thereafter.[5] Rates were also set for wind and solar thermoelectric, which were valid up to a given installed capacity: 150 MW for PVs, 200 MW for solar thermal, and 13,000 MW for wind.

RD 661/2007 was enacted 3 years later in 2007. On this occasion, the regulation included both a floor and a cap rate for solar, thermoelectric, and wind, as well as a review mechanism based on the consumer price index. For wind power, the cap and floor were €85 and €71.2 per MWh, respectively, for the first 20 years, whereas for solar thermoelectric the rates were €344 and €254.

As discussed in the next section, there followed a dramatic increase in the number of PV and wind facilities. RD 1578/2008, enacted in 2008, established new incentives for PV facilities, with different rates for rooftop- and ground-based plants. It also set limits on total capacity to limit the total value of incentives. In 2009, RD 6/2009 implemented a procedure to better determine plans for new facilities to monitor the development of renewables. In 2010, RD 1614/2010 limited the operating hours of solar thermoelectric and wind plants in an attempt to restrict the amount paid out in incentives for renewables (see Section 4).

Two key years for the development of renewables were 2012 and 2013. As discussed in Section 4, the sheer quantity of economic incentives threatened to undermine the financial viability of the power system, given that despite a continuous increase in electricity prices for domestic and industrial consumers, the increase in tariffs fully failed to offset the total increase in costs.

In 2012, Royal Decree Law (RDL) 1/2012 suspended economic incentives for new renewable plants, offering them the pool market price instead. Later the same year, RD 13/2012, among other measures, reduced remuneration for distribution and transmission, as well as capacity payments and *interrumpibilidad* (the service whereby supply can be cut to certain high-consuming clients). In addition, Law 15/2012 established a tax of 7% on power generation and created new taxes on nuclear and hydro generation.

In 2013, RDL 2/2013 abolished economic incentives for special regime facilities and under RDL 9/2013, enacting urgent measures for the financial stability of the electrical system, the government created a new remuneration system for those renewables, based on the concept of reasonable return on investment. The system is based on the "standardization" of the different technologies, so standard capital expenditure for investments is fixed, taking into account a reasonable return on pretax investment (yield on Spanish 10-year government bonds plus 300 basis points). This new method of payment also

---

5. It established values if the producer chose the market rate plus a tariff.

takes the economic life span of the projects into consideration. It is subject to review every 6 years, with an interim review every 3 years.

Law 24/2013 on the Electricity Sector, enacted on December 26, 2013, introduced a large number of major changes and was seen as a reform of the legislation, given that aside from the many changes introduced in RDs, no specific law on the electricity sector had been enacted since 1997.

In terms of developments in renewables, the new law was complemented by Order IET/1045/2014, approving the parameters for remuneration of the different standard investments, having established the life span for economic remuneration (between 20 and 30 years, depending on the technology), standard operation, and maintenance costs, as well as "standard" operating hours.

Other important pieces of legislation include RD 1699/2011 and RD 900/2015, both related to self-consumption; and RD 647/2011 on EVs.

As can be seen, regulatory developments in Spain have lacked permanence, with numerous new pieces of legislation created in response to different regulatory challenges, especially since 2007.

Basic legislation/regulation and energy plans related to renewables

Year	Basic legislation on renewables	Plans for renewables
1980	Law 82/1980 Energy Conservation	—
1986	—	Renewables Plan 1986–1988
1990	—	National Energy Plan 1991–2000 (PEN 1991–2000)
1994	RD 2366/1994	—
1997	Law 54/1997	—
1998	RD 2818/1998	—
1999	—	Plan for Renewable Energy Promotion 2000–2010 (PFER 2000–2010)
2000	RD 1955/2000	—
2004	RD 436/2004	—
2005	—	Renewable Energy Plan 2005–2010 (PER 2005–2010)
2007	RD 661/2007	—
2008	RD 1578/2008	—
2009	RD Law 6/2009	—
2010	RD 1614/2010 RD 1565/2010 RDL 14/2010	National Action Plan for Renewables 2010–2020 (PANER 2010–2020)
2011	RD 1699/2011 RD 647/2011	Renewables Energy Plan 2011–2020 (PER 2011–2020)
2012	RDL 1/2012 RDL 13/2012 Law 15/2012	—
2013	RDL 2/2013 RDL 9/2013 Law 24/2013	—
2014	RD 413/2014 "Order" IET 1045/2014	—
2015	RDL 9/2015 RD 900/2015	—

3   SPANISH RENEWABLE ENERGY PLANS

The development of renewable energy sources in Spain has been governed by a series of national energy plans, backed by the legislation described in Section 2. We shall now examine the planned developments and results, further described in Section 4.

The first energy plan to focus on renewables was the Renewable Energy Plan 1986–1988, which sought to promote the development of this type of energy, although no specific targets were set. The *Plan Energético Nacional* (PEN) *1991–2000* (National Energy Plan), established a target of 3.2% of renewables in primary energy by 2000.

In 1999, a number of specific targets were set in the Renewable Energy Promotion Plan (*Plan de Fomento de Uso de Energías Renovables* or PFER) 2000–2010: 9000 MW for wind, 135 MW for PV, and 200 MW for solar thermoelectric.[6]

Six years later, the government brought out a new renewable energy plan (the *PER 2005–2010*), which targeted 12% of renewables in primary energy, 29.4% of electricity from renewables, and 5.75% of biofuels in transport by 2010. Specifically, 20,150 MW was to be generated from wind power facilities, 500 MW from solar thermal, and 400 MW from PV.

Five years later, a new National Renewable Energy Action Plan (*Plan de Acción Nacional de Energías Renovables* or PANER, *2010–2020*) was implemented. The plan incorporated the targets for renewable energy contained in European Directive 2009/28/EC, that is, 20% of renewable energy sources in final use. This was followed by a new Renewable Energy Plan (*PER 2011–2020*) approved in 2011, which set a target of 20.8% of renewables in final energy by 2020, 39% in electricity generation, and 11% in transport.

Thus since 1986, and more clearly since 1991 (Fig. 7.1), energy policy has been established through a series of plans with specific targets for renewables. These have increased from a 4.5% penetration level in 1990 to a 10% target for 2000, later revised to 12% in primary energy by 2010. For 2010, the target was increased to nearly 30% of renewables in electricity and around 6% in biofuels. The current target is for 21% of renewables in final energy by 2020, with 39% in electricity generation and 11% in transport.

As discussed in the next section, the 2010 targets for installed wind, solar thermal, and PV capacity were clearly surpassed. Nevertheless, a combination of greater-than-planned installed capacity, generous feed-in tariffs initially offered to technologies at an early stage of development, and a drop in demand since 2008 created a very serious economic and financial threat to the stability of the electricity system. Even the European Commission's services warned the Spanish government that the situation in the electricity system was neither economically sound nor sustainable.

---

6. In 2000, the Spanish electricity mix included 20,693 MW of wind energy, 3,921 MW of photovoltaic, and 732 MW of solar thermoelectric.

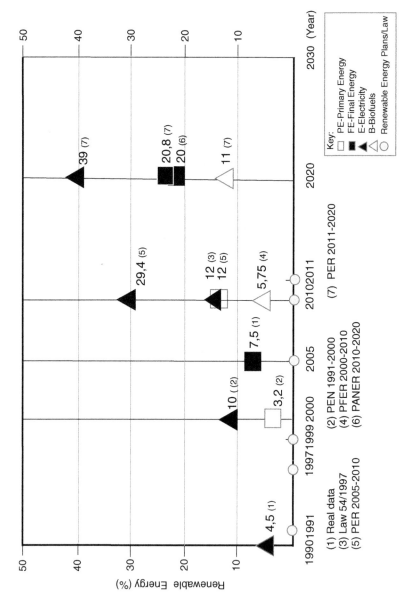

FIGURE 7.1 **Targets for renewables in energy plans.** Note 1: The Renewable Energy Plan 1986–1988, which promoted renewables without setting quantifiable targets, is not included here. (*Source: Original based on various laws and energy directives.*)

4   DEVELOPMENT OF RENEWABLES AND COSTS

Fig. 7.2 provides a more in-depth overview of trends in renewables in Spain, showing the capacity installed each year since 1990.

Fig. 7.3 shows total accumulated installed capacity since 2000.

The legislation described above, particularly RDs 2818/1998, 436/2004, and 661/2007 have had a clear influence on the development of renewable energy, with a logical delay between enactment of the legislation and construction and commissioning of facilities. This is particularly clear in the case of PV, as can be seen in Fig. 7.4.

One of the main reasons for the dramatic fall in electricity demand was the economic crisis of 2008, which lasted beyond 2012,[7] longer than initially expected. In 2014, primary energy demand stood at 118 Mtoe and final energy at 83 Mtoe. Both figures were below the level for 2000. The situation was similar for electricity, with demand of 258 TWh in 2014, comparable to the figure for 2005.[8]

As discussed above, a strong, rapid development in renewable technologies in the Spanish electricity mix had serious consequences, including an increase in regulated costs, to the point of becoming almost uncontrollable.

Regulated costs are calculated as the sum of feed-in tariffs for renewables (RES) and cogeneration, electricity transmission, and distribution costs, the annuity of the tariff deficit and extrapeninsular compensation for the higher cost of supplying power on the Spanish islands, the Balearics, and the Canaries. These costs have increased dramatically since 2006–2007, as can be seen in Fig. 7.5, showing different types of regulated costs, and Fig. 7.6, showing the specific remuneration of the various renewables and other energy sources incentivized under the "special regime."

In 2007, special regime feed-in tariffs came to around €2.1 billion. Six years later, in 2013 they had soared to €9 billion. As a result, regulated costs, which in 2007 totaled around €5 billion, rose to €20 billion by 2013 (Álvarez Pelegry, 2015).

Despite the changing context, the amount of electricity produced from renewables has risen steadily, as Fig. 7.7 shows. In 2015, renewables[9]—mainly large hydro, wind, and solar—accounted for 46.4% of electricity demand, with 66.8% carbon-free, including nuclear.[10]

---

7. By the first quarter of 2012, the situation had become particularly serious. As the Minister for the Economy and Competitiveness later reported "In the context in which we were operating in the first quarter of 2012, the economic situation was worsening by the moment. The situation was far more serious even than we had thought at the beginning of the year, when the picture was already quite bleak" (Guindos, 2016).
8. An analysis of the energy situation during the crisis in Spain can be found in Álvarez Pelegry (2015).
9. Including cogeneration.
10. In Germany, the figures were 26.2% and 41%, respectively (Álvarez Pelegry E. et al., 2016).

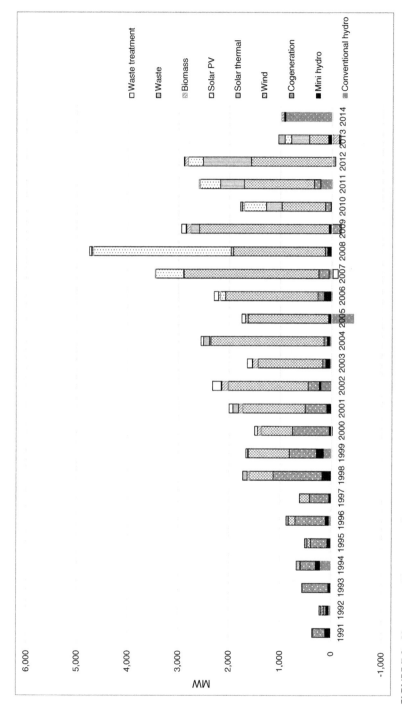

FIGURE 7.2 New renewable power and cogeneration by year (MW). (*Source: Corrales et al. (2016) from REE, CNMC, and Eurostat.*)

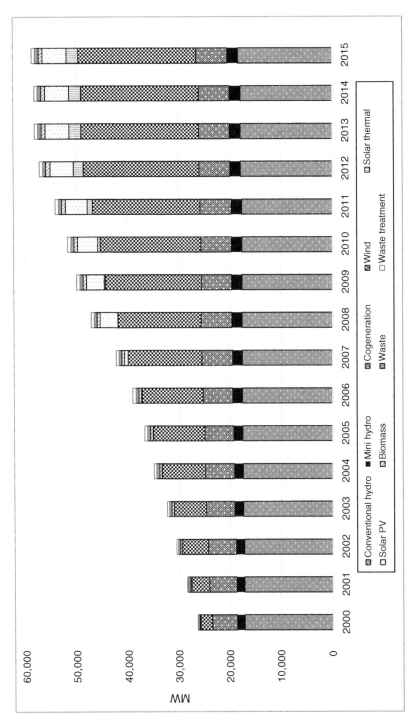

FIGURE 7.3 Development of accumulated renewable power and cogeneration by year (MW). *(Source: Corrales et al. (2016) from REE, CNMC, and Eurostat.)*

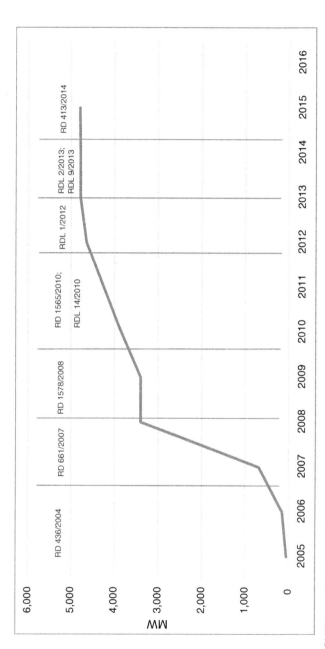

FIGURE 7.4 Development of photovoltaic (PV) power in Spain (2005–2014). *(Source: Corrales et al. (2016) from CNMC, REE, and Mir (2011).)*

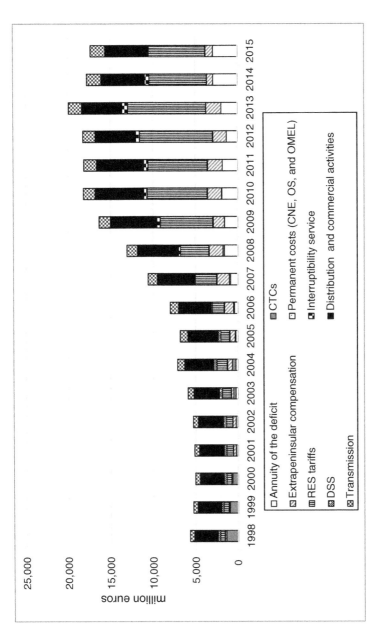

FIGURE 7.5 Regulated costs of the Spanish electricity system. This graph does not include payments for capacity. Sum of feed-in tariffs for renewables (RES) tariffs include RES and cogenerations feed-in tariffs. DSS mainly includes costs of diversification and security of supply. Permanent costs include market and system operator costs and the National Energy Commission Costs (CNE, today CNMC). *CTC, Costes de transición a la competencia. (Source: Corrales et al. (2016) from CNMC and CNE.)*

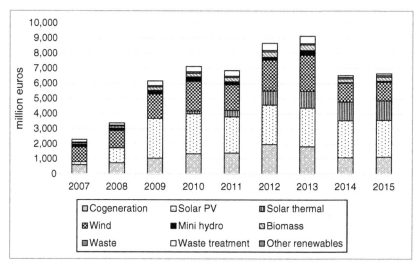

FIGURE 7.6   Development of payments for renewable power production, including cogeneration. Figures for hydro refer to minihydro. *(Source: Corrales et al. (2016) from Foro de la Industria Nuclear (2016) and CNMC.)*

Although this description has focused mainly on renewable technologies in power production, renewable energies can also be found in final energy use, especially solar thermal, biomass, geothermal, and in transport.

Based on this review, one conclusion seems clear: the huge cost of developing renewables led to a real problem with the total cost of the system. To some extent, this has resulted in a sense of caution and a restrictive environment vis-à-vis new developments, which, if improperly managed, might lead to a recurrence of the same problems.

Electricity prices for domestic consumers increased dramatically. The cost per MWh for a domestic consumer using 2500–5000 kWh per year rose from €140 in 2007 to €218 per MWh in 2016 (Eurostat, 2016).

Higher prices are mainly a result of environmental, social, and industrial policies, with a very high percentage, in some years not far from 50%, of the final price related to costs and taxes in these areas. This situation needs to be corrected with intelligent regulation to avoid distortions for the effective development of prosumers. Increased costs for domestic consumers in turn promote measures of energy saving and efficiency and make PV prosumption more attractive in conjunction with a drop in the price of such installations.

5   DER: HOUSEHOLD PROSUMERS AND PV UTILITY SCALE

This section discusses domestic self-consumption, another topic that has sparked debate, and some misunderstanding. To a great extent, this is an issue that depends on regulation, technological improvements, and economics.

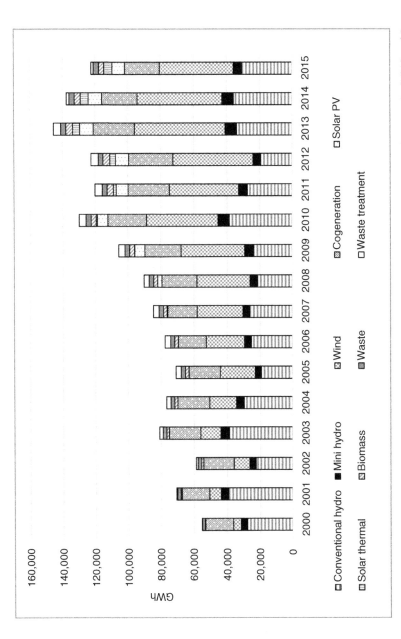

FIGURE 7.7 **Renewable energy produced in the Spanish electricity system (GWh).** *(Source: Corrales et al. (2016) from REE, CNMC, and Eurostat.)*

Our analysis focuses on small household PV installations, with a reference at the end of the section to self-consumption in small industrial factories and at a "utility scale."

Aside from the Spanish regulation itself, it is important to note that the European Commission actively promotes "prosumption" (i.e., self-consumers, including residential prosumers, community/cooperative energy, and commercial and public prosumers) (Sajn, 2016).

RD 1699/2011 transposes European Directive 2009/28/EC and regulates grid connection by small facilities. However, it is Law 24/2013, governing the electricity sector, which makes a clear distinction between producers and consumers. The law includes the concept of self-consumption, which is defined in Article 9 as "the consumption of electricity from generating facilities connected within a consumer's network or through a direct power line associated with a consumer."[11]

RD 900/2015 regulates the administrative, technical, and economic conditions for self-consumption and production with self-consumption. The decree provides for two types of self-consumption, both defined in Law 24/2013. These are: Type 1, facilities of under 100 kW (Article 9.1 a); and Type 2 (Article 9.1 b and 9.1 c of the law). In Type 1 facilities, there is only one "legal" subject, namely the consumer; in Type 2 there are two: the producer and the consumer.

One controversial issue of the current regulation on self-consumption (RD 900/2015) is the payment of charges for the electricity produced from facilities in the 10–100 kW range. Installations with a capacity of under 10 kW are exempt from this charge.[12]

Under the RD, facilities of under 100 kW connected to a sub-1kV grid are exempt from the requirement to obtain prior administrative authorization and administrative authorization for construction (Additional Provision No. 5). Distribution companies are required to report to the National Markets and Competition Commission (CNMC) on the number of contracts for grid access and the volume of contracted power.

To conclude this section, it should be noted that there is no clear figure for installed capacity in small self-sufficiency facilities (households and small industrial facilities), with different sources estimating the figure at between 16 and 50 MW (Barrero, 2016).

---

11. Under Law 54/1997, a producer could produce energy for third parties or for its own consumption. Under RD 1955/2000, producers and self-producers had third-party access to the distribution grid. RD 661/2007 recognized self-consumption and required execution of a standard contract covering energy allocation.

12. There is, therefore, no "sun tax" on small households as some publications claim. Some authors consider the economic impact of this charge not to be very important. It may extend the simple payback period by around 1 year, and in any case self-consumption may save money for the customer with possible returns of over 10% (Romero, 2016a).

5.1 Economic Assessment of a Domestic PV Installation

It is important to assess the economic feasibility of self-consumption. This requires calculating the investment and generating costs of own facilities and estimating some financial parameters to assess the attractiveness of the investment.

It may also be of interest to compare the production cost of the electricity
generated by the facilities themselves and consumed by the owner of the facility
with the price paid to the electricity supplier.

Larrea et al. (2016) made an analysis for two types of consumers (3000 and
5000 kWh per year) in two locations (north and south of Spain). In each case,
they analyzed two scenarios: a consumer–producer with solar PV panels off-
grid and grid connected. The off-grid facility requires electrical storage and, for
economic reasons, a power generator set.

Depending on consumption demands (3000 or 5000 kWh per year) and
location (north and south), the hypothetical consumers would have to invest
between €17,000 and €34,000 to operate off-grid. The calculations showed
that the Levelized Cost of Electricity (LCOE) ranged from €582 to €862 per
MWh, rising to €1548 per MWh where the facilities incorporate batteries
but not a diesel generator. These investments make little economic sense,
given that the simple payback period would be more than 26 years, and there
would be no positive Internal Rate of Return (IRR). If this consumer was
connected to the grid, the cost for the consumer (price) would be around
€215 per MWh.

For a grid-connected consumer with PV facilities, they examined two possible basic scenarios: 100% self-sufficiency and 30% self-sufficiency with no
electrical storage.[13]

Fig. 7.8, intended for illustration purposes, shows a daily curve of PV production and domestic demand for 2 months of the year. The case displayed here
is for 100% self-sufficiency, 3000 kWh per year demand, located in the north of
Spain. As can be seen, the rate of self-consumption ranges from 35% to 58%.

In economic terms, assuming a Weighted Average Cost of Capital (WACC),
of 5.6%, the LCOE ranges from €101 to €141 per MWh and the investment
required from around €1000–8000. The simple payback periods range from
10 years in the south to 15 years in the north for a demand of 3000 kWh per year
and a self-sufficiency rate of 30%.[14]

In this case, however, the LCOE is not the best criterion for comparing electricity costs. Instead, a concept should be employed that is related to the cost of

---

13. The self-sufficiency ratio represents the percentage of energy produced by the household relative to total demand on an annual basis. The self-consumption ratio represents the energy consumed
by the household from the energy it produces itself. A self-consumption ratio of 100% does not
mean that consumers can operate in isolation from the grid, as production and demand profiles differ. The difference may be covered, in this case, by buying electricity "from the system."
14. Depending on the scenario, the Internal Rate of Return (IRR) calculated on the basis of the cash
flow generated by the savings ranges from negative to 9%.

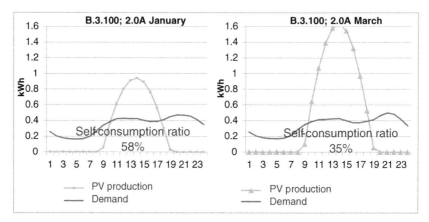

FIGURE 7.8 **Average daily production and consumption (kWh) of a domestic consumer.** Grid-connected 2.88-kWp PV plant, with a self-sufficiency ratio of 100%. *(Source: Larrea et al. (2016) from REE.)*

the electricity produced and consumed.[15] As the household cannot benefit from the total production, but only from the kWh it can consume instantly—if it has no storage system—for a ratio of 30% self-sufficiency, the costs of electricity produced and consumed are practically the same. For 100% self-sufficiency, there is a difference in cost between the electricity produced and the electricity utilized and consumed, where the cost is less beneficial for the consumer.

If the electricity taken up to the grid is remunerated, at 30% self-sufficiency the LCOE is similar to a case in which electricity is not remunerated. However, when the self-sufficiency ratio is 100%, the LCOE decreases by between 19% and 27%, depending on the site (irradiance), yet the return on investment for the household is still very low or even "negative," with a very long payback period.[16]

What conclusions can be drawn from these comparisons?

In the case of off-grid facilities, the grid and electricity system, if available, are a very attractive alternative for the consumer. Amongst grid-connected consumers, self-sufficiency may be attractive in terms of profitability and simple payback period, although potential barriers include the need for consumers to have the financial capability to invest and enough space on their roofs to house the facilities (from 6 to 46 m², depending on installed power and location).

---

15. This cost should not be confused with the Levelized Cost of Electricity (LCOE). The LCOE is the cost of the electricity produced whether or not it is consumed by the producer.

16. Small photovoltaic (PV) linked to domestic consumers needs to be viewed against the context of PV facilities in industry, which already offer a reasonable rate of return and payback period, that is, in the range of 260–600 kW, even when electricity system charges have to be paid. For further information, see the interview with Romero (2016b).

In this case, the grid is of key importance, and necessary for receiving supply and taking up surplus electricity production, as illustrated in Fig. 7.8; a value therefore has to be assigned to the grid.

In this respect, a mechanism needs to be implemented to ensure that self-sufficient consumers contribute to the overall cost of the system. This mechanism should be designed on a technical basis and with political consensus. If this is done correctly, it is then very likely that self-sufficiency will cease to be a political issue, and its development will be mainly determined by a combination of economics, technology, and marketing.

Likewise to develop prosumption, it is also important to consider the process of administrative authorization, construction of the installation, and the space required. In these issues, specialized companies, financing instruments, and new business models might be of help.

In their chapter, Mountain & Harris analyze the economics of electricity supplied by the grid relative to the combination of grid and rooftop PV. They compare market offers and calculate capital cost for various PV sizes in Victoria, Australia, considering capital subsidies and a price for the excess production exported to the grid. These hypotheses and others—such as 2% real cost of capital—are different to the cases analyzed in this chapter. Among their conclusions, they find that payback[17] periods are long with a median of 12 years in Victoria.

From a regulatory perspective and looking at the tariff structure, Haro et al. provide a detailed analysis of the structure of electricity tariffs for domestic consumers. They highlight the importance of the regulated component of the electricity price, which also includes "policy charges". After examining the current make-up of the domestic tariff (PVPC), and taking into account that by 2019 smart meters will have been fully rolled out (thus providing for hourly measures) they advocate an "efficient tariff" to replace the current system of payment of "policy charges." They propose specific methodology to compute hourly network tariffs to incentivize efficient self-generation, storage, and EV deployment.

5.2   Utility-Scale PVs

As mentioned at the beginning of this section, any discussion of this issue should also include the economic advantages of PVs in utility-scale facilities. Álvarez Pelegry and López Cardenete, (2016) report that the advantage of utility-scale solar power plants over rooftop installations has been demonstrated in a study by the Brattle Group, which concluded that a 300-MW solar PV plant has approximately half the cost per kWh of the same capacity divided among 5-kW home installations (in Xcel Energy Colorado System) and performed

---

17. That is to say, the relationship between bill reductions and the outlays needed to deliver those reductions.

consistently better in cost terms in all the scenarios considered in the study (Tsuchida et al., 2015).

Low prices of utility-scale PV projects can also be seen in other regions, such as Latin America. In Argentina under the RenovAr program, 20-year Power Purchase Agreement (PPA) was awarded in 2016 to four PV projects of 100 MW each (La Puna and Cauchari 1, 2, and 3) with prices of $59/60 per MWh[18] (Morrone, 2016).

Low prices are also being obtained in tenders in Chile—where prices fell from $128.90 per MWh in 2013 to $47.60 per MWh in 2016—and in recent tenders in Mexico, where average prices ranged from $44 to $32 per MWh (Relancio, 2016). The figures are corroborated by IEA PVPS (2016), which reports prices ranging from $29.10 per MWh in Latin America and the Caribbean to $67 per MWh in India.

Sadamori (2016) also reports long-term contract prices for new renewable power to be commissioned over 2016–2019 that range from $30 per MWh in Chile and the United Arab Emirates to $65–70 per MWh in the United States, $55–94 per MWh in India, and $87 per MWh in Germany.

The current reduction in costs is mainly due to a drop in the cost of PV modules from a total of around $2.7 per W in 2010 to $0.7–0.8 per W in 2016. The cost of the module represents 40% of the total in 2016, as compared to 70% in 2010, (Relancio, 2016). This trend can be explained not only with single figures, but with price ranges, hence the change for small-scale systems that went from $5.8–6.5 per W in 2006 to $1.8–3 per W in 2015 (IEA PVPS, 2016).

According to Sola González (2016), the drivers of low prices are: (1) large-scale projects, (2) low labor costs, (3) high irradiation, (4) access to low cost financing, (5) stable economy, and (6) long-term PPAs. Elecnor (2016) states that some prices and some projects are not sustainable or are not under construction. However, it recognizes that if the increase in PV volume continues, the "real" costs without dumping or subsidies may fall even further.

Given these developments in PV, it is not surprising that Masson (2016), predicts that global utility-scale capacity will increase from 54 GW in 2016 to 59 GW in 2020, and rooftop from 18 to 31 GW. Overall, therefore, the relative percentage penetration of rooftop PV will increase compared to utility scale, but in terms of actual capacity, utility scale will still be twice that of rooftop in 2020.

Based on this review, it is clear that distributed PV generation will play a more prominent role in the electricity system of the future, both in residential or commercial "prosumption" and at the utility-scale level, as described by Webb et al. Electricity storage will certainly represent a greater breakthrough and will probably have major impacts at the utility scale and at the residential level.

---

18. This level is similar to the 9 projects awarded in the NOA region for 530.8 MW in 19 projects in Mendoza, San Juan, and San Luis.

Adequate and competitive storage may have a strong impact at the PV utility scale, affecting the electricity grid and also fossil fuel generation (including gas), which at medium/high oil prices, may prove uncompetitive in the future.

6 ELECTRIC VEHICLES

Developments in EVs—in terms of both a potential increase in electricity demand and consumption patterns—could have significant implications for the electricity system.

There are three important issues that need to be addressed with regard to EVs. The first is the extent to which they can substitute conventional diesel and gasoline vehicles and how they will compete with alternative energies, that is, compressed natural gas, biofuels, liquefied petroleum gas, and, in a more distant future, hydrogen. The second issue is the effect of EV penetration on power demand and on the shape of the daily demand curve.

The third relevant issue is the impact of replacing diesel vehicles on the emission of greenhouse gases, such as CO_2 and NO_x, and on particle emissions.

With regard to the first point, the penetration of EVs will be mainly related to economic fundamentals, in terms of Total Cost of Ownership (TCO). This indicator is in turn closely related to the prices of vehicles and fuels, and in EVs, to the current and future price of batteries. However, TCO is not the only factor to be considered. Other items, such as subsides, incentives, the regulatory framework, as well as societal issues, also need to be taken into account. And last but not least, the recharging infrastructure is also a key factor.

Based on a hypothesis of specific consumption, prices of vehicles and fuels and cost of the recharging infrastructure, Álvarez Pelegry and Menéndez (2016) have performed a comparative analysis for Spain, focusing on the Basque region. The study calculated the following TCOs (in € per 100 km) for different energy types: gas oil, 18.5; battery electric vehicle, 24.7; compressed natural gas, 20.6; and liquefied petroleum products, 18.2.

In relation to the current and future prices of batteries, which clearly influence the TCO, elsewhere in this volume, Weber & Wilson discuss developments in automotive battery technology. They quote a number of sources, including Nykvist and Nilsson (2015), who give a wide range of prices per kWh (approximately $200–450 per kWh for 2010 and $150–250 per kWh for 2025–2030).[19]

Other assumptions about Li-ion batteries from the DOE indicate a target PHEV battery cost of $125 per kWh by 2020.[20] McKinsey and Company (2014) place the figure at nearly $200 per kWh in 2020 and $163 per kWh in 2025.

---

19. Increasing demand will improve economies of scale. Nykvist and Nilsson (2015) show that an increase in cumulative production from 100 to 1,000 MWh–10,000 MWh has improved the denominated learning rate by 12%–14%. Standardization will further reinforce economies of scale.
20. From $1000 per kWh in 2008 to $268 per kWh in 2015.

Others estimate a price of \$215 per kWh in 2030[21] (Element Energy, 2012). Fraunhofer (2013) examines costs and energy density for 2020, finding a considerable range of prices (€100–300 per kWh) and energy densities (150–300 Wh per kg) (all approximate figures).

Álvarez Pelegry and Menéndez (2016) assume a price of \$383 per kWh in 2015 and \$197 per kWh in 2020 (McKinsey, 2014) and a 25% share of TCO in battery costs (Hensley et al., 2009). The TCO results for the different energy alternatives without incentives are quite similar for the period 2020–2025: in the range of €17–20 per 100 km. These figures are also similar to the TCO for current EVs, with incentives/subsides.

It can thus be concluded that assumptions on penetration based on the relative TCO "prices" of battery electric vehicle, gas oil, compressed natural gas, and liquefied petroleum products are not very robust. Other elements, therefore, need to be considered to make reliable assumptions on the future development of EVs.

In Spain, EVs enjoy considerable policy support at different levels. The central government's MOVEA plan finances the implementation of fast and semifast recharging points on public roads. At the regional and local levels, as well as specific subsidies, a range of other measures has been implemented in cities. These include exemption from parking fees for EVs and priority use of dedicated bus lanes, as well as restrictions on the use of conventional fuels in downtown areas.[22]

Based on these data and considerations, the most conservative hypothesis would involve a continuation of the current rate of growth in new EVs to 2020 (around 5000 vehicles per year[23]), giving a total figure of around 40,000 in 2020. However, given the various promotional mechanisms (economic and others) and the expected decrease in the TCO, a more optimistic forecast, such as that used by the Gobierno de España (2016),[24] would suggest a figure of around 150,000 in 2020. After 2025, the EV market is likely to develop faster, with stronger penetration, resulting in an S-shaped growth in EVs.

With regard to the second of these points, and based on the hypothesis established in CNMC (2015), the effect of EV penetration on electricity demand

---

21. Based on Eurobat (2015). Half of the reduction in cost might come from cell improvements, while the remainder would reflect lowering packaging costs. In terms of improvements in materials, research into cell materials should continue to focus on bringing new cheaper materials in battery cells. Manufacturers are also continuing to develop lower-cost options for battery casings and optimize battery management system design.

22. One development that is likely to have an influence in large cities is the recent announcement by the mayors of Paris, Mexico, and Madrid of their intention to eliminate diesel use in their cities by 2025.

23. In June 2016 there were 18,187 EVs in Spain, of which 37% were passenger cars. Of the total, 87% were battery driven and 11% plug-in hybrids. Further, 2577 EVs were registered in the first half of 2016, representing a 14% increase in sales in 2015 (Gobierno de España, 2016).

24. It should be noted that this figure is less than that used to estimate the effect of EV penetration.

would be an increase of 1.5 TWh[25] in projected electricity demand to 2020[26] [estimated at 273–285 TWh for that year (CNMC, 2015)]. This would mean an increase of 30 MW during demand trough hours and 300 MW at the winter peak.[27]

Álvarez Pelegry and Menéndez (2016), basing themselves on Gobierno de España (2016) assume the same growth rate for 2020–2025 as in the previous 5 years. This assumption is in part based on the fact that TCO is not expected to decrease significantly until 2025. However, it is important to note that this figure may change as a result of technological and/or economic developments. After 2025, TCOs of EVs may become more attractive than diesel or gasoline or even GNC and GLP, and thus an annual growth rate of 5% for the period 2025–2030 has been considered.

One element that will certainly aid EV penetration, and help avoid incremental power capacity is RD 647/2011, which establishes a supertrough tariff (between 1 and 7 a.m.) for EV electricity consumption. This currently stands at €0.886 per MWh for the variable term (IDAE, 2016). Considering an average of 23,400 km per year and a specific consumption of 20 kWh per 100 km, the electricity tariff would be €95.22 per MWh, including payment for a capacity of 3.7 KW. This tariff obviously incentivizes the use of EVs and helps avoid the need to build new capacity for peak demand.

With regard to the third relevant issue—the impact of replacing diesel vehicles with EVs—Álvarez Pelegry and Menéndez (2016) have calculated "savings" in fuel expenses and emissions reductions, basing themselves on figures for a number of journeys in the Basque Country. If the beneficial effects of clean air are taken into account, the introduction of EVs appears to make economic and environmental sense.

If governments and politicians take this factor into account, it is clear that the incorporation of EVs may be one of the key elements in transport innovation and an eventual transformation of power grids.

7   CONCLUSIONS

In Spain, generous regulatory support for renewables, introduced within the framework of the government that plans to encourage renewable energy, had grave economic ramifications. The result was an increase in electricity prices for consumers that made self-consumption attractive, while distorting the effective development of prosumption.

---

25. The incremental consumption of 1.5 TWh may be optimistic. Other sources estimate that there will be 100,000–150,000 electric vehicles in 2020. Taking an assumed power consumption of 0.18 kWh per km for recharging EVs, the figure may be in the range of 0.4–0.6 TWh.

26. This is based on a stock of 500,000 electric vehicles in that year, whereas the government plans to achieve 30% of this figure, that is, 150,000 vehicles.

27. Installed capacity in Spain is 108,328 MW, and peak demand in 2015 was 40,324 MW.

The increased costs to the electricity system caused by that situation have engendered wariness among legislators toward any new developments.

At the same time, technological and economic developments in the field of PVs have led to a faster growth in PV generation than initially expected. This trend is likely to continue, not only for self-consumption at the household and small business levels, but also at the utility scale.

Developments in the area of EVs, both technological (improvements in vehicle range) and economic (reduced battery cost) should help bring the total cost of EV ownership more in line with other alternatives. At this time, the main barriers to large-scale development of this form of transport are economic, including the lack of a sufficient charging infrastructure.

Prosumption and greater penetration of EVs will influence the makeup of the power grids and electricity industry of the future. Specific regulations already exist in Spain in both areas, providing a legal framework for further development. Intelligent and efficient regulation must take into account the total cost of the system, as well as provide technically and economically efficient solutions.

To sum up, the power industry and grids of the future will be different than those of the present model—a message that resonates throughout this volume— and economic policy and regulations are likely to play a decisive role in this transformation. However, it is singularly difficult to predict the exact pace of change, or whether it will be evolutionary or disruptive in nature, as this will depend on a combination of regulatory, economic, and technological factors.

REFERENCES

Álvarez Pelegry, E., 2015. Tres Retos Para la Energía en España: Competitividad, Seguridad, y Crecimiento. ICADE Revista Cuatrimestral de las Facultades de Derecho y Ciencias Económicas y Empresariales de la Universidad Pontificia de Comillas, Madrid.

Álvarez Pelegry, E., López Cardenete, J.L. 2016. Unpublished Manuscript. "Distributed generation and energy transition. Green Energy Law.

Álvarez Pelegry, E., Menéndez Sanchez., J., 2016. Penetración de energías alternativas en el transporte de pasajeros por carretera. El caso del País Vasco. Inversiones, y reducción de emisiones. Propuestas y recomendaciones. Unpublished manuscript.

Barrero, F.A., 2016. La sombra de un impuesto que nadie paga. Energías Renovables, No. 155.

CNMC, 2015. Informe sobre la propuesta de planificación de la red de transporte de energía eléctrica 2015-2020. Available from: https://www.cnmc.es/Portals/0/Ficheros/Energia/Informes/20150618_Informe%20propuesta%20planificacion%202015-2020.pdf

Corrales, J., Dardati, E., Elejalde, R., Fuentes, F., Larrea, M., 2016. Políticas de fomento de las energías renovables. Análisis de la efectividad de las medidas adoptadas: España vs. Chile. Unpublished Manuscript.

Elecnor, 2016. Fotovoltaica 2.0: la nueva oportunidad del sector. III Foro Solar Español, 29–30 November 2016, Madrid.

Element Energy, 2012. Cost and performance of batteries for EVs: Final Report for Committee on Climate Change. Available from: https://www.theccc.org.uk/archive/aws/IA&S/CCC%20battery%20cost_%20Element%20Energy%20report_March2012_Public.pdf

Eurobat E-Mobility Battery R&D Road Map, 2015. Battery technology for vehicle applications.

Eurostat, 2016. Electricity prices for domestic consumers bi-annual data. Available from: http://ec.europa.eu/

Fraunhofer Institute for Systems and Innovation Research, 2013. Technology roadmap energy storage for electric mobility 2030.

Gobierno de España, 2016. Marco Acción Nacional de energías alternativas en el transporte. Available from: http://www.minetad.gob.es/industria/es-ES/Servicios/Documents/Marco-Accion-Nacional-energias-alternativas-transporte.aspx.pdf

Guindos, L., 2016. España Amenazada: De Cómo Evitamos el Rescate y la Economía Recuperó el Crecimiento. In: Península, E. Madrid

Hensley, R., Knupfer, S., Pinner, D., 2009. Electrifying Cars: How Three Industries Will Evolve. McKinsey Quaterly.

IDAE, 2016. Informe de precios energéticos regulados. Available from: www.idae.es

IEA PVPS, 2016. Survey Report of selected IEA countries between 1992 and 2015, T1–30.

Larrea, M., Castro, U., Álvarez, E., 2016. Instalaciones fotovoltaicas aisladas y conectadas a la red eléctrica. Un análisis técnico-económico. Unpublished manuscript.

Masson, G., 2016. Keys for Strategies & PV Industry Development. III Foro Solar Español, 29–30 November 2016, Madrid.

McKinsey and Company, 2014. Evolution. Electric vehicles in Europe: gearing up for a new phase? Amsterdam Roundtable.

Mir, P., 2011. La Regulación Fotovoltaica y Solar Termoeléctrica en España. Economía de la Generación Solar Eléctrica. In: Civitas, E. Madrid.

Morrone, M., 2016. Energías Renovables en Argentina. El rol del Estado Nacional en el desarrollo de la energía solar. III Foro Solar Español, 29–30 November 2016, Madrid.

Nykvist, B., Nilsson, M., 2015. Rapidly falling costs of battery packs for electric vehicles. Nat. Clim. Change 5, 329–332.

Relancio, C., 2016. Eficiencia de las subastas en la asignación de precio y en la garantía de capacidad. III Foro Solar Español, 29–30 November 2016, Madrid.

Romero, F., 2016a. Valoración y recomendaciones relativas a un RD pésimo. Energías Renovables, No. 152.

Romero, F., 2016b. Entrevista. Energías Renovables, No. 155, pp. 28–29.

Sadamori, K., 2016. Medium and long term perspectives for PV. III Foro Solar Español, 29–30 November 2016, Madrid.

Sajn, N., 2016. Electricity prosumers. European Parliamentary Research Service (Briefing, November 2016).

Sola González, E., 2016. EDF energies nouvelles. III Foro Solar Español, 29–30 November 2016, Madrid.

Tsuchida, B., Sergici, S., Mudge, B., Gorman, W., Fox-Penner, P., Schoene, J., 2015. Comparative Generation Costs of Utility Scale and Residential. Scale PV in Xcel Energy Coloradós Service Area. The Brattle Group.

FURTHER READING

Álvarez Pelegry, E., Castro Legarza, U., 2014. Generación Distribuida y Autoconsumo. Análisis Regulatorio. Cuadernos Orkestra, Bilbao.

Álvarez Pelegry, E., Ortiz, I., 2016. La Transición Energética en Alemania (Energiewende). Política, Transformación Energética y Desarrollo Industrial. Cuaderno de Orkestra. Available from: http://www.orkestra.deusto.es/images/investigacion/publicaciones/cuadernos/La_transici%C3%B3n_energ%C3%A9tica_en_Alemania_Energiewende_-_Versi%C3%B3n_web.pdf

Díaz Mendoza, A.C., Larrea Basterra, M., Kamp, B., Álvarez Pelegry, E., 2016. Costes de la energía y competitividad industrial. Cuadernos Orkestra. Available from: http://www.orkestra.deusto.es/es/investigacion/publicaciones/cuadernos-orkestra/928-precios-de-la-energia-y-competitividad-industrial

IEA, 2016. Global EV Outlook 2016. Beyond one million electric cars.

Macías, E., 2016. Entrevista. Energías Renovables, No. 155.

McKinsey and Company and Bloomberg New Energy Finance, 2016. An integrated perspective on the future of mobility.

Wietschel, M. et al. Market Evolution Scenarios for Electric Vehicles. Summary. Fraunhofer Institute for Systems and Innovation Research ISI.

Zsolosz, H., 2016. Autoconsumo. Ya no hace falta viajar a Nevada a preguntar por Tesla. Energías Renovables, No. 155.

Chapter 8

Quintessential Innovation for Transformation of the Power Sector

John Cooper
Prsenl, Austin, TX, United States; MaaS Energy, Austin, TX, United States

1 INTRODUCTION

In the controlled environment of regulations and monopolies, electric utilities have grown accustomed to an evolutionary change environment managed through top-down rules and processes. Used to managing and controlling change, incumbents now face changing conditions that risk the loss of that control and unanticipated negative outcomes as described in other chapters of this volume. Fundamental changes over the past decade now challenge old assumptions and stress these foundations, begging the question of how utilities may respond and maintain their important place in the economy and society. This chapter argues that *innovation is the key to answering the challenges of ongoing and accelerating change and bringing utilities into synergy with other stakeholders in the emerging energy ecosystem.*

A first step in tackling these twin challenges is to acknowledge new realities associated with change only now becoming apparent. The first reality is that change, not status quo, is the new cultural norm: that is, tomorrow is likely to be less like today. We should start expecting disruption, rather than continuing to be surprised by it. But most organizational processes inside utilities still remain structured around execution of routines designed to maintain quality of service and manage growing complexity. The second reality is that the pace of change accelerates in today's digital world: that is, if you are not moving forward and faster to keep up, you are falling behind. Again, most large bureaucratic organizations are challenged to become sufficiently flexible and adaptable to change so rapidly.

Understanding the problem—new realities associated with change—leads to a discussion on the principal consequence, rapid disruption and destabilization, and the potential of digital technologies (especially data analytics) and

Innovation and Disruption at the Grid's Edge. http://dx.doi.org/10.1016/B978-0-12-811758-3.00008-5

new business models to answer those challenges. With a foundation in data analytics, utilities are able to shift to a new approach to change that marries innovation with business transformation—not as a one-time project, but as a new organizational foundation that anticipates and answers ongoing, accelerating change with innovation, flexibility, and resilience.

This chapter connects the dots between (1) the challenges of change—most notably, Decarbonization, decentralized energy resources (DER) Integration, and Grid Modernization; (2) a new driving vision, the coming Energy Internet Economy; and (3) the innovation and business transformation strategies required to achieve that vision.

The rest of the chapter is structured as follows. Section 2 examines the twin challenges of adapting culture to ongoing change and business models to accelerating change. Section 3 introduces the Dynamic Innovation Cycle as a tool to understand individual and organizational adjustments to embrace, rather than fight, dynamic change. Section 4 examines data analytics as the foundation of a platform that supports strategies in response to serial disruptions and introduces the Quintessential Innovation (Q^2i) platform, designed to make innovation the foundation of sustainable business transformation. Section 5 examines the emerging Energy Innovation Market featuring the rising energy prosumer, the emergence of platforms and apps enabled by the Q^2i approach, and the growing importance of market data followed by the chapter's conclusions.

2   TWIN CHALLENGES: NONSTOP, EVER FASTER CHANGE

Electric utilities, and more and more, new energy economy stakeholders as well, face a twofold challenge. First, adaptations to climate change drive *more rapid decarbonization* of the energy supply chain, starting with coal now, and moving on to petroleum and natural gas in the coming decades. Filling in the gaps left by receding fossil fuels are innovative forms of clean energy and DER, driving the *personalization of energy* and creating new markets of *rising prosumers*. Second, change is becoming dynamic—changes driven by technology advances occur at an increasing pace, putting increasing pressure on incumbent companies to adapt. Changes just now becoming apparent can be expected to continue to unfold and to accelerate over time. This dynamic state of constant and increasing change is unprecedented in an industry historically defined by stability and long-term capital investments in infrastructure.

All energy consuming and producing individuals and organizations will need to find ways to accommodate ongoing change that moves faster over time. This *Pace Problem* will hit conservative institutions like governments and electric utilities especially hard. The advancing disparity between rapid external change and slower, more methodical internal change creates a compounding problem that signals the need for a paradigm shift. To address these change challenges and keep pace with still more changes in the future, innovation in all its forms will become the means for fundamental shifts in our approaches to energy.

Consider, for instance, business practices in the grid ecosystem and the Internet economy with regard to innovation. Where the grid ecosystem moves slowly to ensure system reliability and regulatory equity, the Internet economy has embraced the concept of *emergence*, a signature quality that leverages innovation to address constant change. Emergence and innovation open the entire system to new industry players who seemingly introduce transformative concepts overnight, growing large with market acceptance, recently described as Big Bang Disruption.[1] Perhaps the best evidence that such disruption has become commonplace in the Internet economy is the rapid dominance of platforms provided by companies like Google and Amazon, which emerged and steadily grew to ubiquitous influence over the consumer economy with a common business model: *harnessing data to drive innovation.*

One way for the electric utility industry to accelerate adaptation to innovation will be to take it in small doses. In this way, *Iterative Innovation* may become the key to situational data analytics, grid optimization, power continuity, DER resilience, sustainability and fossil fuel transition, and a shared future of *Personal Energy.* Given the collapsing timeframe to address transformation challenges, effective and sustainable business and utility transformation will depend upon collaboration among utilities associated with innovation.

As introduced above, driving this innovation imperative is an increasingly dynamic energy environment characterized by new technologies and business models introduced at a dizzying rate, well beyond the capacity of today's utility organizational culture and regulatory conservatism to keep up. A new paradigm that not only integrates technology in creative ways, but also goes beyond technology to embrace cultural, organizational, and institutional transformation, is needed to make the nascent *energy Internet—the eNet—*more like the Internet and less like the grid. This chapter will outline the connections between innovation and various aspects of change, leading to a formula for market and industry transformation, and importantly, a formula that is still possible in a collapsing timeframe.

Embracing change while preserving legacy investments that still have value won't be easy, not by a long shot. Conventional approaches, like extending the current top-down grid management paradigm with technological enhancements or system redesign, will encounter fundamental challenges from ongoing environmental changes, including:

- the rise of third-party competitive energy efficiency and site-based energy alternatives to grid power;
- consequent flat or declining utility rate-based revenues from diminishing kWh sales;

---

1. Big Bang Disruption, Larry Downes and Paul Nunes, Harvard Business Review, March 2013; see also https://www.accenture.com/us-en/~/media/Accenture/Conversion-Assets/DotCom/Documents/Global/PDF/Industries_18/Accenture-big-bang-disruption-strategy-age-devastating-innovation

- the rising cost of grid modernization and a managed transition to a new paradigm and related resource needs; and
- nontechnology transition challenges (organizational change, community energy with third-party stakeholders, new business models, regulatory reform and changes in energy policy, etc.).

These are the elements that are increasingly understood to drive the need for *business transformation*, a broader term that embraces technology changes as a component, but looks more holistically at new business models and an evolving energy ecosystem. But to address these broader issues successfully and in a timely fashion, such business transformation won't be enough: a *transformation framework* that embraces innovation will be needed.

To summarize, *ongoing change* and *the pace problem* require *innovation*, *business transformation*, and a *transformation framework* to achieve the qualities of adaptability, reliability, and resilience needed in the future. For instance, a typical utility's grid topology must be transformed from a one-way "step down" distribution network into something closer to the Internet's highly redundant mesh. Certainly a variety of new technologies and business practices would be needed to accomplish such grid transformation. But as complex as grid redesign and technology integration would be while maintaining reliability and low energy prices, there's one more thing standing in the way of technology and business model upgrades: cultural acceptance.

For transformation success, individuals and organizations must be willing to accept innovation as the new cultural norm. The imperative of innovation is highlighted throughout this book, but especially so in the chapter by Woodhouse and Bradbury. Section 3 addresses the cultural change that will be enabled by a *Dynamic Innovation Cycle* and remaining sections will describe how organizations can turn innovation into a core competency.

3   MANAGING THE NEW REALITY OF DYNAMIC CHANGE

The human element, generally overlooked in discussions of innovation, transformation, and technology, is fundamental to any successful organizational change program. Recurring and incessant, the ongoing change we'll face going forward will be driven by two fundamental themes:

- *Climate Change* as a growing, cataclysmic threat to global societal and humanity; and
- *Personal Energy* that empowers energy consumers and prosumers—the emergent, individual, and collective response to that threat.

Climate Change holds the potential of unavoidable catastrophe, absent serious mitigation as seen in the Paris Agreement,[2] ratified on October 5, 2016, and

---

2. http://unfccc.int/paris_agreement/items/9444.php

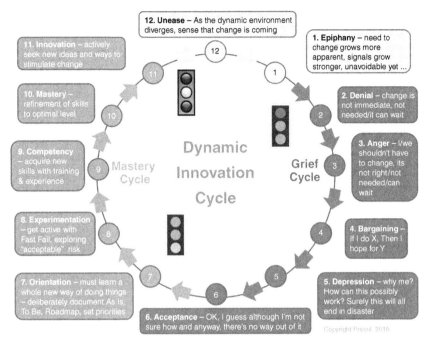

FIGURE 8.1   The Dynamic Innovation Cycle. *(Source: Prsenl.)*

effective 1 month later. As coal recedes as a source of electricity, natural gas and petroleum must follow for global temperature rise to be contained. The shift from fossil fuels to a clean energy economy promises an increasingly dynamic energy ecosystem that is ever more dependent on ubiquitous, reliable electricity. Such business transformation will include a shift from selling electricity as a commodity to viewing clean energy as a personal service. *Personal Energy* as a service will require every individual, and organizations of all types and sizes, to adapt to ongoing change. The acceptance of Personal Energy will grow with improved technologies and new business models and widespread consumer adoption will contribute to an improved toolset for addressing change and disruption.

Based on innovation, *Personal Energy* will require multiple passes through the *Dynamic Innovation Cycle*, as in Fig. 8.1. With each pass, it is hoped, personal or organizational familiarity with the pattern and developing skill sets will make the process more seamless, even automatic. The Dynamic Innovation Cycle is especially appropriate given that the urge to deny climate change is strong, the path to mastery in Personal Energy is difficult to discern, and Energy Innovation is at the heart of any solution. This 12-step cycle addresses rapid and episodic disruption, the fundamental challenge facing individuals and more particular to this chapter, those inside utilities and other organizations that must grapple with significant change and transformation.

1. *Epiphany.* The beginning step of the 12-step Dynamic Innovation Cycle is a realization (i.e., an epiphany) that myriad warning signs form a pattern, which provides growing clarity that disruptive change is on the horizon.

Grief Cycle

The Five Stages of Grief, identified and made famous by Dr. Elizabeth Kubler-Ross,[3] were originally developed to explain the behavior of patients facing a terminal diagnosis. This model has been borrowed and expanded to explain coping with grief in many guises. In this case, we borrow it to explain the grief of accepting disruptive changes in routines. Persistent change is debilitating, as are predictions of calamity. Both apply here. Dynamic change means that these five stages will reoccur over and over, so we call it the *Grief Cycle*.

2. *Denial.* Refusal to accept facts and reality is a common defense in the face of distressing conclusions and impacts. But delays that accompany denial compound neglected problems, delay change, constrain options, and raise costs. Less time on denial and delay means more time and resources for adaptation.

3. *Anger.* Loss of status quo, routines, even real economic loss generates anger, a natural defense mechanism humans experience when stressed or threatened. While anger can serve a constructive purpose in some cases, unfocused or blind anger is a destructive and damaging reaction to change.

4. *Bargaining.* With no improvement and exhaustion from anger, focus shifts to bargaining in the form of If/Then propositions… "If I can just have X (some more time to adjust, relief from my stress, etc.), Then I will be willing to reform, make changes, accept (my fate, this change, etc.)."

5. *Depression.* With previous gambits to avoid change proven consistently ineffective, Depression sets in and it seems as if one is at the bottom of a deep dark hole and there's no way out. Helplessness gives way to paralyzing despair. "I am screwed, I am a victim and there is nothing I can do" …

6. *Acceptance.* With other options played out, acceptance of change ushers in a new world order to replace the old. But acceptance does nothing to obviate the loss of the old and the vacuum it creates. Something must be done. As we all know, "Nature abhors a vacuum."

Mastery Cycle

Acceptance marks an end to the Grief Cycle, but it also opens a new beginning for the *Mastery Cycle*, which fills the vacuum with a new paradigm comprising the steps an individual or organization takes on gaining maturity and growing

---

3. https://en.wikipedia.org/wiki/K%C3%BCbler-Ross_model

into a new paradigm. Progress through this cycle is subjective, with some moving through its natural steps much more rapidly than others.

7. *Orientation.* Stripped of old rules, the individual or organization is at first on shaky ground in the new, unfamiliar paradigm. It's time to go back to school, in a way, to learn the new ways and new vocabulary and grow into the new paradigm.

8. *Experimentation.* In this exciting, but awkward stage, newfound skills are executed haltingly, mistakes and missteps abound. Experimentation provides room to learn lessons more thoroughly, to apply the general to the specifics of any situation. In organizations, experimentation takes form in Trials, providing opportunities for individual stakeholders to engage with utilities and corporations themselves mastering the stages of change, innovation, and adaptation.

9. *Competency.* As skills are mastered, those seeking to change gain a handle on what is necessary. They acquire their own toolbox, and master how to use their favorite tools for different situations.

10. *Mastery.* As hidden talents are developed, skills mastered, and experience logged, mastery begins to emerge within the new paradigm. In time, habits and routines even turn the formerly odd and new into the commonplace and normal.

11. *Innovation.* Moving beyond the commonplace, the master can begin to play with the rules; invent new and better ways; and innovate to create a unique personal style. The Innovation Stage is exciting: sharing, experimenting, turning new technologies, products, services, and business models into novel approaches.

Unease

12. *Unease.* Having gone from mastery to innovation, one hopes to spend time in this exciting stage to reap the benefits of all that hard work. But hints appear that show the environment is changing and that the conditions and rules that support mastery are disappearing. With a threat to one's security and well-being growing, unease becomes the first signal that the cycle is completing itself and soon it will be time to repeat the process.

A key point to remember about this dynamic cycle is that it is not zero sum: each turn around the cycle imparts new skills and wisdom to the individual, organization, or society. With each turn, patterns become more obvious, stages more recognizable, and indeed, innovation skills more apparent, eventually, becoming habits. Experience brings pattern recognition, anticipation, and rapid progress through the Grief Cycle, perhaps even skipping it altogether in zeal to get to the Mastery Cycle and Orientation, where progress begins. Those who master this cycle become natural community leaders, their common enemy, the Grief Stages, which slow everyone down and inhibit successful adaptation to change.

4 QUINTESSENTIAL INNOVATION (Q²i)

4.1 Data Analytics and Disruption

For utilities to execute the necessary changes to the grid ecosystem at the pace of change that consumers have come to demand in all other areas of the economy, they will need to emulate the success of innovative Internet companies with situational data analytics as the universal, fundamental core of the platforms that will emerge to foster innovation. Data analytics (of both operational and market data) and emerging innovative ideas will guide the automation required to transform the grid. For stakeholders to manage increasing complexity, such as 100% distributed solar energy, more and more innovation will be necessary. The current operating culture of utilities holds system reliability as a primary function, stressing routines over innovation. Situational data analytics supports strategies to both overcome conventional change inhibitions and stimulate homegrown innovation as it grows within the organization.

Incorporating data-driven innovation as a foundational platform, as shown in Fig. 8.2, is the secret to overcoming the Pace Problem. Such a platform will position innovation to drive waves of transformation like a virus through progressive maturity cycles. *Iterative innovation*, as described below, enables stakeholders (i.e., utilities, consumers, and vendors) to adapt to a new pace, tools, stakeholder roles, and business models.

Utilities enjoy such benefits as:

- grid modernization;
- distribution-based power;
- distribution feeder balancing;
- positive energy integration;
- strategic positioning of DER; and
- deferral of distribution system upgrades.

In turn, consumers see:

- a growth path to Prosumer;
- decarbonization;
- increasing innovation;
- ways to support grid modernization; and
- new revenue from positive energy.

Finally, vendors enjoy:

- new avenues to partnering with utilities;
- effective customer programs; and
- access to utility marketing efficiencies.

For utilities to embrace a new business model that enables expected technological changes, new operational models must be mastered, as further described below and elsewhere in this volume. But that's not all: emerging markets and market

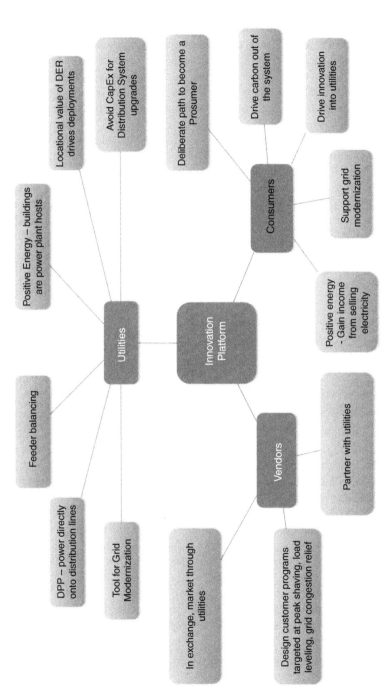

FIGURE 8.2 Innovation Platform. (*Source: Prsenl.*)

data must join operations planning and protocols. Such an integrated operations/ market platform will blend (1) utility IT and OT data with (2) third-party data (e.g., weather data), as well as (3) data about energy consumers and prosumers.

In short, the concept described henceforth, the *Quintessential Innovation Platform (Q²i)*, becomes the means to harness data and situational analytics to foster data-driven iterative innovation. Q^2i requires situational analytics software to provide a stable open data architecture that effectively connects the scores of incumbent software products used in regular utility operations.

The iterative innovation framework becomes the foundational key to a sustained transformation of the electric grid into a dynamic system that manages nearly infinite complexity by making flexibility and adaptability core elements of integrated system and market operations. Q^2i describes five-stage innovation in two vertical and horizontal integrated platforms: the vertical platform enables a five-stage *iterative data innovation cycle* from data to innovation; and the horizontal platform enables a five-phase *iterative business transformation cycle* across progressive maturity stages, from planning to innovation. The sections below show how the Q^2i platform enables an Energy App Economy to drive the pace required to keep up with an ever more dynamic economy and society.

4.2 The Iterative Data Innovation Cycle

The Iterative Data Innovation Cycle progresses in stages as the situation dictates, from Data Acquisition to Data Analytics to Insight to Action to Innovation and then back to repeat the cycle at a higher plane. In a dynamic world, the utility never really completes or exits this process, but makes becoming ever more innovative its principal core competency and business driver (Fig. 8.3).

1. *Data Acquisition.* Iterative Data Innovation begins with investment in a data acquisition system. A utility can't transform without access to critical operational, third-party and market data.
2. *Data Analytics.* Data analytics combines multiple types of data and provides ready access to utility managers via well-designed user interfaces.
3. *Insight.* Adding value through experienced operational insights is the essence of situational data analytics and key to transformation.
4. *Action.* Shifting from good ideas and insights to putting them to the test requires action and integration of new technologies, helped by third-party participation.
5. *Innovation.* Analysis and insights from trials and pilots confirms or refutes hypotheses, underscoring the need for more data and new trials to hone understanding of new approaches and new capabilities.

4.3 The Iterative Business Transformation Cycle

As developed at Prsenl and shown in Fig. 8.4, a corresponding *Iterative Business Transformation Cycle* with its multiple iterations and experience moves across

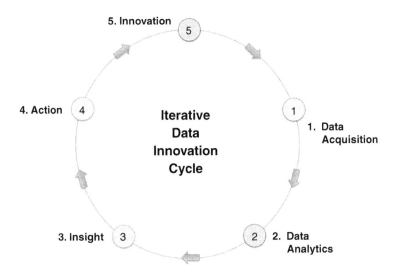

FIGURE 8.3 The Iterative Data Innovation Cycle. *(Source: Prsenl.)*

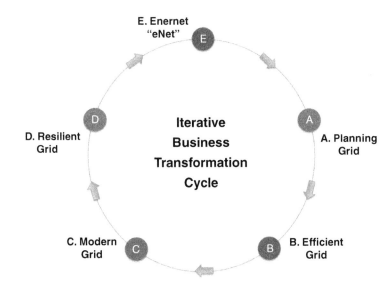

FIGURE 8.4 The Iterative Business Transformation Cycle. *(Source: Prsenl.)*

five maturity stages, from Planning (system planning) to Efficiency (efficient business processes) to Modernization (modernized grid technology integration) to Resiliency (system resiliency and DER integration) to Innovation (innovative transactive energy economy). The organization uses each step of the Iterative Data Innovation Cycle inside each maturity stage to become consistently more innovative as it adopts a variety of change activities moving through progressions. In both cycles, the utility uses steady and consistent iterations to mature and attain its vision of becoming an innovative, resilient company.

1. *Planning.* The essence of business transformation is sound planning; shifting from traditional department-level planning to holistic, integrated strategic planning.
2. *Efficiency.* The most rapid and easiest place to begin a business transformation is to identify and eliminate wasteful processes to become more efficient, typically in terms of process innovation/reform.
3. *Modernization.* More capital-intensive than efficiency, grid modernization channels Smart Grid ideas to embrace improved communications and enhance the acquisition of sensor data, enabling improved operations and greater flexibility. IT/OT integration and technology investment invite the collaboration of tech vendors to reimagine utility operations for new purposes.
4. *Resiliency.* Closely tied to DER integration, system resiliency considers system flexibility and adaptability. The secret to blending DER and centralized systems is to open the system to new ideas and participants by leveraging new DER capabilities to address operational conundrums like resiliency.
5. *Innovation.* The ultimate goal of a flexible, adaptive system is to promote, rather than react to change. Visions, such as a peer-to-peer transactive energy system and an energy Internet (i.e., eNet) suggest an evolved grid that operates as an efficient market enabler of innumerable energy consumers, producers, and prosumers to meet the goals of system resiliency, low cost, and highly reliable power, and low-to no-carbon emissions in a more open, efficient, and sustainable manner.

4.4 Quintessential Innovation for Electric Utilities: The Q^2i Platform and Innovation Apps

The emerging Energy Innovation Market (described in more detail below and also throughout this book whose main theme is innovation) will draw on lessons learned from the emergence of the App Economy introduced by companies like Google, Apple, and Amazon. Platforms and Apps represent the ultimate consumer vehicle to incorporate innovations and enjoy improved customer value. The genius of the platform model is to invest in technology that supports near constant innovation offered in the form of modular services designed for specific market niches and specific problems, creating solutions that provide specific value (i.e., Apps).

The Q^2i platform, shown in Fig. 8.5, is designed to enable this modular approach to accelerate problem solving by inviting utilities and third parties to share, and profit from, their innovations in the form of Apps. A current challenge for utilities suffering with the Pace Problem is the adherence to a

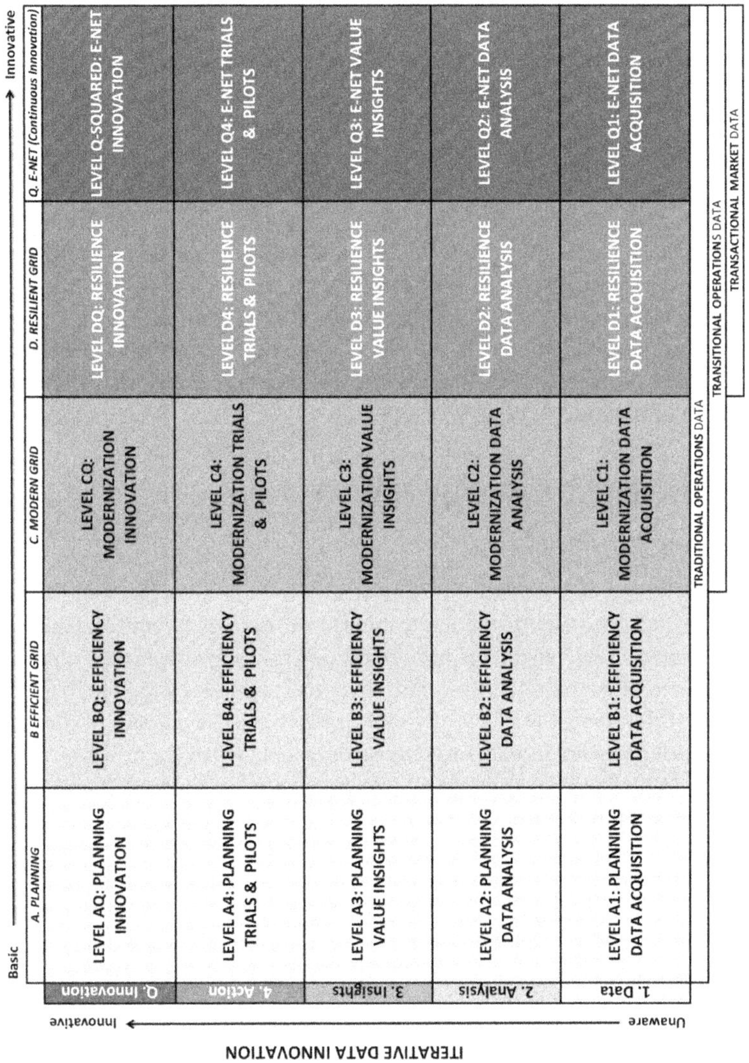

ITERATIVE BUSINESS TRANSFORMATION INNOVATION

Basic ——————————————→ Innovative

	A. PLANNING	B EFFICIENT GRID	C. MODERN GRID	D. RESILIENT GRID	Q. E-NET (Continuous Innovation)
Q. Innovation	LEVEL AQ: PLANNING INNOVATION	LEVEL BQ: EFFICIENCY INNOVATION	LEVEL CQ: MODERNIZATION INNOVATION	LEVEL DQ: RESILIENCE INNOVATION	LEVEL Q-SQUARED: E-NET INNOVATION
4. Action	LEVEL A4: PLANNING TRIALS & PILOTS	LEVEL B4: EFFICIENCY TRIALS & PILOTS	LEVEL C4: MODERNIZATION TRIALS & PILOTS	LEVEL D4: RESILIENCE TRIALS & PILOTS	LEVEL Q4: E-NET TRIALS & PILOTS
3. Insights	LEVEL A3: PLANNING VALUE INSIGHTS	LEVEL B3: EFFICIENCY VALUE INSIGHTS	LEVEL C3: MODERNIZATION VALUE INSIGHTS	LEVEL D3: RESILIENCE VALUE INSIGHTS	LEVEL Q3: E-NET VALUE INSIGHTS
2. Analysis	LEVEL A2: PLANNING DATA ANALYSIS	LEVEL B2: EFFICIENCY DATA ANALYSIS	LEVEL C2: MODERNIZATION DATA ANALYSIS	LEVEL D2: RESILIENCE DATA ANALYSIS	LEVEL Q2: E-NET DATA ANALYSIS
1. Data	LEVEL A1: PLANNING DATA ACQUISITION	LEVEL B1: EFFICIENCY DATA ACQUISITION	LEVEL C1: MODERNIZATION DATA ACQUISITION	LEVEL D1: RESILIENCE DATA ACQUISITION	LEVEL Q1: E-NET DATA ACQUISITION

TRADITIONAL OPERATIONS DATA

TRANSITIONAL OPERATIONS DATA

TRANSACTIONAL MARKET DATA

ITERATIVE DATA INNOVATION

Innovative ←——————————————— Unaware

FIGURE 8.5 Quintessential Innovation: Q²i Platform and Apps Model for *Utilities*. (*Source: Prsenl.*)

service territory definition of the provisioning of commodity kWhs. Locked into relatively small markets, utilities have a small base on which to apply their innovations.

The degree to which utilities are able to share their innovations in the form of apps will provide three essential solutions to the challenges introduced at the start of this chapter.

1. *Time and Resources to Implement Change.* The requirements to implement solutions will be dramatically reduced if utilities are able to adopt vetted, modular, specific solutions in the form of affordable apps, rather than develop and implement more expensive, slower custom solutions over and over again.

2. *Monetization of Innovation.* The ability to participate in the app revenue stream provides utilities with an alternative to rate-based revenue, a key objective of business transformation projects, now widely accepted among utility executives.

3. *Motivation to Innovate.* With the potential revenue component to offset the costs/risks associated with solution development, a utility will be encouraged to be ever more innovative.

5 THE ENERGY INNOVATION MARKET

5.1 Rising Prosumers

Consumers have a new relationship with electricity, driven in part by increasing outages, but also by new technologies that open up new avenues of independence. Electric utility ratepayers have traditionally been a dependent class with little access to information, little control over their electricity, and perhaps consequently, little interest in electricity except maybe in the amount of the utility bill they must pay each month, or on those occasions when the grid experiences an outage. In the developed world to date, outages have been short and rare, at least, that is, until recently.

Extended outages—days or weeks without electricity—have become more common over the past decade where extreme weather makes the grid vulnerable to disruption. And those consumers who can have begun to take steps to ensure power continuity in the event of a future loss of power. This newfound energy independence is not just tied to outages and storm-prone areas. Other consumers are looking into new approaches to energy, spending time on the web to learn how to conserve power by making changes to their home or building, for instance.

As energy consumers grow in power consciousness, they may begin to look at various appliances that consume power and replace them with Energy Star versions that consume much less (e.g., new refrigerators). Many are intrigued by the rows of now-affordable LEDs at the local hardware store, not to mention the growing interest in onsite power—the solar panels they see on neighboring rooftops—and electric vehicles. There's a word for those who consume as well as produce power—*prosumers.*[4]

To mark this transformation from dependent ratepayers to consumers to prosumers, let us use the term *Rising Prosumers*, showing that ratepayers are on unique, individual journeys to energy independence. Rising Prosumers make up a new class of energy consumers, rapidly growing into a significant market segment with different expectations, different perspectives, and a growing appetite to take control of their energy destiny, perhaps not only to avoid outages or save money, but also to align their consumption with their personal values associated with climate change.

The Rising Prosumer represents new market needs, as more and more maturing consumers, actively engaged in seeking a more sophisticated perception of energy value, move beyond the commodity focus on monopoly dependency and $/kWh. As consumers shift to ever greater energy independence they represent a key challenge for electric utilities, namely, to redefine the nature of consumer engagement, a difficult task for monopoly utilities who have traditionally seen marketing as relatively one-dimensional, principally distributing information one-way out to their ratepayers via bill stuffers (analog), and websites and social networking applications (digital).

A recent study by Accenture[5] examined the average time a consumer engages with a utility: "In 2016, the average customer of a regulated U.S. utility spent about 8 minutes interacting with their utility through digital channels and about 11 minutes with a representative. Half of all customers did not digitally interact with their energy provider at all."[6]

5.2 Marketing Data: The Missing Link

As we move from monopolies to markets, the emerging *Personal Energy* sector will be comprised of multiple stakeholders, including:

- consumers maturing to become *rising prosumers* and associated communities;
- new DER vendors—solar power providers and their counterparts in the fields of Energy Efficiency, Demand Response, Energy Storage, EV Charging, and Microgrids;
- incumbent utilities, retail electric providers, and grid companies and associations; and
- governments at all levels.

---

4. Alvin Toffler introduced this term in his bestseller *Future Shock*, published in 1970, then elaborated on it in *The Third Wave* in 1980.
5. The New Energy Consumer: Thriving in the Energy Ecosystem, new customer plays to help utilities transform amid infinite disruption, Accenture, 2016.
6. https://www.greentechmedia.com/articles/read/customers-spend-8-minutes-a-year-interacting-online-with-their-utility

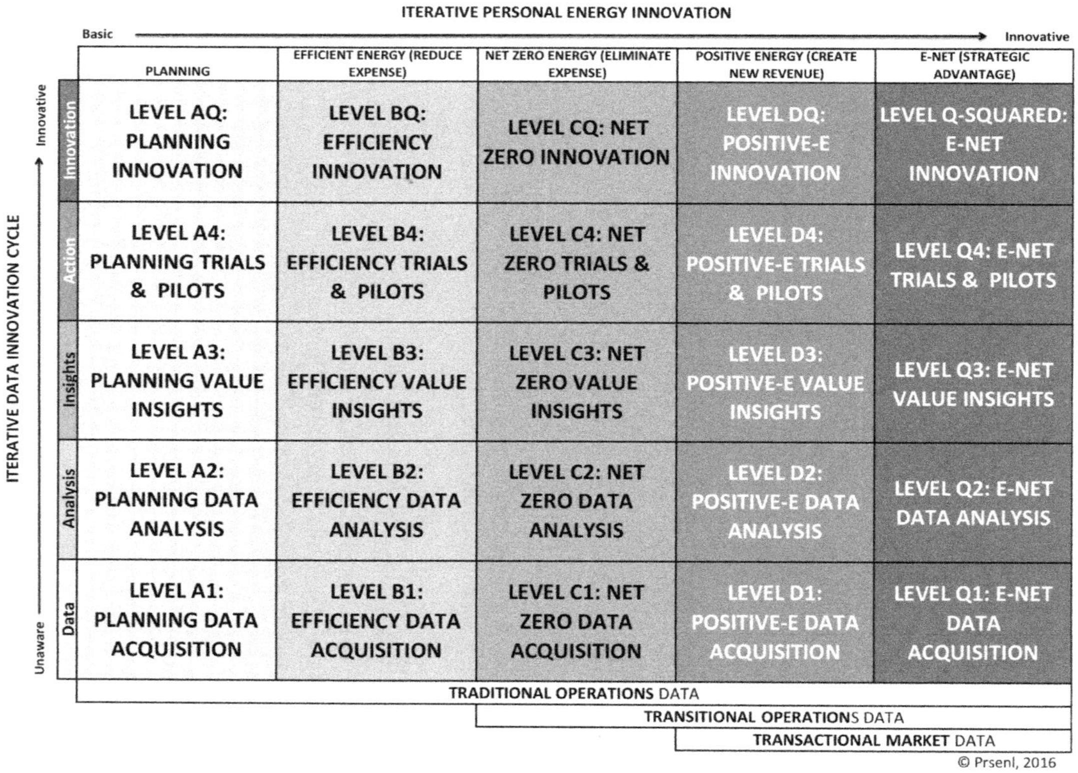

ITERATIVE PERSONAL ENERGY INNOVATION

Basic ──────────────────────────────────────→ Innovative

ITERATIVE DATA INNOVATION CYCLE	PLANNING	EFFICIENT ENERGY (REDUCE EXPENSE)	NET ZERO ENERGY (ELIMINATE EXPENSE)	POSITIVE ENERGY (CREATE NEW REVENUE)	E-NET (STRATEGIC ADVANTAGE)
Innovation	LEVEL AQ: PLANNING INNOVATION	LEVEL BQ: EFFICIENCY INNOVATION	LEVEL CQ: NET ZERO INNOVATION	LEVEL DQ: POSITIVE-E INNOVATION	LEVEL Q-SQUARED: E-NET INNOVATION
Action	LEVEL A4: PLANNING TRIALS & PILOTS	LEVEL B4: EFFICIENCY TRIALS & PILOTS	LEVEL C4: NET ZERO TRIALS & PILOTS	LEVEL D4: POSITIVE-E TRIALS & PILOTS	LEVEL Q4: E-NET TRIALS & PILOTS
Insights	LEVEL A3: PLANNING VALUE INSIGHTS	LEVEL B3: EFFICIENCY VALUE INSIGHTS	LEVEL C3: NET ZERO VALUE INSIGHTS	LEVEL D3: POSITIVE-E VALUE INSIGHTS	LEVEL Q3: E-NET VALUE INSIGHTS
Analysis	LEVEL A2: PLANNING DATA ANALYSIS	LEVEL B2: EFFICIENCY DATA ANALYSIS	LEVEL C2: NET ZERO DATA ANALYSIS	LEVEL D2: POSITIVE-E DATA ANALYSIS	LEVEL Q2: E-NET DATA ANALYSIS
Data	LEVEL A1: PLANNING DATA ACQUISITION	LEVEL B1: EFFICIENCY DATA ACQUISITION	LEVEL C1: NET ZERO DATA ACQUISITION	LEVEL D1: POSITIVE-E DATA ACQUISITION	LEVEL Q1: E-NET DATA ACQUISITION

Unaware ──────────────────────────────────────→ Innovative

TRADITIONAL OPERATIONS DATA

TRANSITIONAL OPERATIONS DATA

TRANSACTIONAL MARKET DATA

© Prsenl, 2016

FIGURE 8.6   Quintessential Innovation: the Q²i Platform applied to *Rising Prosumers*. (*Source: Prsenl.*)

As monopolies transition into markets with multiple stakeholders and relationships, it will be increasingly important to understand how the market functions and for that, much more marketing data will be needed. By focusing on data and a framework to (1) manage complexity; (2) channel innovation into transformation; and (3) accelerate the pace of change, the Q^2i platform, now adapted for use by Rising Prosumers (Fig. 8.6) becomes the essential component of any company seeking to get more out of data analytics. By providing a practical means for a managed transition at the Rising Prosumer level, Q^2i enables medium and long-term goals for the consumer and producer sides of the marketplace to come together. Industry transformation seeks a shared, repeatable format; Q^2i helps avoid both the redundancy inherent in our fractured utility industry and the challenge of the Pace Problem.

6 CONCLUSIONS

This chapter began with a two-pronged challenge: ongoing and accelerating change and the threat of disruption to the historically stable electric industry. The remainder of the chapter engaged a progression through the challenge and proposed a framework for innovation as an answer to the challenge, as described below.

- *Dynamic Change* will impact all stakeholders in the energy ecosystem.
- The *Pace Problem* challenges bureaucracies and incumbents most of all.
- *Energy Innovation* is positioned as the answer to a more dynamic energy economy, providing greater value for customers and greater flexibility and adaptability for organizations of all kinds.
- The *Dynamic Innovation Cycle* helps individuals and organizations understand the means to adapt to ongoing, accelerating change and disruption.
- *Platforms* are a proven mechanism for making innovation a core competency of a new business model.
- The *Quintessential Innovation (Q^2i) platform* comprises two parts: the *Iterative Data Innovation Cycle* and the *Iterative Business Transformation Cycle*.
- These two cycles combine to provide an individual or organization a practical means to mature into an acceptance of innovation as a core competency.
- The Q^2i platform is adaptable to be used with apps to address the Pace Problem and to integrate the consumer and producer sides of the emerging *Energy Innovation Market*.

As shown in this chapter and throughout this book, the dynamic nature of change will continue to confound individuals and companies in our decarbonizing, digitally driven energy world. For those who seek to thrive in this new business climate, innovation offers a path to a new paradigm, but the path is loaded with constraints and obstacles, as shown in many chapters herein. This chapter suggests a particular strategy, a managed transition framework that leverages platforms and apps, which is a proven mechanism outside the energy industry. Adopting best practice in this manner points to new directions and new pathways to manage complexity and open up to new stakeholders in the evolving energy economy.

Part II

Enabling Future Innovations

9. Bringing DER Into the
   Mainstream: Regulations,
   Innovation, and Disruption
   on the Grid's Edge   167
10. Public Policy Issues
    Associated With Feed-In
    Tariffs and Net Metering:
    An Australian Perspective   187

11. We Don't Need a New
    Business Model: "It Ain't
    Broke and It Don't
    Need Fixin"   207
12. Toward Dynamic Network
    Tariffs: A Proposal for Spain 221
13. Internet of Things and the
    Economics of Microgrids   241

Chapter 9

Bringing DER Into the Mainstream: Regulations, Innovation, and Disruption on the Grid's Edge

Jim Baak
Vote Solar, Oakland, CA, United States

1 INTRODUCTION

Thomas Edison and Nikola Tesla were pioneers and fierce rivals. Edison, a proponent of direct current, electrocuted live animals to demonstrate the dangers of Tesla's alternating current for transmitting electricity. But there is one thing they both had in common—they both saw a future where renewable energy would dominate.

In 1931, Edison described his vision of the future this way:

We are like tenant farmers chopping down the fence around our house for fuel when we should be using Nature's inexhaustible sources of energy—sun, wind and tide. … I'd put my money on the sun and solar energy. What a source of power! I hope we don't have to wait until oil and coal run out before we tackle that.[1]

Likewise, on the issue of harnessing the sun's power, Tesla said:

A far better way, however, to obtain power would be to avail ourselves of the sun's rays, which beat the earth incessantly and supply energy at a maximum rate of over four million horsepower per square mile. Although the average energy received per square mile in any locality during the year is only a small fraction of that amount, yet an inexhaustible source of power would be opened up by the discovery of some efficient method of utilizing the energy of the rays.[2]

---

1. In conversation with Henry Ford and Harvey Firestone (1931); as quoted in *Uncommon Friends: Life With Thomas Edison, Henry Ford, Harvey Firestone, Alexis Carrel & Charles Lindbergh* (1987) by James Newton, p. 31.
2. The Problem of Increasing Human Energy, *Century*, illustrated magazine, by Nikola Tesla, June 1900, http://www.tfcbooks.com/tesla/1900-06-00.htm

Innovation and Disruption at the Grid's Edge. http://dx.doi.org/10.1016/B978-0-12-811758-3.00009-7

While they both recognized the potential for renewable energy, they could not have foreseen the success of distributed solar PV, let alone advances in energy storage or the expanded role consumers would play in this idyllic future.

Nearly 100 years after Tesla and Edison made these prognostications, the utility industry is facing perhaps the most significant transition since the two energy pioneers debated the virtues and pitfalls of alternating versus direct current. Driven by climate change policies, advances in technology, significant cost reductions, and the emergence of the prosumer—as discussed in companion chapters in this volume—states like California, Hawaii, New York, and a host of others are forced to recognize and begin planning for significant growth potential for distributed energy resources.

Like the divisive Edison/Tesla debate over how best to transmit electricity, today there are clear divisions among the various stakeholder groups about what the future of the utility industry should look like. Given the great deal of uncertainty surrounding the future role of utilities and the resulting impact on their financial well-being, utilities are trying to protect their vested interests while at the same time complying with regulatory mandates and state policy imperatives.

DER providers are also trying to carve out a profitable niche in the burgeoning prosumer energy space and create innovative products and services to differentiate themselves from one another and from utilities. Yet these same providers view utilities as potential customers and are caught between advocating on behalf of consumers and fostering good relationships with utilities.

Prosumer motives vary from a desire for energy independence from utilities, to a commitment to addressing environmental and climate change challenges, to strong interest in clean energy and technologies, or a desire to realize economic benefits DER may provide. Generally speaking, their interests focus on benefits they accrue rather than supporting the reliable and cost-effective operation of the grid, in part because they have no visibility into, and likely little interest in a macroview of utility operations. To the degree there are financial incentives, programs, or mechanisms made available, they may participate in supporting distribution grid operations. Aggregators and service providers are the likely vehicle for residential and small business consumers for participating in distribution services, while larger commercial, industrial, or agricultural customers may choose to participate individually.

Finally, regulators must try to balance these diverse and often divergent interests, all while looking out for disadvantaged customers and those customers who have little desire to change their relationship with utility providers. Aside from installing PV on their roof or buying an electric vehicle (EV), many consumers are, at this stage, likely unaware of DER options available to them now or in the near future, such as the potential role they play as prosumers or the impact their energy choices have on grid planning, operations, and ultimately the financial health of utilities.

Section 2 of this chapter looks at the various stakeholder interests and drivers of DER. Section 3 provides a high-level comparison of the approaches to planning for DER in California and New York. Section 4 is an in-depth look at the foundational steps necessary to bring about a transition to grid-edge technologies. Section 5 provides an overview of the current regulatory framework and how it must change to support a future with significant amounts of DER, followed by the chapter's conclusions.

2 CHALLENGES AND OPPORTUNITIES OF HIGH LEVELS OF DER

For the most part, DER exists today under a patchwork of disparate utility programs, nurtured by public policies and regulations focused on avoiding large capital investments, keeping rates low or in response to customer demands. Programs have varied widely from state to state and utility to utility, depending on regional grid needs or policy goals.

For example, California has one of the most successful energy efficiency programs in the country, holding average residential energy consumption steady for decades, as well as the largest rooftop solar PV programs in the nation, along with a growing base of EVs—largely a result of state policies and regulations. Similar variations are seen in other parts of the world, as described in other chapters of this book.

On the other hand, demand response programs have been relatively weak and piecemeal in the Golden State. In comparison, demand response programs have been more successful and prolific in the East and Northeast. The common thread has been for regulators and utilities to treat these resources separately via special tariffs or programs as adjuncts to conventional grid resources.

However, as DER adoption grows, regulators are beginning to realize that the future of planning and investment is in fact largely driven by what happens at the grid's edge, namely the intersection of the distribution network and customers' premises. One need only look at states with high solar PV penetration levels, such as Hawaii (Box 9.1), to see the impact of not planning for DER.

Spurred by more aggressive public policy goals aimed at mitigating the impacts of climate change as in California[3] and Hawaii,[4] or in response to resiliency concerns postnatural disaster in the case of New York,[5] some regulators and utilities are beginning to view DER as collective resources to support the grid and allow customers more control over their energy choices. In large part,

---

3. For the latest Orders, Rulings, Staff Reports, and other documents for the California Public Utilities Commission, see: http://www.cpuc.ca.gov/General.aspx?id=5071

4. For the latest Orders, Rulings, Staff Reports, and other documents from the Hawaii Public Utilities Commission, see: http://dms.puc.hawaii.gov/dms/DocketDetails?docket_id=84+3+ICM4+LSDB9+PC_Docket59+26+A1001001A14H14A84843E4191418+A14H14A84843E4191 41+14+1873&docket_page=4

5. For the latest Orders, Rulings, Staff Reports, and other documents for the New York Public Services Commission, see: http://www3.dps.ny.gov/W/PSCWeb.nsf/All/C12C0A18F55877E785257E 6F005D533E?OpenDocument#Orders

---

BOX 9.1 Smart Inverters Help Stabilize Hawaii's Grid

With a combination of high electric rates and plentiful sunshine, Hawaii is home to the highest penetration of rooftop solar in the country. Having such a high density of residential PV, however, was affecting the distribution grid of the state's largest utility, Hawaiian Electric Company (HECO). Voltage and frequency fluctuations would cause existing PV systems to trip off and on, causing huge spikes in demand. As a result, customers applying to interconnect new systems languished in long queues while the utility attempted to fix the problem.

In February 2015, HECO and microinverter manufacturer Enphase Energy worked out a solution. Enphase remotely deployed new software settings for 800,000 microinverters on 51,000 residential solar systems. The fix involved remotely adjusting the frequency and low voltage ride-through settings, reducing nuisance tripping and helping stabilize the entire grid. By treating the microinverter as a grid resource, HECO was able to not only eliminate the huge interconnection backlog, but also accommodate more PV systems on circuits with high PV penetrations.

---

these policies have helped accelerate the decline in prices for technologies like solar PV, resulting in rapid consumer uptake of behind-the-meter solar.

There are several drivers spurring the current evolution of DER:

- Public Policy Goals
- Utility/Grid Needs
- Market Forces

These drivers, described below, and their respective constituencies, often have conflicting motives and needs. As acknowledged in the introductory and concluding statements in this volume by leading regulators from Australia, California, New York, and Europe, the challenge regulators and policymakers now face is how to align these drivers to create value for all market participants. As will be discussed later, a major barrier to achieving this objective is the current regulatory and financial framework that has existed for the past 100 years or longer.

2.1   Public Policy Drivers

Regulated utilities have always been subject to public policy priorities. As evidence, one must only look at the litany of utility rates and tariffs, which may include rates for low-income customers, economic development, direct access, community choice aggregation, green energy, distributed generation, and a host of other policy-inspired fees, charges, and tariffs. Specific rate components are often used to fund energy commissions or generic public benefits programs. In the case of DER, more and more utility commissions are seeing the potential opportunity to avoid greenhouse gas emitting resources or even traditional capital investment in the grid.

2.2   Utility/Grid Needs Drivers

DER presents both challenges and opportunities for distribution utilities. Smart inverters, whether associated with PV, stationary storage, or even EVs, can provide voltage regulation, volt-VAR optimization, frequency response, and a host of other services. Energy storage is seen as a potential source for fast-ramping to address the diurnal profile of solar PV, and along with EVs, can absorb excess energy at times of high solar production, reducing or eliminating overgeneration issues in spring or fall months. Solar PV provides midday energy and capacity, which can help defer capacity upgrades on the distribution and even transmission grid. Particularly when combined in optimized portfolios, DER can be strategically deployed in a manner that enhances grid reliability and at lower costs than traditional utility capital investments.

Deploying DER, from a utility perspective however, is challenging and complex. Most distribution grids were designed for one-way power flow under a utility command and control structure. And while utilities have had little control over energy consumption outside of tariffs and demand-side management programs, the growing popularity of solar PV and EVs and the potential for consumer-oriented energy storage technologies makes managing the grid far more uncertain and complex.

These concerns have been highlighted by the tremendous success of the California Solar Initiative (CSI) program along with the State's aggressive Renewable Portfolio Standard (RPS). As of the fourth quarter of 2016, the CSI program has resulted in 1813 megawatts (MW) of installed behind-the-meter solar PV, with another 126 MW in application.[6] Through the end of 2015, California had a total of 5498 MW of installed in-state central station solar PV, plus an additional 1292 MW of solar thermal generation[7]—largely a result of the State's aggressive RPS policies. The growth in solar has changed the net load profile for the state so much that daytime solar production at times now exceeds demand, resulting in negative prices for solar energy in the California Independent System Operator's (CAISO) markets. CAISO has estimated that, absent operational changes and other mitigations, this situation will continue to worsen, as illustrated in the now-infamous "California duck curve" (Fig. 9.1).

On the distribution grid, the success of behind-the-meter solar PV could significantly increase bidirectional energy flows on the grid, impacting the operation of protection devices, possibly increasing wear and tear on equipment, such as voltage regulators. Some of these impacts can be mitigated or avoided using intelligent inverters deployed with PV and energy storage, while some may require grid modernization upgrades. The degree to which such upgrades are necessary, and which can be avoided via intelligent deployment of DER that support grid operations, is the subject of significant debate.

---

6. Source: https://www.californiasolarstatistics.ca.gov/reports/monthly_stats/
7. Source: http://www.energy.ca.gov/almanac/electricity_data/electric_generation_capacity.html

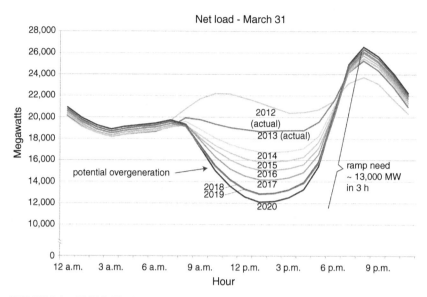

FIGURE 9.1   CAISO "duck curve." *(Source: Licensed without endorsement with permission from the CAISO.)*

As will be discussed in Section 5 of this chapter, utilities, and in particular investor-owned utilities (IOUs) have a vested interest in remaining viable and continuing to generate value for their shareholders. Under the current regulatory construct, utilities can't earn value for shareholders from third-party or customer-owned DER as they can from traditional grid infrastructure capital investments. The widespread deployment of DER within their borders is an existential threat to these utilities, as significant capital investment in both resources and infrastructure shift from utilities to the grid edge—all the way to the customer or prosumer premises.

Regulators are also interested in making sure utilities remain viable so they can serve vulnerable customer segments and those consumers who simply want to remain full-service utility customers. The challenge for regulators is balancing the needs of a diverse group of stakeholders with sometimes diametrically opposing priorities and needs. A particularly vexing problem is determining a new role for utilities and a regulatory framework that both supports traditional consumers, prosumers, third-party DER technology and service providers, the utilities themselves, and their shareholders.

2.3   Market Drivers

What's unique about DER is the degree to which market forces are driving technological adoption. To be sure, technological advances and significant price declines have driven consumer adoption of lighting, space conditioning, and appliances since the very early days of electric utilities. Most of those advances, however, resulted in increased energy consumption as a result of automation and improvements in consumer comfort and convenience. Consumers were, and largely remain passive takers of utility services and subject to the utility rate-making framework.

The current crop of DER, which includes distributed generation, energy storage, and EVs—further covered in other chapters—not only serve customer needs, but also have the ability to support grid operations. Solar energy provides consumers with the option to reduce their carbon footprint while reducing energy costs and acting as a hedge against future rate increases. The current crop of energy storage is largely aimed at arbitraging demand charges, with a segment of residential customers viewing it as a way to avoid consumption of fossil generation. Enabled by the Internet of Things—devices connected to the Internet and controllable via computers and smartphones, consumers have more control over how and when they consume or even produce energy. The newly empowered prosumer has emerged, requiring utilities and regulators to rethink the traditional utility–consumer paradigm.

Importantly, third-party owned distributed resources have the potential to significantly lower energy sales and displace traditional utility grid investments. While energy efficiency and other demand-side management programs have been a staple of utility offerings since the first energy crisis in the early 1970s, they have had a relatively minor impact on utility revenue and profits. Often regulators have put in place performance-based or earnings adjustment mechanisms to offset lost sales from energy efficiency program requirements.

Prudency reviews have ensured, or attempted to ensure, utility investments are necessary, just, and reasonable, avoiding overinvestment in the grid and helping keep rates more affordable for consumers. While not a perfect system, the regulatory compact has ensured monopoly utilities have the ability to earn a guaranteed, fair rate of return for their shareholders or owners, while providing a strong measure of protection for consumers against unreasonably high rates.[8]

Regulators must try to implement public policy goals while balancing the needs of consumers and utilities. In the past, this has mainly focused on reducing customers' energy use and bills and avoiding unnecessary grid investments by the utilities. However, with the challenges presented by climate change and the policies enacted to mitigate these impacts, along with incentives and the rapid decline in prices for solar PV and EVs and the increasing availability and affordability of DER, balancing these needs has become more difficult.

---

8. See, for example, Dr. Karl McDermott, *Cost of Service Regulation in the Investor-Owned Electric Utility Industry, A History of Adaptation*, Edison Electric Institute, June 2012.

3 CALIFORNIA AND NEW YORK—A TALE OF TWO REGULATORY APPROACHES

California and New York are the two highest profile examples in the United States for addressing DER planning and deployment. Both recognize the potential benefits and challenges DER presents and have begun to develop regulations to facilitate the transition to a more decentralized future.

New York began with a vision for reinventing the utility sector, an effort they called Reforming the Energy Vision (REV). The vision laid out a new role for utilities as Distribution System Platform Providers (DSPPs). These DSPPs are responsible for optimizing the distribution system utilizing DER as well as providing aggregated energy and ancillary services to the bulk system operator. The New York Public Service Commission (NYPSC) views the accurate pricing of DER as essential to achieving these objectives.

The ultimate objective of the NY REV process is to create a market where DER owners or aggregators are able to offer services into a distribution level market, much like wholesale markets operated by Independent System Operators (ISOs) or Regional Transmission Operators (RTOs), such as the New York ISO (NYISO) and CAISO. In addition to bilateral transactions, these markets offer day-ahead and hour-ahead energy markets, ancillary services markets, and sometimes capacity markets. By establishing a distribution services market, the NYPSC hopes to achieve efficiencies inherent in market structures and create opportunities to monetize the wider range of attributes DER can provide.

In contrast, California's approach has focused more on the technical aspects of identifying ideal locations for DER and determining the locational net benefits rather then developing market mechanisms to accelerate deployment of DER. The focus is largely on deferring capital investments at specific locations on the grid rather than addressing autonomous DER growth driven by consumer needs. Sourcing has focused on competitive procurement by the utilities based on granular locational net benefits rather than creating market structures to address these needs.

Unlike New York, California has thus far avoided the issue of identifying a new structure or role for utilities. In late 2016, the California Public Utilities Commission (CPUC) did issue a decision that created a pilot utility incentive mechanisms that address, at least in part, the issue of compensating utilities for procuring third-party DER rather than making traditional capital investments, which under the current regulatory framework earn value for their shareholders.[9] Aside from this limited incentive pilot program, however, the CPUC has yet to

---

9. In D.16-12-036 (Decision Addressing Competitive Solicitation Framework and Utility Regulatory Incentive Pilot, December 15, 2016), the CPUC authorized a pilot program for CA's largest regulated utilities, allowing them to earn a 4% pretax incentive based on the annual payment for distributed energy resources that defer investments in the distribution grid. The pilot, set to last for 17 months, will evaluate the effectiveness of this type of incentive and inform future discussions on incentive mechanisms or potential changes to the regulatory framework.

consider changes to the regulatory framework, such as the creation of market mechanisms or the formation of a distribution system operator structure.

Thus the focus in California is more on building the foundation for a modern grid via creating the modeling, data, processes, and regulations that identify optimal locations and locational net benefits, rather than addressing the evolution of the utility's role and the regulatory and economic framework necessary to sustain DER growth. That's not to say California regulators and policymakers are not aware of the need to address these important issues, but rather the focus has been more short-term and pragmatic, emphasizing building the foundational components that are necessary for an eventual market structure and regulatory framework.

To varying degrees, California utilities are generally ahead of utilities in New York from the standpoint of deploying advanced grid analytics and planning, and in particular smart meter technologies. Access to detailed subhourly load data and system data should give utilities in California the ability to be more precise in determining highly granular locational net benefits for DER. However, lacking a longer term vision for how this data might inform more accurate and efficient sourcing mechanisms and identifying the role utilities and consumers will each play in a future distributed grid has caused a misalignment between the IOUs' financial interests and State policy goals driving DER deployment.

Concerns about this misalignment have surfaced in other proceedings, including IOU general rate cases and EV infrastructure deployment proposals. Parties to the proceedings have raised concerns about overinvestment in grid infrastructure, or "gold-plating" the grid by utilities to compensate for lost opportunities to rate base capital investments that could be deferred or avoided via procurement of third-party DER. Utilities have countered that the investments in grid infrastructure are necessary for integrating DER and maintaining reliability, but the questions surrounding the future role of utilities and their compensation structure have created uncertainty and calls for increased scrutiny of utility spending.

4   GETTING THE MOST OUT OF DER

Deploying significant amounts of DER while maintaining reliable grid operations requires a significantly higher level of situational and operational awareness and planning. Data, from consumption/production by consumers to highly granular locational grid conditions and even current and forecasted weather conditions, are foundational to successfully operating a grid with high levels of DER penetration. In this section, I will describe how California is addressing these data needs in the near term, as well as future data considerations.

4.1   Need for Granular Geographical and Temporal Data

In the past, DER included mainly utility energy efficiency and demand response programs along with a smattering of customer-owned combined heat and power systems or on-site generation. Each resource was treated separately via tariffs, programs, or bilateral agreements, but generally speaking was optimized to meet bulk power system needs or accommodate customer needs. With the dramatic increase in distributed solar PV, growth in customer demand for EVs, and early interest in distributed energy storage, the trend in DER is shifting much more toward meeting customer wants with less regard for impact on the grid.

At low levels of penetration, utilities can manage DER integration via the interconnection process. However, as DER becomes more and more prevalent, utilities will have a much more difficult time managing the grid and meeting customer expectations. Since DER can be deployed throughout the distribution grid, utilities need to understand loads and system conditions at a highly granular level, both geographically and temporally.

PG&E, for example, has modeled the DER integration capacity, or hosting capacity, of over 3,000 circuits, with over 102,000 line segments and 500,000 nodes.[10] In addition, they have installed roughly 5 million smart meters that collect subhourly load data on electric customers in their service territory. They have modeled all of their three-phase line segments, but have not yet modeled the single-phase line segments, which comprise roughly one-third of their total circuit miles and which are mainly radial lines serving residential customers.

Modeling the single-phase line segments would increase precision, and is ultimately needed for determining distribution marginal costs for eventual distribution markets and for matching detailed customer load data with operational data for the line segment on which the customer is interconnected. The California utilities do have plans to evaluate the single-phase line segments to determine hosting capacity, which will provide more granular cost data to inform future procurement and sourcing options and markets.

Rather than reacting to DER deployed solely to meet customer needs, utilities are attempting to leverage advanced distribution planning tools to identify where DER can provide the greatest benefits to the grid and which attributes and combinations of DER can provide the greatest value to the grid. Secondarily, understanding locational conditions and needs better enables utilities to understand how to support customer-driven DER by designing programs or incentives to mitigate any potential negative grid impacts such deployment might cause. Since not all customers who install DER will be interested or able to provide grid services, it's important to understand where and in what quantities this DER growth may occur and mitigate the impacts to the extent possible.

For example, if a customer located on a circuit that has little available capacity wanted to purchase one or more EVs, the utility could either charge the customer a fee to upgrade the circuit to accommodate this new load, or instead offer them programs or incentives to make energy efficiency improvements, install rooftop solar or energy storage, or participate in a demand response program to avoid making costly system upgrades. Particularly in areas where there

---

10. Source: Pacific Gas and Electric Company (2015), p. 23.

are GHG reduction, air quality improvement or customer choice policies driving DER deployment, it is important to understand and mitigate costly upgrades that would otherwise make DER less economic to customers.

4.2   Customer Load and Demographic Data

Utilities around the country have installed, or are in the process of installing advanced "smart meters" throughout their service territory in an effort to reduce operational costs, improve metering/billing accuracy and grid reliability, and eventually enable customers to respond to signals to better manage their usage and allow the utility to better manage the grid. Obtaining subhourly customer load data from these smart meters is an important first step in understanding current usage patterns and predicting potential future patterns.

Forward-thinking utilities are now using forecasting models that include economic and weather data to forecast loads at the subcircuit level. This highly granular analysis enables the utility to better understand not only usage trends, but also DER growth potential and output. These data are essential for determining the locational value of DER and designing sourcing mechanisms to deploy DER in a manner that supports the efficient and reliable operation of the grid.

Historical and forecast weather data are necessary not only for understanding and predicting heating, cooling, and lighting loads, but also for predicting solar PV output. Larger utility service territories may cover a wide range of climatological and geographical conditions, increasing the importance of locational forecasting.

For example, PG&E's service territory in California covers cool coastal regions where air conditioning is less common, to the much warmer Central Valley, and into the Sierra Mountains. Even within the San Francisco Bay Area, where the company is headquartered, temperatures can vary by as much as 25–30°F from the coast to the inland valleys. Understanding weather impacts on loads historically and in near-real time at the subcircuit level enables utilities to understand what the grid needs are and which forms of DER and in which combinations can best meet those needs.

4.3   Hosting Capacity Analysis

The first step in determining the value of DER is to identify the optimal locations where they can provide the greatest benefits to the grid. Using data collected from smart meters and other sensors on the grid, along with the load, economic, and forecasting data described previously, utilities perform an analysis of the circuit DER hosting capacity, sometimes referred to as an Integration Capacity Analysis (ICA).[11] This is done using a power flow analysis that is

---

11. See, for example, Electric Power Research Institute, *Distribution Feeder Hosting Capacity: What Matters When Planning for DER?*, April 2015.

FIGURE 9.2 Sample Pacific Gas and Electric Company Integration Capacity Analysis (ICA) map.

performed for each circuit using data representing load profiles, weather conditions, DER operational profiles, seasons, and other factors. The analysis is used to predict the available unused capacity along each circuit, showing how much additional DER can be accommodated on the circuit.

Utilities in California then publish this information in graphical form using geographic information systems "heat maps."[12] Circuits with ample available hosting capacity are colored green, ones with more constraints are yellow, while highly constrained circuits are depicted in red. Although details in early versions of the maps vary from one utility to the next, clicking on the circuit on the map brings up information, such as circuit voltage, capacity, projected peak load, and existing and planned distributed generation megawatts.[13] This information allows DER project developers to identify locations where capacity exists to install more solar PV or other DER.

Fig. 9.2 shows a screenshot from PG&E's ICA map[14] for a portion of Oakland, California. Clicking on one of the line segments in the map brings up information on the circuit, shown in Fig. 9.3.

Once the utilities complete the analysis on all of their circuits, more detailed information is expected to be included. This will include the timing, duration,

12. See http://www.cpuc.ca.gov/General.aspx?id=5071, Integration Capacity Analysis (ICA) Maps. Users must register with each utility to access maps.
13. As of this writing, only PG&E included this level of detail. SCE and SDG&E maps contain information about existing and queued distributed generation megawatts.
14. See https://www.pge.com/en_US/for-our-business-partners/energy-supply/solar-photovoltaic-and-renewable-auction-mechanism-program-map/solar-photovoltaic-and-renewable-auction-mechanism-program-map.page

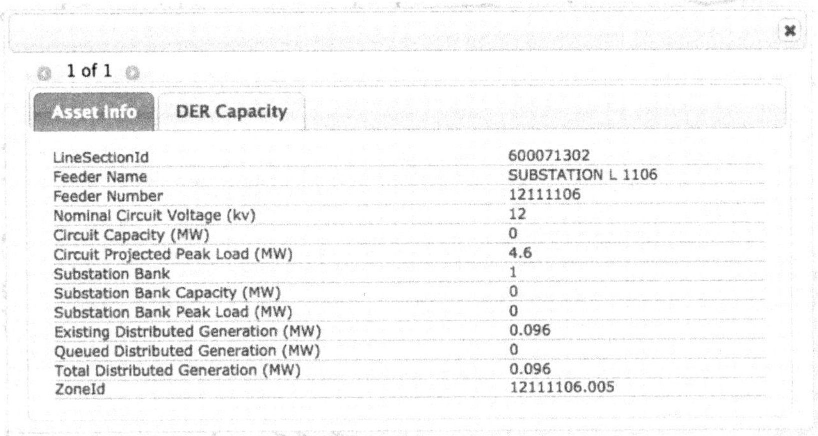

LineSectionId	600071302
Feeder Name	SUBSTATION L 1106
Feeder Number	12111106
Nominal Circuit Voltage (kv)	12
Circuit Capacity (MW)	0
Circuit Projected Peak Load (MW)	4.6
Substation Bank	1
Substation Bank Capacity (MW)	0
Substation Bank Peak Load (MW)	0
Existing Distributed Generation (MW)	0.096
Queued Distributed Generation (MW)	0
Total Distributed Generation (MW)	0.096
ZoneId	12111106.005

FIGURE 9.3 Sample Pacific Gas and Electric Company ICA circuit data.

and nature of the constraint that exists, such as a thermal overload, power quality or voltage violation, protective device conflict, or safety and reliability issue. These more detailed maps will provide valuable tools for third parties to identify high priority areas to target for DER development. The additional data describing the specific constraints will help them tailor their solutions to meet specific grid needs. This type of targeted deployment, tailored to meet specific grid needs, helps grid operators maximize the benefits of DER for all customers while keeping costs low.

4.4 Locational Net Benefits Analysis

While the ICA identifies locations on the grid where constraints exist and includes criterion of what may be causing the constraint, the Locational Net Benefits Analysis (LNBA) attempts to determine the value for DER at specific locations along the grid. In simple terms, the LNBA is an accounting of the difference between the benefits and costs of integrating DER at specific locations on the distribution grid.

California Public Utilities Code Chapter 4, Article 3, Section 769 establishes the requirement for utilities to evaluate locational benefits of DER:

> *Evaluate locational benefits and costs of distributed resources located on the distribution system. This evaluation shall be based on reductions or increases in local generation capacity needs, avoided or increased investments in distribution infrastructure, safety benefits, reliability benefits, and any other savings the distributed resources provide to the electrical grid or costs to ratepayers of the electrical corporation.*[15]

15. State of California Public Utilities Code Division 1—Public Utilities Act, Chapter 4—Regulation of Public Utilities, Article 3—Equipment, Practices, and Facilities, § 769; effective January 1, 2015.

To understand the benefits of DER, the utilities first identified the potential value components. These include the more obvious benefits, such as avoided distribution capacity, avoided line losses, ancillary services, increased reliability and resiliency, along with societal benefits, greenhouse gas reduction, and renewable energy integration benefits. Table 9.1 shows the various components from PG&E's Distribution Resources Plan that are used to determine locational net benefits.

The California Public Utilities Commission directed the State's IOUs to use an avoided cost calculator, modified to capture distribution system impacts, as the basis for determining the avoided costs on which the locational benefits are to be derived.[16] The calculator was originally developed for evaluating system level avoided costs, but has been modified to accommodate distribution level impacts.

To maximize the net benefits for all customers, it is essential to get as complete and accurate accounting as possible of the benefits and costs of DER at specific locations along the grid. This includes the value to not only the distribution system, but also to the bulk grid and to customers, both with and without DER. Utilities, such as PG&E have the ability to identify highly granular data—from customer smart meters and potentially intelligent inverters, to grid conditions and historical data at the three-phase, and soon single-phase circuit levels, to very locational weather and demographic historical and forecast data. This will enable very accurate evaluation of the locational net benefits, assuming the utilities include all of the appropriate value components mentioned previously.

There is one very important additional consideration, however. To realize the full value DER can provide, utilities must evaluate optimal combinations, or portfolios, of DER. For example, rooftop solar alone may provide capacity value and potentially voltage or frequency support. But a rooftop PV system that combines energy storage and/or an EV can provide ramping support, avoid overgeneration at times of peak solar production and low loads, or help stabilize the grid in other ways. Therefore, utilities should consider the potential impact and value of DER portfolios instead of or in addition to individual DER. This step is also important for developing appropriate incentives and sourcing mechanisms to influence the design and deployment of third-party DER.

4.5 DER Sourcing

Understanding the more precise value of DER is essential for creating new sourcing options and for transitioning from current, cruder methods of compensating DER, including Net Energy Metering tariffs. At low levels of penetration that now exist in most jurisdictions, NEM is an important tool for encouraging

---

16. Assigned Commissioner's Ruling on Guidance for Public Utilities Code Section 769—Distribution Resource Planning (R.14-08-013), February 6, 2015, and Assigned Commissioner's Ruling (1) Refining Integration Capacity and Locational Net Benefit Analysis Methodologies and Requirements; and (2) Authorizing Demonstration Projects A and B (R.14-08-013), May 2, 2016.

TABLE 9.1 Consolidated Components For PG&E's Locational Impact Analysis

#	Component	PG&E definition
1	Subtransmission, substation, and feeder capital and operating expenditures (distribution capacity)	Avoided or increased costs incurred to increase capacity on subtransmission, substation, and/or distribution feeders to ensure system can accommodate forecast load growth
2	Distribution voltage and power quality capital and operating expenditures	Avoided or increased costs incurred to ensure power delivered is within required operating specifications (i.e., voltage, fluctuations, etc.)
3	Distribution reliability and resiliency capital and operating expenditures	Avoided or increased costs incurred to proactively prevent, mitigate, and respond to routine outages (reliability) and major outages (resiliency)
4	Transmission capital and operating expenditures	Avoided or increased costs incurred to increase capacity on transmission line and/or substations to ensure system can accommodate forecast load growth
5a	System or local area RA	Avoided or increased costs incurred to procure RA capacity to meet system or CAISO-identified LCR
5b	Flexible RA	Avoided or increased costs incurred to procure flexible RA capacity
6a	Generation energy and GHG	Avoided or increased costs incurred to procure electrical energy and associated cost of GHG emissions on behalf of utility customers
6b	Energy losses	Avoided or increased costs to deliver procured electrical energy to utility customers due to losses on the T&D system
6c	Ancillary services	Avoided or increased costs to procure ancillary services on behalf of utility customers
6d	RPS	Avoided or increased costs incurred to procure RPS eligible energy on behalf of utility customers as required to meet the utility's RPS requirements
7	Renewables integration costs	Avoided or increased generation-related costs not already captured under other components (e.g., ancillary services and flexible RA capacity) associated with integrating variable renewable resources
8	Any societal avoided costs which can be clearly linked to the deployment of DERs	Decreased or increased costs to the public which do not have any nexus to utility costs or rates
9	Any avoided public safety costs which can be clearly linked to the deployment of DERs	Decreased or increased safety-related costs which are not captured in any other component

LCR, Local capacity requirement; RPS, Renewable Portfolio Standard.
Source: Pacific Gas and Electric Company's Electric Distribution Resources Plan, July 1, 2015, Table 2-12, p. 65.

distributed PV, helping such technologies to achieve greater economies of scale. It also works quite well for most utilities today given the general lack of sophisticated grid analysis tools and data.

However, as distributed PV reaches higher levels of penetration and other forms of DER become more widely available, and as utilities deploy advanced grid analysis and planning tools, more precise, locationally and temporally, granular values of DER are necessary for cost-effectively deploying DER. To facilitate the transition from more crude sourcing options and incentives requires a more complete accounting of the full value DER can provide.

Most utility sourcing today is done through pricing (via rates, tariffs, and incentives), programs (such as mandatory levels of energy efficiency or RPS requirements), and procurement. As was previously mentioned, each distributed resource type is typically treated in siloed fashion for pricing and programmatic sourcing options. The same is generally true for procurement, where the utility issues a Request for Offers (RFO) for a specific resource type and megawatt level to meet an identified need at a specific point in time.

What's lacking in the sourcing mix in California is a transition to market-based sourcing structures. The NY REV process specifically identifies markets as the end-state goal of the process, though data and analytics have not yet been developed sufficiently to support such mechanisms. The DSPP framework established by the NYPSC changes the role of the utilities to one that supports and enables markets to operate efficiently and effectively.

California's approach of relying on Purchased Power Agreements (PPAs) instead of transitioning to markets places greater emphasis on contingency planning—what to do if the DER fails to materialize in the quantities and at the times needed, if at all. While penalties for nonperformance can be incorporated into PPAs between utilities and DER providers, this does little to address very real reliability concerns. The typical fallback if the DER fails to show up at all is for the utility to go forward with grid upgrades that would have otherwise been built, or to screen and restrict DER providers to a narrow list of "preferred providers" that could hinder competition and technological innovation.

By their very nature, markets create opportunities for multiple suppliers to compete and encourage innovation by providing price transparency. This of course assumes timely, accurate, and detailed data is accessible to all market participants in a manner that enables bidding of optimal resource attributes to meet the identified needs.

From a pragmatic standpoint, however, markets may take longer to develop and require careful thought to protect against gaming the system. Enron's actions in California during the deregulated era at the beginning of the century highlight the potential dangers of poorly designed markets and lax oversight. Given the significant potential disruption to the utilities' operational and financial modus operandi, it may make sense to start with more familiar sourcing mechanisms, such as RFOs for distribution investment deferral, tariffs, and programs, modified to incorporate data from the ICA and LNBA evaluations.

One example of a more conventional tariff structure that could support DER deployment is an options payment tariff for voltage support. Much like a demand response tariff, where customers who agree to reduce load when called upon during critical peak periods are paid a monthly fee, and then are compensated for any actual demand reduction they make in response to the utility's signal. In the case of DER like solar PV with smart inverters or energy storage, the DER customer would be paid an options fee that allows the utility to activate voltage regulation capabilities of the inverter. The customer would also be compensated for the actual voltage support provided, as metered by the inverter or a utility meter, to compensate for the reduced energy output of the inverter.

The steps both New York and California are taking to identify hosting or integration capacity, optimal locations, and to identify and assign precise locational values for DER are of course fundamental to creating more effective sourcing mechanisms, whether they be tariffs, programs, competitive solicitations, or market mechanisms. A planned approach that incorporates several sourcing mechanisms while building toward a market structure will likely prove most successful.

5 ALIGNING UTILITY FINANCIAL MOTIVES WITH DER POLICY GOALS

A fundamental barrier standing in the way of widespread DER deployment is the existing regulatory framework and IOU revenue structure that has existed for over a 100 years. Most utilities earn a guaranteed rate of return that is calculated on a combination of their energy sales and capital investments, although in some states earnings have been "decoupled" from sales.

In California, an example of a decoupled regulatory framework, IOUs, earn a rate of return that is largely based on the depreciated value of approved capital assets. This gives the utilities an inherent bias toward owning assets rather than procuring services via PPAs. Yet DER is typically owned by customers or third parties, creating a misalignment between the utilities financial interests and the policy objectives enacted by the state with regard to DER. As a result, there exists a situation where utilities are being asked to defer capital investments, on which they would otherwise earn a rate of return, and instead procure third-party owned DER, which provides no value to shareholders. The Commission has attempted to address this in the short term via DER incentive pilot programs, but has left open the larger question of what role the utilities will play and how will they earn value for their shareholders as more and more DER are deployed throughout the grid.

This situation creates an additional regulatory burden to ensure utilities don't overinvest in grid upgrades, correctly value the benefits DER can provide, or delay the deployment of advanced grid planning tools and data platforms in an effort to forestall DER deployment. This presents regulators with a difficult task during the early stages of transitioning to the grid of the future, as utilities submit annual distribution plans and begin a new cycle of general rate cases.

Regulators must ensure capital investments support higher levels of customer and third-party owned DER and be wary of approving traditional investments in the distribution grid that may later become stranded or which foreclose opportunities for deploying DER. Utility commissions may lack the technical expertise to evaluate whether a specific utility investment could be deferred or displaced by DER, or fully understand the capabilities and limitations of DER as alternatives to traditional grid investments.

Smart inverters, for example, can provide many of the voltage regulation capabilities that might otherwise require utilities to install capacitors or other equipment to control voltage. Substation, transformer, or conductor upgrades, unless done to replace aging unreliable equipment, could be deferred with the addition of DER in the right locations. Even some of the communications and control equipment that utilities are considering to allow them to directly control dispatching of DER could be duplicative of the inherent capabilities DER possesses.

If utilities overinvest in the grid, questions arise about who should pay for these investments—shareholders, all utility customers, or just those customers who've installed DER. Regulators must be very careful in balancing the interests of vulnerable customers, those who choose not to participate as prosumers, and those who do. Allocating grid modernization costs to non-DER customers will elicit complaints about one group of customers subsidizing another group of customers. Conversely, burdening DER customers with stranded cost recovery will likely result in higher fixed costs, negating some or much of the economic benefits of DER and having a chilling effect on DER deployment. Yet burdening shareholders with the costs of potential stranded investments, which were approved in regulatory proceedings, is also unfair.

Regulators must therefore carefully evaluate all utility infrastructure requests and determine whether the investments are truly necessary to support the reliable and cost-effective operation of the grid, or whether they could be deferred or displaced with well-placed and well-designed portfolios of DER. It is therefore imperative that regulators create the proper incentives, motivations, and regulatory framework for utilities to consider DER as alternatives to traditional grid investments without concern for the impact on their financial well-being.

6 CONCLUSIONS

The utility world has changed dramatically since the time of Edison and Tesla, yet in many regards the utility industry remains largely familiar to the time when alternating current and direct current were first being debated. But an industry that fueled the second industrial revolution cannot remain immune for long from the advances it helped spawn. Technological advances in solar power, energy storage, vehicle propulsion, communication systems, and data processing and analytics have converged to produce a new world of distributed energy resources to complement the Internet of Things. These resources are empowering a new breed of prosumers, who have the ability to control how and when they use, generate, or even store electrical energy, fundamentally changing their relationship with utilities that once served as their sole provider of electric service. In this Darwinian world, utilities and regulators must also adapt to meet the evolving demands and desires of consumers, or face extinction like species and industries of the past.

Fortunately, states like California and New York are blazing new trails for others to follow. New York, with its strong vision of the future and focus on building a new role for utilities and driving toward distribution markets, has created a degree of regulatory certainty that allows each party to focus on their respective role. California's more pragmatic approach, with its focus on planning tools, data, modeling, and analysis, is establishing the fundamentals and building a solid foundation upon which DER technology and service providers, markets, and sourcing options will be built. Though neither approach is without its flaws, regulators in other jurisdictions can take the best approaches from each state to use as a template upon which to build a framework best suited for their situation.

Regulators must recognize the needs and drivers of the various stakeholder groups to understand how to create a framework to support DER growth in a way that balances these varied and sometimes conflicting interests. They must continue to protect vulnerable customers and those that choose not to participate in this new energy future, ensuring that the grid operates reliably and that rates remain reasonable and fair for all customers. They must recognize the inherent conflict that exists between the existing utility financial structure and newly enacted public policy goals that could potentially erode shareholder value. Most importantly, regulators must create a clear vision for the future of the grid and establish some measure of regulatory certainty. Ultimately, each of the participants in this new energy paradigm—utilities, DER technology and service providers, prosumers and traditional utility consumers—must succeed in order for this new framework to succeed. It's imperative that regulators show strong leadership to help bring this about in the most efficient, cost-effective, and reliable manner possible.

REFERENCE

Pacific Gas and Electric Company, 2015. Application of Pacific Gas and Electric Company for Adoption of its Electric Distribution Resources Plan Pursuant to Public Utilities Code Section 769 (U39E), Application Number 15-07-006, Filed July 1, 2015.

FURTHER READING

Advanced Energy Economy Institute, 2015. Toward a 21st century electricity system for California. A Joint Utility and Advanced Energy Industry Working Group Position Paper, August 11, 2015.
California Public Utilities Commission, 2014. Order Instituting Rulemaking Regarding Policies, Procedures and Rules for Development of Distribution Resources Plans Pursuant to Public Utilities Code Section 769, August 14, 2014.

Cleveland, F., 2016. California DER Regulatory Requirements: Smart Inverter Working Group (SIWG) and the Distribution Resources Planning (DRP), presentation to the Western Interstate Energy Board, May 9, 2016, Distributed Energy Resource Interconnection Timelines and Advanced Inverter Deployment.

Electric Power Research Institute, 2015. The Integrated Grid: A Benefit-Cost Framework. Electric Power Research Institute, Palo Alto, CA.

Greentech Media, 2015. Evolution of the grid edge: pathways to transformation, January 2015. Available from: https://www.greentechmedia.com/research/report/evolution-of-the-grid-edge-pathways-to-transformation

MIT Energy Initiative, 2016. Utility of the Future. An MIT Energy Initiative Response to an Industry in Transition.

National Renewable Energy Laboratory/Southern California Edison Company, 2016. NREL/SCE High Penetration PV Integration Project: FY13 Annual Report (NREL/TP-5D00-61269), Barry A. Mather, National Renewable Energy Laboratory; Sunil Shah, Southern California Edison; Benjamin L. Norris and John H. Dise, Clean Power Research; Li Yu, Dominic Paradis, and Farid Katiraei, Quanta Technology; Richard Seguin, David Costyk, Jeremy Woyak, Jaesung Jung, Kevin Russell, and Robert Broadwater, Electrical Distribution Design, Inc.; June 2014.

New York Department of Public Services, 2016. Staff Report and Recommendations in the Value of Distributed Energy Resources. Proceeding 15-E-0751, October 27, 2016.

Osterhus, T., Ozog, M., Stevie, R., 2016. Distributed Marginal Price (DMP) Methodology Applied to the Value of Solar. Integral Analytics, Cincinnati, OH.

San Diego Gas & Electric Company, 2015. Application of San Diego Gas & Electric Company (U902E) For Approval of Distribution Resource Plan, Application Number 15-07-003, Filed July 1, 2015.

Seguin, R., Woyak, J., Costyk, D., Hambrick, J., Mather, B., 2016. High-Penetration PV Integration Handbook for Distribution Engineers (NREL/TP-5D00-63114). National Renewable Energy Laboratory, Golden, CO.

Southern California Edison Company, 2015. Application of Southern California Edison Company (U 338-E) for Approval of Its Distribution Resources Plan, Application Number 15-07-002, Filed July 1, 2015.

St. John, J., 2014. Distributed Marginal Price: The New Metric for the Grid Edge? Pinpointing the Value of the Distributed, Intelligent Grid with Integral Analytics, Greentech Media, August 21, 2014.

State of New York Public Services Commission, 2016. CASE 14-M-0101: Proceeding on Motion of the Commission in Regard to Reforming the Energy Vision, Order Establishing the Benefit Cost Analysis Framework, January 21, 2016.

Woychik, E., 2015. Electric Utility Adaption to Disruptive Change: Dashboards for Success and Profitability by 2020?, Presentation to CRRI 34th Annual Eastern Conference, May 15, 2015.

Chapter 10

Public Policy Issues Associated With Feed-In Tariffs and Net Metering: An Australian Perspective

Darryl Biggar* and Joe Dimasi**

*Australian Competition and Consumer Commission, Melbourne, VIC, Australia;
**Independent Competition and Regulatory Commission, Canberra City, ACT, Australia

1 INTRODUCTION

As many chapters in this book have emphasized, the electric power sector is experiencing a fundamental transformation. The rapid uptake of distributed energy resources, energy management solutions, and energy storage has opened up new opportunities for electricity customers, particularly customers connected to distribution networks and on the grid's edge. Around the world there is an increasing interest in allowing small consumers to produce their own electricity, and not just to sell that electricity back to the grid, but to actively trade that electricity with others, through peer-to-peer trading, open platforms, community-shared solar schemes, and so-called group or virtual net metering.

Many electricity consumers believe they have a right to self-produce electricity if they are able to do so. It is only a small further step to seek to trade that self-produced electricity with neighbors. However, these proposals raise a host of deep and complex issues for policymakers, which are increasingly being raised in Australia,[1] and in many other countries around the world.[2]

---

1. For example, Langham et al. (2013), and the Local Generation Network Credits rule change: http://www.aemc.gov.au/Rule-Changes/Local-Generation-Network-Credits. There is also an interest in virtual net metering in some specific locations: "Byron shire to be the first in Australia to pilot virtual net metering," RenewEconomy, March 23, 2015; "Thinking outside the square: virtual net metering could reduce power bills," www.ergon.com.au, June 25, 2015.

2. For example, in the United Kingdom, www.openutility.com and, in New Zealand, www.p2power.co.nz. Many states in the United States, such as Vermont, Massachusetts, and California, allow for group or virtual net metering for community shared solar schemes. See the chapter by Jones et al. in this book; US DoE (2012); "Making projects happen with group net metering policies," Renewable Energy World, August 7, 2012; and "Vermont Group Net Metering Information & Guidelines," December 2010, http://energizevermont.org/wp-content/uploads/2010/06/Group-Net-Metering-Info-Guidelines_Final-1.pdf

Innovation and Disruption at the Grid's Edge. http://dx.doi.org/10.1016/B978-0-12-811758-3.00010-3
187

As shown in this chapter, the associated policy issues of net versus gross metering, virtual net metering, and peer-to-peer trade in electricity come down to questions of tariff design. Historically the tariffs paid by smaller electricity consumers have not been efficient or cost reflective. The proliferation of on-site generation and storage options places these historic tariffs under considerable strain. The problem is exacerbated when end-customers seek to trade their self-produced or stored electricity with neighbors. A decision will have to be made: either to abandon the historic structure of the network tariffs, or to halt the proliferation of embedded generation and storage and prevent peer-to-peer trade in electricity. The resolution of that tension will determine the course of the electricity industry over the next few decades.

In developed countries, such as Australia, almost all electrical loads (customers) have the right to connect to the grid at their desired location and to purchase electricity at the prevailing retail tariff.[3] This chapter starts with the assumption that, subject to meeting technical requirements, embedded generation has a similar legal and physical option to connect to the distribution network and to sell its output at some rate, such as the prevailing wholesale spot price, the local feed-in tariff, or the local retail rate.

This assumption is important. It implies that the underlying public policy issue is not the *ability* to trade the output of the embedded generation; after all, as long as the overall power system is in balance, every unit sold by the embedded generation must be, in effect, purchased by a consumer somewhere. Rather, the primary issue is the *price* paid for the output of that embedded generation: both the level and structure of that price, and its level and structure relative to the price paid by load. It is these pricing issues that make the handling of embedded generation and peer-to-peer trading politically tricky.

This chapter takes as given that the overall public policy objective—as set out in Australia's National Electricity Objective[4]—is the efficient use of and investment in the electricity sector. This chapter also assumes that all external environmental effects, including the cost of carbon pollution and/or the use of environmental water, are internalized through existing mechanisms, such as taxes, fees, or carbon-pricing mechanisms. This assumption allows us to treat all forms of generation on a level playing field.

This chapter consists of three sections in addition to this introduction. The next section briefly sets out the theoretically efficient tariff design and the implications for net metering and peer-to-peer trade. Section 3 then looks at how differences in the retail tariffs for loads and for embedded generators affect the incentives of end-customers to either combine generation with load (as in net metering) or to separate generation from load (as in gross

3. Remote loads may have to pay the cost of connection assets; unusual loads may have to pay for network strengthening.
4. http://www.aemc.gov.au/Energy-Rules/National-electricity-rules

metering). In practice new embedded generation in Australia is paid a tariff well below retail tariffs, creating strong incentives for net metering. Section 4 looks at the problems that arise as a result and the various options for solving those problems: requiring separate metering for embedded generation, on the one hand, or extending net metering to neighbors, on the other. Section 5 concludes.

2   EFFICIENT TARIFFS FOR GENERATION AND LOAD IN THEORY

As noted in the introduction, the public policy issues associated with virtual net metering and peer-to-peer trading in electricity come down to questions of tariff design. In particular, these public policy issues depend critically on the design of two tariffs: the tariff paid for the output of the embedded generation, and the tariff paid for electricity consumed by loads.[5]

In particular, the public policy issues arise from:

- the *efficiency* (*cost reflectivity*) of each of these two tariffs; and
- the *difference* between the tariff for generation and load.

It is useful to be clear at the outset what theory suggests is the optimal tariff. Economic theory is quite clear that first best pricing for generation and load involves tariffs that are fully cost reflective and for which there is *no difference* in the tariff paid to generation and the tariff paid by load. Efficient tariffs would involve every load being charged (and every generator being paid) the short-run marginal cost of the *production* and the *delivery* of a unit of electricity to the location of that customer.

As is well known from wholesale electricity market pricing theory, the marginal cost of production of electricity depends on the marginal cost of the "marginal" generator or load, and varies continuously according to the level of demand relative to the stock of available generation. At off-peak times, when there is plenty of surplus capacity, the marginal cost of producing electricity is low. At peak times, when additional peaking generation must be brought on-line, the marginal cost of generation can be very high.

The marginal cost of delivery of electricity depends on the presence of congestion on the transmission and distribution network. At times when congestion is absent the marginal cost of delivery of electricity is very low, and limited only to the electrical losses on the system. At times and places where congestion is

---

5. As noted in footnote 4, the "tariff paid by load" is not, strictly speaking, the tariff paid by the end-customer, but rather the tariff paid by the retailer that serves the end-customer. The retailer may repackage that tariff in different ways. However, this distinction is not important for the argument in this chapter.

present, the marginal cost of delivery may be very high, leading to substantial price differences in the marginal price of electricity at different locations.

In theory, the marginal cost of production and the marginal cost of delivery can be determined jointly through a market process, which yields a distinct price at each location or node on the network and at each point in time. These prices are known as *locational marginal prices*[6] or nodal prices.

The mechanism to determine such prices is well known, and is now quite routine at the high-voltage level in liberalized electricity markets around the world. This process requires a centralized market operator who accepts bids and offers from market participants and who seeks to maximize the value of trade, subject to the physical constraints of the physical power system. Research is underway in extending these principles to the lower-voltage distribution network level.[7]

This theory is clear that at any given location on the network, and at any point in time, the price paid for power injected into the network at that location should be equal to the price paid for power withdrawn from the network at that location. In such a market there would be no restrictions on net metering or peer-to-peer trade: any market participant could, in principle, trade electricity with any other market participant, subject to the constraint that all trades would be mediated via the market operator, and all market participants would be paid or would receive their own local locational marginal price. As already noted, these processes are commonplace in the high-voltage or transmission component of liberalized electricity markets around the world.

Locational marginal prices are dynamic and may vary widely from one location to another. Many end-customers will want to be protected from the volatility in the wholesale prices. In the Australian market, this function is performed by retailers. Retailers typically use hedge products to convert volatile wholesale prices into the time-averaged retail prices that end-customers seem to desire. As locational marginal prices are extended down to the distribution network level it will be important to ensure that retailers have the hedging tools they need to be able to convert the volatile wholesale prices into the retail contracts that consumers want, including flat or time-averaged retail prices if that is what consumers desire.[8]

---

6. To be clear, we are not referring here to the Locational Marginal Prices, which are currently determined at the transmission network level in many wholesale electric power markets around the world (such as PJM). Here we are referring to hypothetical locational marginal prices, which would be, in principle, determined in a future electricity distribution network.

7. For example, Yuan et al. (2016). NARUC (2016), p. 63: "With widespread adoption of DER and integration with utility distribution system planning efforts, the availability of hosting capacity analyses can also be paired with development of distribution locational marginal prices to drive economic siting of DER, much the same way that transmission planning and locational marginal prices identifies areas in need of additional resources to relieve congestion."

8. Biggar and Reeves (2016).

3   RETAIL TARIFFS FOR GENERATION AND LOAD IN PRACTICE, AND THEIR IMPLICATIONS

In practice, of course, existing tariffs paid by loads, and existing tariffs paid to embedded generators, depart very significantly from this theoretical ideal.[9]

3.1   Existing Retail Tariffs for Loads

Retail tariffs paid by loads depart from the marginal cost pricing ideal in two main ways: the presence of substantial fixed costs that are typically recovered through the variable (per kWh) retail charge; and the ubiquitous practice of time averaging and geographic averaging of retail rates.

Historically several, largely fixed, costs have been recovered through variable, energy-based (per kWh) surcharges on tariffs paid by load. There are several such fixed costs:

- the fixed costs of the transmission and distribution networks;
- the costs of running the wholesale market (the cost of the system operator);
- retailer costs; and
- the costs of government schemes to promote renewable generation, such as the Renewable Energy Target.

These costs are summarized in Fig. 10.1 from a report by the public utility regulator in the Australian state of Victoria.

In addition, retail tariffs for load in Australia have been highly time averaged and geographically averaged. Furthermore (although perhaps less significantly) retail tariffs typically include sales taxes or value-added taxes, which are not paid on self-produced goods and services.

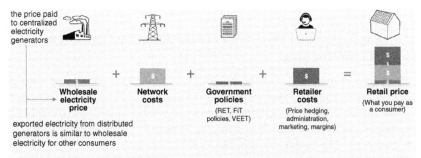

Source: ESC, based on Australian Energy Market Commission *2015 Residential Electricity Price Trends*, December. The illustration of the extent of retail price breakdowns is broadly representative of AEMC data.

FIGURE 10.1   Composition of the retail price of electricity (Victoria, Australia). *FiT*, Feed-in tariff; *RET*, Renewable Energy Target; *VEET*, Victorian Energy Efficiency Target.

---

9. This issue is also addressed in the chapter by Haro et al. in this book.

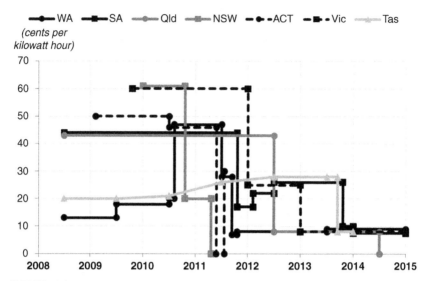

FIGURE 10.2   **Evolution of feed-in tariffs 2008–2015 Australian states.** *ACT*, Australian Capital Territory; *NSW*, New South Wales; *QLD*, Queensland; *SA*, Southern Australia; *Tas*, Tasmania; *Vic*, Victoria; *WA*, Western Australia.

When it comes to tariffs for embedded generation in Australia, a few such generators sell their output in the existing wholesale spot market. However, most small-scale embedded generators in Australia are paid a simple time-averaged feed-in tariff. As set out in Fig. 10.2, although some of these feed-in tariffs started out much higher than the retail tariff, over the past few years the feed-in tariff offered to new solar PV customers has declined rapidly[10] and now typically reflects the avoided cost of wholesale generation (on the order of 6–8 c/kWh).[11]

3.2   Implications of Differences in the Tariff for Generation and the Tariff for Load

Differences in the price paid for the output of embedded generation and the price paid for electricity consumption at the same location and the same point in time give rise to strong and often highly undesirable incentives. The precise incentives that arise depend on whether the generation price is above or below the load price. The simple mathematics of these results are set out in Box 10.1.

---

10. The premium feed-in tariffs for some of the historic schemes are ending in late 2016 and early 2017 (TEC, 2016).

11. The current minimum, maximum, and median feed-in rates (c/kWh) paid to the majority of households and small businesses that are not on premium rates is as follows: VIC (5,10,5), SA (6.8,12.6,8), NSW (0,7.5,0), and QLD (0,10,6). Source: www.markintell.com

BOX 10.1 Impact of Differences in Tariffs on Incentives

Let's suppose that, at a given site, the load at a point in time is Q^L and the genera-tion on the same site is Q^G. The price paid by the load is assumed to be P^L and the price paid for the output of the generation on the same site is P^G. The total amount paid by this customer is therefore:

$$P^L Q^L - P^G Q^G$$

Let's assume that the system operator cannot monitor the output of the em-bedded generation and the on-site load directly. Instead, the system operator can only easily monitor the *net* consumption at the site, which we will label $N = Q^L - Q^G$. The end-customer is assumed to be able to manipulate the appar-ent on-site generation and load subject to the constraints that the net consump-tion is given by N and the on-site generation and load are both required to be positive. The total amount paid by the customer can be written in two other equivalent ways:

$$P^L Q^L - P^G Q^G = (P^L - P^G)Q^L + P^G(Q^L - Q^G) = (P^L - P^G)Q^L + P^G N$$
$$P^L Q^L - P^G Q^G = (P^L - P^G)Q^G + P^L(Q^L - Q^G) = (P^L - P^G)Q^G + P^L N$$

In the case where generation price is equal to the load price $P^L = P^G$, the total amount paid is just the common price multiplied by the net consumption $P^G N = P^L N$. In this case the customer cannot manipulate the apparent generation or consumption (while holding the net consumption fixed) in such a way as to increase or reduce his/her payment.

However, when the generation price is different to the load price the customer may be able to manipulate his/her payment by altering the apparent local genera-tion and load. There are two cases to deal with:

- When the generation price is less than the load price ($P^L > P^G$), the previous equations above show that the customer would like to make both the volume of generation Q^G and the volume of load Q^L as *low* as possible (while hold-ing the net consumption N constant). This can be achieved by using on-site generation to offset on-site load. If the customer is a net importer of power ($N > 0$), the best the customer can do is to make the on-site generation appear to be zero $Q^G = 0$, in which case the customer pays the load price P^L for any increase or decrease in the net consumption of power (from the second equa-tion). On the other hand, if the customer is a net exporter of power ($N < 0$), the best the customer can do is to make the on-site load appear to be zero $Q^L = 0$ in which case the customer pays the lower generation price P^G for any increase or decrease in the net export of power (from the first equation).
- On the other hand, when the generation price is above the load price ($P^L < P^G$), the customer would like to make both the volume of generation Q^G and the volume of load Q^L as high as possible (while holding the net consumption N constant). This could perhaps be achieved by separate metering of generation and load (i.e., gross metering) and by diverting load through the generation meter (in the manner suggested in the text). In principle there is no upper limit on the revenue the customer can receive by arbitraging in this manner.

Let's consider first the typical case in which the price paid for the output of embedded generation is less than the price paid by load. For example, a retail load in Australia may pay a price of around 27 c/kWh. However, a generator on the same site may only be paid the avoided wholesale cost (around 6 c/kWh).

In this case the owner of the embedded generator has a strong incentive to sell *directly to load* in the vicinity, rather than sell its output to the wholesale market at the feed-in tariff (Box 10.1). This can be achieved by:

- using the output of the local generation to offset local load (which is referred to below as a form of net metering),
- constructing a physical duplicate network to link up the generation with neighboring local loads (even where such duplication is inefficient), or
- establishing a separate local embedded network combining generation and load.

To summarize, when the generation price is lower than the load price, it can be said that the generation has a strong incentive to *combine* with the load as much as possible. These actions can undermine legitimate public policy objectives, as explained further below.

Now let's consider the opposite case in which the price paid for the output of embedded generation is greater than the price paid by load at a point in time. This might be the case for customers on a "premium" feed-in tariff,[12] or for generators that are paid the time-varying wholesale spot price, at times of high wholesale prices.

In this case there are strong incentives to *separate* generation from load. This may involve:

- Gross metering, that is, the separate metering of the output of the generation from the load (as set out below).
- Breaking up a microgrid or embedded network into its constituent parts, even where it is inefficient to do so.
- Diverting load so as to appear as though it is generation. One possible way this could be achieved is for the load to be used to charge batteries (paying the lower load price), and for the output of the batteries on discharge to be used to supplement the output of the embedded generation (receiving the higher generation price). This will be referred to as "load masquerading as generation." In principle, there is no upper limit on the extent to which the customer could increase revenue by arbitraging in this way.

These results are summarized in Table 10.1.

---

12. For example, customers on the New South Wales Solar Bonus Scheme, who were paid as much as 60 c/kWh the output of rooftop solar PV (TEC, 2016, p. 4).

TABLE 10.1 Differences in Prices for Generation and Load Affects Incentives of End-Customers

Case	Incentive
Generation price below load price	Generation seeks to *combine* with load, using net metering, colocation, or the creation of a duplicate physical network (such as an embedded network or microgrid)
Generation price above load price	Generation seeks to *separate* from load using gross metering, separate location, or diversion of load to masquerade as increased generation

4   CURRENT PROBLEMS AND POSSIBLE FUTURE DIRECTIONS

4.1   The Status Quo and its Problems

In Australia the price paid for the output of embedded generation (the feed-in tariff) is typically much *lower* than the price paid by retail load.

In addition, for reasons discussed further below, in practice it is very difficult for an electricity network or retailer to identify and separately meter the output of embedded generation against the will of the site owner. Typically the broader network will only be able to meter the *net consumption* or the *net exports* of each site.

The status quo in Australia therefore involves the following three elements:

1. Retail tariffs for load that are both inflated and highly time averaged.
2. A price for net exports that is time-averaged and much lower than the price paid for net imports.[13]
3. Difficulty (or impossibility) of separately identifying and metering embedded generation.

This combination of elements gives rise to several, material public policy issues relating to the use of and investment in embedded generation:

- There are inefficient incentives for both the use of and investment in embedded generation. End-customers do not have the correct incentive to use their embedded generation at times when it is most highly valued, and they do not have an incentive to choose the right type of generation (specifically, generation that can best respond to wholesale market conditions). In particular, the private incentive to install solar PV may be much larger than the social benefit of installing solar PV generation.[14]
- There are inefficient incentives regarding the sizing of embedded generation. End-customers have an incentive to install embedded generation sufficient

---

13. This approach (where there is, in effect, a different price for net exports as for net imports) is also known as "net purchase and sale." Wikipedia: Net Metering.
14. For example, Wood et al. (2015) and AEMC (2014).

to offset only their own load (unless they can combine with other customers to form an embedded network, as noted in the next bullet point). This may result in inefficient sizing or scaling of embedded generation.

- There are inefficient incentives regarding the establishment of embedded networks; end-customers have an incentive to create separate local networks combining load and generation, even if doing so involves duplication of existing network infrastructure.
- There are inefficient incentives regarding investment in storage. As the price paid for net exports is lower than the price paid for net imports, end-customers with sufficient embedded generation to enable net exports at times have an overly strong incentive to install storage, so as to arbitrage between times of net exports and times of net imports, even if there is little or no social benefit arising from that storage.[15]
- The increasing penetration of embedded generation poses a threat to the revenue stream of network operators. As long as fixed costs are recovered through inflated variable (usage) charges for electricity, increased output of embedded generation reduces the revenue received by the network business without a corresponding reduction in its costs. This may result in higher prices for all customers, including customers without embedded generation, further stimulating investment in energy efficiency and embedded generation, and increasing the risk of a death spiral.[16]

There are also potential arguments that the status quo is inequitable or unfair because not all end-customers have equal access to embedded generation. Some customers, such as those with ample roof space, can enjoy the benefits of reduced electricity charges while others, such as those in apartment buildings, cannot.

4.2 Gross Metering, if it Were Feasible, Could Partially Solve These Problems

What can be done to solve these problems? We noted above that the problems stem from three underlying elements: inefficient tariffs for load, tariffs for generation less than the tariff for load, and the difficulty in separately metering the output of on-site generation from the output of on-site load.

One possible solution, of course, would be to insist—despite the difficulty—that any on-site generation must be separately metered from an on-site load. This is known as *gross metering*.[17] The definitions of gross metering and net metering are set out in Table 10.2. This possibility is illustrated as "solution 3" in Fig. 10.3.

---

15. The chapter by Mountain and Harris in this text also emphasizes that solar PV customers may have overly strong incentives to invest in storage.

16. This is mentioned in the chapter by Orton, Nelson, Pierce, and Chappel in this text.

17. In the United States, gross metering is also referred to as the "buy–sell" or "buy all/sell all" approach (Borlick and Wood, 2014; NARUC, 2016, p. 132).

TABLE 10.2 Definitions of Gross and Net Metering

Gross metering	The on-site generation and the on-site load are separately identified and metered, from the first unit of output or consumption in the same way as would arise if the generation and load were located on separate sites
Net metering	The on-site generation is required to be offset against on-site load at the same point in time. Net imports to the site are charged at the standard retail tariff for that point in time. If there are net exports from the site, the amount paid depends on the form of net metering which is discussed further below

Status Quo: Net metering for collocated generation

1. Retail tariffs are inflated and time averaged	2. Price paid by retail load is well above the price paid for the output of embedded generation	3. Output of embedded generation can offset local load

Problems:

- Inefficient incentives to use and invest in embedded generation.
- Erosion of revenue stream for networks
- Inefficient generation scale decisions

- Inefficient incentives to invest in storage
- Inefficient incentive to bypass existing networks

Solution 1: Cost-reflective tariffs

Solution 2: Extend net metering to neighbors or beyond

Solution 3: Enforce separate metering of all embedded generation

Locational Marginal Pricing

Pros:

- Resolves all of these problems including efficient use of and investment in embedded generation, investment in storage, bypass, and so on

Cons:

- Remains largely untried

Virtual net metering and extensions

Pros:

- Partially resolves inefficient incentives for scale, storage, and bypass

Cons:

- Exacerbates inefficient incentives for use and investment in embedded generation and further threatens revenue stream of networks

- Ignores network congestion

Gross metering

Pros:

- Can improve incentives to use and invest in embedded generation

- Can improve incentives to invest in storage and/or to bypass existing networks

- Maintains revenue stream for networks

Cons:

- May not be feasible or practical

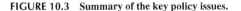

FIGURE 10.3   Summary of the key policy issues.

To many end-customers a requirement that the output of embedded generation be separately metered and paid a lower price seems wrong. Many people have a strong sense that, where it is feasible, they should be allowed to produce on-site to meet their own needs, whether in electricity or in other services. When the Queensland Government considered adopting gross metering for solar PV installations, the response from some solar PV owners was one of disbelief. The Policy Director of the Clean Energy Council observed:

What the Queensland Competition Authority has proposed is the equivalent of telling people they can't just use the lemons growing on the lemon tree in their backyard – they have to sell the produce to a wholesaler for next to nothing, and then buy the lemons back at a premium from the supermarket.[18]

In fact, as emphasized above, there are circumstances where a government might consider telling customers they cannot directly consume the lemons they grow in their backyard. As long as the retail tariff for load is artificially inflated and time averaged, a requirement for separate metering for on-site generation has several potential benefits in partially addressing the problems identified above. Specifically, gross metering potentially:

- allows for improved price signals to embedded generation (while preserving the existing flat retail tariff for loads), potentially increasing the efficiency of both the use of and investment in embedded generation;
- eliminates the incentive to artificially size on-site generation, allowing for more efficient larger-scale generating installations where it is feasible;
- eliminates the incentives to create artificial embedded networks and to inefficiently bypass the existing distribution networks;
- eliminates the incentive to install inefficient storage facilities to arbitrage between times of net exports and times of net imports; and
- preserves the contribution to fixed charges embedded in the retail tariff.

On the other hand, as noted earlier, gross metering may simply be infeasible or impossible to enforce. There is no sensible distinction between an increase in on-site generation and a reduction in on-site load. Any attempt at defining on-site generation for the purposes of gross metering will result in artificial boundary issues, which are difficult to police.

For example, what constitutes a "generator" for which separate metering is required? Must a generator always export power? If gross metering is imposed, manufacturers may seek to install small-scale generation or storage inside consumption devices. Consider for example a pool pump or an air-conditioning unit that normally draws from the grid, but which has the capability to switch to an internal battery at peak times. Should such devices be labeled as generators or loads? The distinction is entirely artificial.

---

18. Solar Choice (Rebecca Boyle), "A gross Solar Feed-in Tariff for Queensland?," 2012, http://www.solarchoice.net.au/blog/news/a-gross-solar-feed-in-tariff-for-queensland-200912/

Even more importantly, without costly detailed on-site inspection, a network or retailer can not know whether or not a customer has installed an on-site generator.[19] Historically solar PV installations have tended to be more visible than other forms of on-site generation, but even these could be hidden in the future[20] and, in any case, it would make no sense to target one form of embedded generation over others. Even if an on-site generator could be identified, it is likely to be extremely difficult to inspect and verify the way the generator is physically wired, to ensure that the full output of the generator passes through the meter, rather than being used to offset on-site load.

There is also the difficult question of how to handle storage, which is both a consumer and a producer of electrical energy. Should battery storage be charged the load rate when it is charging, and the generation rate when it is discharging? Would this require two separate meters? How would this be enforced?

Although gross metering has some economic merit over the status quo, it is not clear that it is feasible in practice.

4.3   Extended Forms of Net Metering Could Also Partially Solve These Problems, but Each has Problems of its Own

Rather than a separate metering of generation and load, perhaps the solution is to allow further integration of generation and load, allowing generation not just to offset on-site load, but also loads located on neighboring sites. This possibility is illustrated as "solution 2" in Fig. 10.3.

At one level, allowing local generation to offset neighboring loads does not seem unreasonable. After all, this generation is already physically connected to the neighboring loads through the local distribution network. These arguments seem particularly compelling where the end-customer owns the neighboring buildings,[21] or where the neighboring buildings don't have the capability to install embedded generation of their own.

The arguments may also seem reasonable where allowing such trade allows for the efficient sizing of an embedded generation facility. If a customer is allowed to install solar panels on their own roof to offset their own load, should they not be allowed to receive the same (or larger) benefit by paying for a share of the larger, more cost-effective, community-scale facility at the end of the street?

These questions amount to asking the question: with whom should an embedded generator be allowed to trade "directly" (without going through the market operator). Should an embedded generator be only allowed to trade directly

---

19. End-customers in the United States who have installed solar PV without permission from their local utility have been referred to as the "guerrilla solar movement." http://www.motherearthnews.com/renewable-energy/guerrilla-solar.
20. Solar PV could be embedded in the roof itself, in the paint, or in glass. http://www.scientificamerican.com/article/im-getting-my-roof-redone-and-heard-about-solar-shingles/
21. In this case the definition of what constitutes "on-site" generation seems entirely artificial.

TABLE 10.3 Definitions Net Metering Used in This Chapter

On-site net metering	The on-site generation is only allowed to be offset against on-site load, as long as there are no net exports from the site. Net imports to the site are charged at the standard retail tariff. If there are net exports from the site, the amount of the net export is paid the feed-in tariff. This form of net metering is standard in Australia
Virtual net metering	On-site generation is allowed to be offset against on-site load *and* other designated loads in a designated neighborhood at the same point in time, such as the participants in a community-shared solar scheme. Other net exports are paid the feed-in tariff
Local or community net metering	On-site generation is allowed to be offset against all load in a designated site, campus or neighborhood at the same point in time, such as a shopping center or housing subdivision
Global net metering	On-site generation is allowed to be offset against all loads at the same point in time

with on-site load? With designated neighboring loads? With all neighboring loads? Or, with all loads? These questions can be thought of as different forms of net metering. There are at least four different net metering options, summarized in Table 10.3.[22]

In the United States the most common form of net metering allows a customer to "bank" credits for exported generation and use those credits to offset the customer's own consumption at later points in time.[23] Another version of net metering sees the customer paid "the average retail utility energy rate" for these export credits. Using this terminology above, these are both forms of global net metering, with slight variations in the tariff paid for net exports, although with limits on the total amount of exports that can earn the retail tariff rate.[24]

Some states, such as California, allow the owner of a multimeter property to use credits for net exports from one customer to offset the energy bills of other tenants in the same building. The California Public Utility Commission refers to this as virtual net metering.[25]

---

22. Langham et al. (2013) set out a different typology based in part on the identity of the generator and loads. They distinguish Single Entity VNM, Third Party VNM, Community Group VNM, and Retail Aggregation VNM.

23. There remains the question as to what to do with any excess credits at the end of the billing year. The treatment of these excess credits varies across utilities.

24. As the net exports can only be used to offset the customer's own net imports, if the customers total annual net exports exceed the total annual net imports, additional credits for exported energy are effectively worthless.

25. CPUC, Virtual Net Metering, http://www.cpuc.ca.gov/General.aspx?id=5408

These extended forms of net metering could partially solve some of the problems identified above. Specifically, extended forms of net metering could potentially:

- Reduce the incentive for inefficient network bypass, by eliminating the incentive to build a physical network between generation and load.
- Reduce the incentive to create artificial embedded networks, instead allowing the creation of "virtual" embedded networks.
- Reduce the incentive to inefficiently scale the embedded generation. If the end-customer is able to find sufficient load to offset the local generation, the local generation can be efficiently sized.
- Reduce the incentive to install inefficient storage. If the end-customer is able to find enough load to offset the local generation, there may be little or no output that is paid the net export price, thereby reducing the incentive to install inefficient amounts of storage.

In addition, extended forms of net metering could be argued to reduce the apparent inequity or unfairness in access to embedded generation opportunities. Extended forms of net metering allow all customers (or at least all customers in the proximity of an embedded generation facility) to share in some of the benefits of that facility, whether or not they are able to install a generating facility of their own.

However, allowing for extended forms of net metering exacerbates the problems identified above. Specifically, extending the scope of net metering will:

- not improve (and potentially worsen) the price signals on the embedded generation, which leads to inefficient usage and investment decisions in embedded generation[26]; and
- exacerbate the problem of erosion of the revenue streams of network operators by reducing the contribution to various fixed costs, including the fixed costs of network provision.[27]

These problems have led in the United States and Canada to the placing of limits on the volume of embedded generation that can qualify for net metering, either on an individual site (kW) or collectively (by the number of subscribers or the total volume of embedded generation).[28]

---

26. NARUC (2016), p. 130: Net metering "does not account for time or locational differences in the costs or value of energy. Of course, the timing and location question is not attributable specifically to [net metering], but is a feature of traditional monthly billing systems with or without customer generation. Still, the matter becomes more complex when both consumption and production are involved."

27. NARUC (2016), p. 130: "Traditional electric rates carry a margin in excess of the direct costs of the measured kWh so that the total costs of the electric utility, including fixed costs and other variable operating costs, can be recovered through that charge. By measuring only net energy, and crediting excess against the total bill, [net metering] reduces not only the energy revenue of the utility but also the margin available for the coverage of other costs." Also: Raskin (2016).

28. Wikipedia: Net Metering. More recently there have been moves to replace net metering with more efficient tariff structures. For example, "New Hampshire consumer advocate floats net metering successor with TOU credits," UtilityDive, October 25, 2016.

There is another flaw with extending the scope of net metering, which is potentially even more serious. As noted at the outset, the efficient price for both generation and load is a price that reflects the short-run marginal cost of producing electricity and delivering that electricity to the location of the customer. As long as the generation and load are located at the same physical site, the price paid by load and the price paid for generation should be the same. However, as long as the generation and load are physically separated there is at least a possibility that the marginal cost of delivering electricity to those two sites is different.

The marginal cost of delivering electricity to location A may differ from the marginal cost of delivering electricity to location B due to either losses or congestion. The further apart the sites are, the greater the likelihood that the electrical losses will differ from one site to the other, and the greater the likelihood that, from time to time, congestion will arise on the network on some path between A and B. Such congestion prevents the carrying out of a simple trade between A and B.

Extending net metering from on-site net metering to neighboring locations effectively sets the generation price at location A equal to the load price at neighboring location B. This is only efficient where there are no losses or network congestion between those two locations. This is possibly a reasonable assumption for small subnetworks, such as within a building, shopping center, campus, or subdivision. It becomes less reasonable the greater the geographic separation between the locations, and therefore the greater the scope for congestion to affect the locations differentially.

The failure of these extended forms of net metering to reflect congestion on the local network is a major drawback. It raises the possibility that local trading of electricity will be incompatible with the physical capability of the local power system. This will at least be inefficient and may, at worst, force the intervention of the system operator to maintain system security.[29]

To summarize, extended forms of net metering are essentially a local form of electricity trading that ignores the physical limits of the local distribution network. Such forms of peer-to-peer trading can only ever be possible over a relatively small geographic area. The larger the area, the greater the likelihood that the resulting trades will be incompatible with the local network, forcing the system operator to intervene. If there is to be large-scale, sustained peer-to-peer trading in electricity, it will only be possible through a mechanism that integrates that trading with the physical limits of the power system. Such mechanisms already exist at the high-voltage (transmission) end of the market in liberalized electricity markets around the world. If peer-to-peer trading in electricity

---

29. According to Langham et al. (2013), virtual net metering programs in the Unites States "attempt to alleviate this issue by limiting the collective installed generator capacity to a proportion of the feeder capacity or voltage related statutory limit."

is to be sustained, it is essential that such mechanisms be extended to include smaller generators and loads.

5   CONCLUSIONS

Around the world there has been substantial interest in facilitating the introduction of small-scale embedded generation into local distribution networks. This process has been eagerly promoted by advocates who see small-scale embedded generation as environmentally sound and as giving a degree of control, autonomy, and choice to the small consumer. It is a small step from these developments to envisage local peer-to-peer trading in electricity, particularly over private embedded networks, community schemes, and microgrids. These schemes are said to further enhance environmental outcomes, and to promote community cohesion.

However, historically retail tariffs have been both high (inflated with a contribution to fixed costs) and time averaged. In this world, end-customers have distorted incentives to invest in on-site generation; the incentive is both too strong, doesn't provide the right usage signals, and doesn't provide the right incentives to choose the right type of on-site generation. Furthermore, investment in embedded generation threatens to undermine the revenue stream of distribution networks, by reducing their contribution to fixed costs. These effects have been observed around the world.

Under the status quo in Australia, end-customers can use the output of their embedded generation to offset their own on-site load, but not load on other sites. In addition, the tariff paid for net exports is almost always much lower than the tariff paid by net imports. As a result there are strong incentives to limit the size of the on-site generation to match the on-site load, even if that means foregoing economies of scale. In addition, where end-customers cannot control the output of the embedded generation (as with solar PV), the customer has a strong incentive to invest in assets to store the electricity produced at times of net exports to reduce consumption at times of net imports, even if those storage assets have no overall social value. Furthermore, end-customers have an incentive to physically connect embedded generation to local loads, in effect setting up a duplicate local distribution network, or to set up a private embedded network, even where the use of the existing public network is more efficient.

Faced with these problems, policymakers may be tempted to move in one of two directions: The first direction is to insist on gross metering of embedded generation. This would, in principle, allow for a much more efficient tariff for the embedded generation, while preserving the contribution to fixed costs embedded in the current load tariffs. However, gross metering is difficult to enforce in both theory and practice. There is no meaningful distinction between an increase in on-site generation and a reduction in on-site load. To our knowledge gross metering has not been imposed as a regulatory requirement (although it has been allowed in situations where the generation tariff is substantially above the retail tariff).

An alternative direction for policymakers is to allow extensions to the principle of net metering, essentially allowing the embedded generation to trade directly with other loads in the neighborhood. This has some apparent benefits: it allows exploitation of economies of scale, reduces incentives for network bypass, and reduces the incentive to set up separate private embedded networks. However, in a classic application of the theory of the second best, it is not possible to say that extending net metering will be desirable overall, as extending net metering makes the problem of bad price signals for embedded generation worse and worsens the erosion of revenue from the contribution to fixed costs.

The extended forms of net metering discussed in this chapter offer the potential for peer-to-peer or direct trading of electricity between end-customers. However, these extended forms of net metering ignore the potential for local network congestion. The theory is quite clear that *direct* trade between two parties is only possible under quite extreme assumptions, including the absence of local network congestion. Otherwise, all trade must be mediated through the system operator. It is conceivable that some subnetworks will be designed so as to not feature any internal network congestion. However in the broader network we do not foresee the possibility of unrestricted direct peer-to-peer trade. Attempts to introduce blockchain-based peer-to-peer trading in electricity networks[30] will not become mainstream, unless they can integrate the physical limits of distribution networks.

Without further tariff reform there will be an on-going tension between embedded generation on the one hand, and public policy objectives of efficiency and fairness on the other. If net metering continues, or is expanded, embedded generation will likely proliferate, but the objectives of efficiency and revenue stability will be undermined. On the other hand, if net metering is prevented (perhaps insisting on gross metering), it will be possible to improve efficiency in the use of embedded generation, but investment in embedded generation is likely to be restricted by regulatory controls. These public policy trade-offs can only be resolved through a move to more cost-reflective tariffs, such as locational marginal pricing. There are some encouraging early signs of the need to reflect congestion in distribution network pricing in Australia,[31] but so far few concrete steps have been taken. The resolution of this tension will set the direction of the electricity industry for the next several decades.

---

30. "Blockchain Transactive Grid Set to Disrupt Energy Trading Market," Engerati, April 11, 2016; "Bitcoin-inspired peer-to-peer solar trading trial kicks off in Perth," RenewEconomy, August 12, 2016; and "Blockchain power trading platform to rival batteries," Australian Financial Review, August 14, 2016. Endesa Energy's Blockchain Lab has launched a challenge for the best innovative ideas making use of blockchain technology in the energy industry.

31. Essential Services Commission (2016) and Australian Energy Markets Commission (2016).

ACKNOWLEDGMENTS

Darryl Biggar is the Special Economic Advisor (Regulatory) for the ACCC and the AER. Joe Dimasi is the Senior Commissioner at the Independent Competition and Regulatory Commission, ACT and the Chair, Tasmanian Economic Regulator. The views expressed in this chapter are those of the authors and do not reflect the views of the ACCC, the AER, the ICRC, or OTTER.

REFERENCES

Australian Energy Markets Commission, 2014. Rule Determination: National Electricity Amendment (Distribution Network Pricing Arrangements) Rule 2014.

Australian Energy Markets Commission, 2016. Distribution Market Model, Approach Paper.

Biggar, D., Reeves, A., 2016. Network pricing for the prosumer future: demand-based tariffs or locational marginal pricing? In: Sioshansi, F. (Ed.), Utility of the Future: The Future of Utilities. Academic Press, Cambridge MA, (Chapter 13).

Borlick, R., Wood, L., 2014. Net Energy Metering: Subsidy Issues and Regulatory Solutions, Issue Brief. Edison Foundation Institute for Electric Innovation, Washington DC.

Department of Energy (US), 2012. A Guide to Community Shared Solar: Utility, Private, and Nonprofit Project Development.

Essential Services Commission (Victoria), 2016. The Network Value of Distributed Generation, State 2 Draft Report.

Langham, E., Cooper, C., Ison, N., 2013. Virtual Net Metering in Australia: Opportunities and Barriers. Report for the Total Environment Centre, Sydney, NSW.

NARUC, 2016. Distributed Energy Resources, Rate Design and Compensation: A Manual prepared by the NARUC Staff Subcommittee on Rate Design.

Raskin, D., 2016. A rose by any other name: response to 'Solar Battle Lines'. Public Utility Fortnightly 154 (3), 16–19.

Total Environment Centre, 2016. Life after FiTs.

Wood, T., Blowers, D., Chisholm, C., 2015. Sundown, sunrise: how Australia can finally get solar power right. Grattan Institute, Melbourne, VIC.

Yuan, Z., Hesamzadeh, M.R., Biggar, D., 2016. Distribution locational marginal pricing by convexified ACOPF and Hierarchical dispatch. IEEE Transactions on Smart Grids (forthcoming).

FURTHER READING

Essential Services Commission (Victoria), 2016. The energy value of distributed generation.

Mountain, B., Szuster, P., 2014. Australia's million solar roofs: disruption on the fringes or the beginning of a new order. In: Sioshansi, F. (Ed.), Distributed Generation and its Implications for the Utility Industry. Academic Press, Cambridge, MA, (Chapter 4).

Chapter 11

We Don't Need a New Business Model: "It Ain't Broke and It Don't Need Fixin"

Clark W. Gellings

Clark Gellings and Associates, LLC, Morgan Hill, CA, United States

1 INTRODUCTION

As described in several of the chapters in this volume, a popular subject these days has become the idea that "we" need a new business model for distributional utilities. Usually this is offered during discussion about the future of investor-owned distribution utilities, but the argument is meant to apply to public power and rural electric cooperatives as well. This dialogue is prompted by the dramatic changes in the cost and performance of distributed energy resources (DERs), including solar photovoltaic (PV) power, energy storage, and hyperefficiency appliances, as well as consumer interest in having control over energy costs and general support for sustainability. Indeed, there is a rapid pace of technological innovation taking place in the DER space.

The natural conclusion by the critics of the existing regulatory paradigm is to say that regulation, as currently practiced, has outlived its usefulness; hence the need for a new regulatory regime, new business models, new tariffs, etc. Others have offered that transactional energy is the ultimate solution and that utilities' usefulness has passed us by. The purpose of this chapter is to argue that there is nothing conceptually wrong with the existing regulatory model based essentially on the rate of return regulation and suggests that, if properly applied, can continue largely as it was conceived. In fact, it will lead to a preferred basis for the distribution utility of the future.

This chapter is organized in seven sections, followed by the chapter's conclusions. Section 2 revisits the driving force behind the discussions about today's distribution utility business model; Section 3 offers an opinion for more changes; Section 4 describes the focus of a new regulation using the existing model; Section 5 discusses the rate of return regulation; Section 6 discusses as to why we do not need to reinvent the wheel; Section 7 discusses how we can

Innovation and Disruption at the Grid's Edge. http://dx.doi.org/10.1016/B978-0-12-811758-3.00011-5

move forward; and Section 8 compares the existing regulatory model to transactive energy (TE).

2 A REPRISE: WHAT HAS PROMPTED THE DISCUSSIONS ABOUT NEW BUSINESS MODELS?

Electricity demand is driven by a complex combination of the desire for the comfort, convenience, and productivity that energy provides, but is tempered by the price, availability, and functionality of competing energy forms (e.g., natural gas) and the technology that converts electricity into services. Fundamental to this demand is the presence of income (in households) and/or economic activity (in business and industry). In each energy application, the technology that converts electricity or other energy forms into lighting, heating, cooling, motive power, or other energy services plays a key role in the effectiveness of the desired energy service, its economy, and the resultant environmental footprint.

Until recently, the relationship of electricity demand to these factors was straightforward. Now with the advent of cost-effective PV power generation, the expanding use of combined heat and power (CHP), and dramatic changes in the end use of electricity, the equation changes. Consumer demand for electricity is now potentially supplied by a combination of grid-supplied energy services and power generated on-site.

Beyond distributed generation, the advent of a myriad of other new and improved technologies also has a substantial impact on the technology landscape and then the overall demand for energy. Recent adoption of plug-in electric vehicles (PEVs) is a prime example of new technologies that can significantly increase the use of electricity. Other new technologies, such as tablets, PCs, home entertainment systems, and heat pumps, are additional examples. Finally, there are a myriad of new entrants into the industry offering everything from purchase power agreements to funding and revenue and cost sharing for innovative distributed power generation, storage, and efficient end-use technology and programs. These changes have led to a consideration of the evolution to a modern power-delivery infrastructure referred to as the Smart Grid. It is defined by the Energy Independence and Security Act of 2007 (EISA, 2007) as a modernization of the electricity-delivery system. Thus it monitors, protects, and automatically optimizes the operation of its interconnected elements, from the central and distributed generator, through the high-voltage network and distribution system, to industrial users and building automation systems, energy storage installations, and end-use consumers, including their thermostats, electric vehicles, appliances, and other household devices.

Fig. 11.1 illustrates the dilemma these changes will cause for the regulated distribution utility. The study it illustrates was based on the 2012 Annual Energy Outlook (AEO, 2012); it highlights how these changes may impact the demand for electricity leading up to 2035 and was responsible for igniting some of the most active debate among utility executives (Electric Power Research

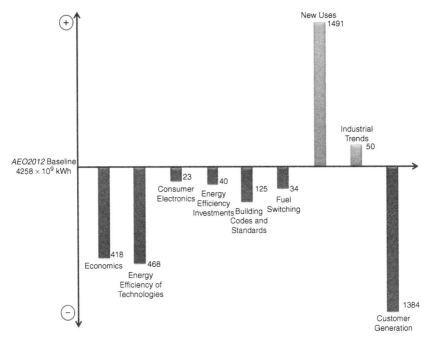

FIGURE 11.1   Factors affecting growth and decline in the demand for electricity-related services by 2035 based on 2012 estimates. *(Source: AEO (2012).)*

Institute, 2013). In this study the net electricity demand from grid-related services is projected to decrease substantially from energy efficiency and customer generation. This despite projected offsets by new uses of electricity, primarily electric vehicles. Most startling is that since the study was published: (1) energy efficiency has accelerated, (2) new uses have only grown modestly, and (3) customer generation has increased well beyond expectations. The bottom line is a further decline in electricity sales from grid-related services.

These changes, further described in accompanying chapters, in technology and the introduction of new entrants offer the possibility that the grid will be characterized by a two-way flow of electricity and information to create an automated, widely distributed energy-delivery network. The grid will enjoy the benefits of distributed computing and communications to deliver real-time information and to enable the near-instantaneous balance of supply and demand at the device level.

3   WILL THERE BE MORE CHANGES?

As a result of these factors among others, there are likely to be more changes in today's power system in the next 10 years than there have been in the last 100. At no time since the first power systems were invented and deployed has there been such a variety of new energy technology available or waiting in the wings. These have the potential to fundamentally and profoundly change the generation, delivery, and utilization of electricity and related electric energy services.

It should be carefully noted that the author does NOT disagree that rapid change is coming with significant implications for the incumbents. What this chapter wishes to explore is whether we need a new business model to deal with these changes.

4   WHAT SHOULD THE NEW REGULATORY FOCUS BE?

The author agrees that we are not yet positioned to leverage the full value of all the innovation that new technologies have brought us. However, we do not need a new business model to do so. We can make the necessary changes within the current regulatory framework to enable them. These changes are as follows:

1. Increase focus on the consumer by building, maintaining, and operating the distribution system to maintain the highest practical reliability and power quality.
2. Support the adoption of DERs by adopting sufficient flexible resources so as not to constrain their proliferation, moving toward an integrated grid, installing and maintaining DER on the customer's premises as part of rate-based assets, developing DER leasing or loan programs, and offering purchase power agreements.
3. Adopting innovative and dynamic rate design options, which align prices to reflect both fixed and variable costs, stimulate the adoption of innovative technology, fully recover capacity costs, and recognize that costs vary by time and season, as described in several chapters in this volume.
4. Facilitate the development and adoption of new uses of electricity, which displace fossil fuels, so as to recognize that electrification is the optimal path toward a low-carbon future. Facilitation must include enhancing R&D on electric end uses, incentive programs, and special tariffs.
5. Develop a full-service strategy to provide installation, operation, and maintenance of all electric technologies, including DERs, efficiency appliances, and other electrical devices. This includes enabling utility investment on customers' premises for generation and storage assets.

Utilities should position themselves as the broker for all things electric in the customer's eyes, "the great rebalancing" described in chapter by MacGill & Smith. The changes now underway will make the marketplace increasingly complex for energy consumers. Without the utility's intervention, consumers will not be able to successfully deal with a number of actors in the marketplace (Gellings and Gudger, 1997). Collectively these service offerings would satisfy the needs originally identified by The Modern Grid Initiative (2007) (MGI). These also include:

- enabling active participation by consumers,
- facilitating participation in alternative service offerings and demand response programs,
- accommodating all generation and storage options,
- enabling new products and services,
- providing power quality for the digital economy, and
- optimizing asset utilization.

Theoretically these services could continue to be provided by the myriad of firms now in the marketplace, as well as others. However, for society to rely solely on some arbitrary organic growth will not insure that potential benefits of this revolution in technology will be fully realized. These needs are ones that could be provided by enabling electric utilities, in partnership with their regulators, to use existing arrangements to supply new and expanded desirable characteristics using the electric power system in its entirety from consumers to central and distributed resources. As alternative business models are considered, it is important to consider how these needs will be met in those arrangements and compare those potential arrangements to that which leverages today's regulated business model.[1]

5 RATE OF RETURN REGULATION

Rate of return regulation has traditionally been used most often used by regulators to determine fair and reasonable prices for electricity. For the utility, fair and reasonable is intended to reflect the utility company's ability to earn a reasonable return on costs incurred in providing electricity service and to avoid monopoly profits. For the customer, it is to restrict the profits the company can make, therefore protecting them from paying more than necessary. In the rate of return regulation, the regulator determines the appropriate amount for the company's rate base, cost of capital, operating expenses, and depreciation (Jamison, 2005).

It uses these determinations to calculate how much revenue the utility needs to recover expenses. The rate of return regulation became popular during the development of the basic utility infrastructure we have today. It allowed for investors to comfortably provide capital to utilities to build power plants and power-delivery systems in exchange for guarantees against certain competition. For example, during the formation of the industry, this became evident in an arrangement, called the "regulatory compact" in which distribution utilities were franchised or allotted a geographic area in which they became the exclusive provider, provided they agreed to serve all consumers, referred to as an obligation to serve, in return for which they were guaranteed a "fair" rate of return.

---

1. Refer to Woodhouse, S., 2016. Decentralized reliability options: market based capacity arrangements. In: Sioshansi, F. (Ed.), Future of Utilities—Utilities of the Future. Academic Press.

There are several inherent advantages to the rate of return regulation. The first is that it is easily sustainable if there is limited competition because prices can easily be adjusted. Second, it provides more comfort to potential investors, as it constrains regulators from enabling too much volatility in price and therefore lowers risk and the cost of capital. In the rate of return regulation, investors can be confident that they have a fair opportunity to receive profit from their investments.

There are, however, four primary disadvantages of the rate of return regulation:

- It provides meager incentives for companies to operate optimally.
- The utility can easily succumb to encouragement to make unnecessary investments so as to inflate its allowed return.
- It requires frequent rate cases during times of high inflation.
- Without careful regulatory oversight, it provides and easy mechanism for companies to shift costs from competitive markets to noncompetitive markets. In short, the rate of return regulation continues to be useful in offsetting the power of natural monopoly. It is mostly applied to Investor-Owned Utilities (IOUs), where there is a divergence of basic interests between the customer wanting to pay little for electricity and investors wanting to earn a handsome return on the utility stocks and bonds it invests in.

Regulation as applied to public infrastructure in the United States first appeared in the case of railroads. During the 1800s railroads expanded, financed by investors eager to earn a premium return. Understandably their investments were targeted at the most profitable new routes, and prospective customers would be charged sometimes usurious rates. At first, railroads expanded in a rather haphazard fashion with inconsistent prices being charged for the same relative service. Both investors and politicians developed the realization that to best serve the public's interest, regulation would be needed. Railroad regulation came forward in the 1880s, leading to formation of the National Association of Regulatory Utility Commissioners (NARUC), still active today, and well known for an association that supports regulators, which oversee electric utility regulation. Following the initial regulation of railroads, consumers and utilities themselves began to clamor for regulation of electric utilities.

The rate of return regulation of electric utilities resulted in oversight of prices for electricity and, in turn their capital expenditures, cost of capital, and operating expenditures. In that regard, they must justify and prioritize investments in the infrastructure for electric power generation, transmission, and distribution. To effectively judge the needs of consumers and to best measure utility performance, regulators measure reliability, customer service, power quality, environmental performance, and benchmark expenses as compared to others.

Admittedly, the rate of return regulation is an imperfect scheme, but it works and can easily be molded to accomplish what is needed for an evolving near-perfect power system.

6 WHY WE DO NOT NEED TO REINVENT THE WHEEL?

In considering the costs necessary to accommodate the needs of tomorrow's power system, there will be expenditures, such as communications, information technology, controls, energy storage, and sensors, which may not be viewed purely as electricity related. However, over the history of the industry, allowed expenditures have often included those only remotely related to the provision of reliable and affordable electric service. Utilities have installed poles for community ball fields, constructed community parking lots, supplied free electricity to county fairs, repaired the electrical service entrance in the city hall building, strung Christmas lights on downtown streets, and performed many other gratuitous tasks, which were built into their expenses and paid for by average ratepayers.

The most substantial nontraditional expenditures are those related to Demand-Side Management (DSM). The most prevalent programs and activities considered to be DSM are: load control; thermal energy storage and dual-fuel heating; innovative rates; energy efficiency; demand response; and electrification (Demand-Side Management, 1984). In these programs and activities, distribution utilities often paid the majority share of expenses, with consumers paying little, if any. In considering future business models, it is not out of the question that this practice could be expanded to investments of communications, information technology, sensors, and the establishment of more complex energy-management systems.

The best DSM examples are energy efficiency and Demand Response. It is now common practice for most utilities to include energy efficiency–type DSM in their portfolios. These are often funded by ratepayer-based programs or other mechanisms to reduce energy consumption, over what it would have been without DSM. Upward trends in energy efficiency program expenditures signal downward pressure on electricity demand. The energy intensity of end uses will reduce further as a result of increased spending by electric utilities on DSM programs and activities. For example, utility spending in the United States and Canada on energy efficiency programs has risen to $9.9 billion in 2014 (Consortium for Energy Efficiency, 2016).

The US Energy Independence and Security Act of 2007 popularized the term Demand Response and generally defined it as including programs and activities that reduce peak demand by the use of dynamic pricing, advanced metering, and enabling technologies. The US Energy Information Administration estimates that over 9 million customers are enrolled in Demand Response Programs (in 2014), yielding an actual peak demand savings of 12,700 MW. The US Federal Energy Regulatory Commission (FERC) has estimated that the potential for Demand Response is such, so as to reduce peak demand in 2019 by as much as 150 GW (NADRP, 2009).

The concepts of using the utility to fund programs that all utility customers my benefit from, albeit the benefits participants, nonparticipants, and ratepayers at large receive may differ, are very important in considering new business models. We managed to do this very effectively with DSM: why can't we do it with DER and other service offerings?

As these changes evolve, the best societal option will increasingly be a truly dynamic power system. This approach enables resources and technologies to be deployed operationally to realize all potential benefits. It requires a robust and modern grid characterized by connectivity, rules enabling interconnection, and innovative rate structures that enhance the value of the power system to all consumers. DER, Demand Response, and DSM will be increasingly important in enabling a marriage between these various technologies and systems, so as to enhance the utilization of the power system and enable increased reliability and enhanced consumer choice, while maximizing the penetration of DERs, including PVs, energy storage and efficient buildings, appliances, and devices (Demand-Side Management, 1984).

The Solar Energy Industries Association (SEIA) recently reported that the United States solar market remains what they refer to as "on track" for a record-breaking year, with 1361 MW installed during the third quarter of 2015 and 4.1 GW installed during the first three quarters of 2015. Much of this is new solar power generation installed on residential buildings (www.seia.org). These installations are often funded by third parties and contracted for by consumers through a variety of purchase power agreements. The lowest cost and most effective power system can no longer be configured solely by combinations of central station power generation knitted to customers by power-delivery systems. Today and increasingly into the future, society's needs for reliable, affordable, and sustainable electricity can best be met by an optimal combination of distributed generation, distributed energy storage, energy efficiency, and new uses of electricity integrated with central generation and bulk system storage.

A more recent report by GTM Research in collaboration with the SEIA reports that a new record was set with the installation of 4.1 GW at an average of 2 MW of solar PV installed per hour through the third quarter of 2016 (GTM Research and SEIA, 2016).

As these distributed solar installations proliferate, the value of DSM will evolve, as will the DSM market participants. DSM incentives targeted toward consumers to influence the pattern and amount of electricity usage will still be valuable, but will change in how they are offered.

7 HOW CAN WE MOVE FORWARD?

One example, can be gleaned from Salt River Project (SRP), an electric utility in the Phoenix, Arizona area, faced with substantial increases in PV system adoption by its consumers, has modified how it compensates customers who produce excess energy. SRP's current tariffs allow consumers with excess PV power generation to sell those kilowatt-hours to SRP at a flat rate, which does not vary by time of day. Rather than pay a flat rate, SRP's plan encourages the use of emerging storage and control technologies so as to reduce the demand on its system during critical periods (www.srpnet.com).

Another example is the City of Austin, Texas. In the so called Pecan Street project, Pecan Street Inc., a research and development organization (www.pecanstreet.org) learned through experimentation that the kilowatt-hours generated from homes with west-facing solar panels were more valuable than those generated from homes with south-facing roofs. The energy generated from west-facing panels occurred later in the day when the sun's arc was westward leaning generating energy that could displace more expensive central generation alternatives. In this example, as DSM evolves, incentives like those for solar energy will need to be tied to the time-varying benefit of load shape changes. If the displaced central generation is more expensive later in the day, then utilities can pay more for the replacement.

Likewise, if the displaced generation is less costly, then the utilities will pay less. As power systems evolve, so will the focus of DSM, but with the same overall objective of maximizing the benefits of electricity to consumers and society. Albeit neither of these examples is of IOUs, they each have their own forms of regulation; SRP has an oversight board and the City of Austin has the city council to deal with.

Any change that derives itself form expansion of distributed technologies can be resolved using the existing regulatory model. The key to enabling necessary changes is for regulators to allow utilities to make investments in DERs, including generation, storage, hyperefficiency tend uses, and the necessary controls and infrastructure to enable their integration with the grid. Fortunately, these concepts are under discussion. For example, in April 2016, Mike Florio, one of the commissioners at the California Public Utilities Commission (CPUC) *floated* the idea that—perhaps—utilities should be compensated for buying DERs from their customers (California Current, 2016). In its April 8, 2016 issue, California Current reported that: California's IOUs could get a 3.5% return on power they buy from distributed resources *if* a *proposal* by Commission Florio goes forward (EEnergy Informer, 2016).

While many of the details of this proposed approach remain to be resolved, finding a regulatory fix to make DERs an attractive option to IOUs, just as investing in poles and wires currently are, would make sense in this context. A similar regulatory "fix" was conceived and implemented in California to make energy efficiency profitable for IOUs. Why not try the same for DERs?

Admittedly, it is a difficult concept to implement, requiring more regulatory intervention. However, as noted by Baak in his chapter, all indications are that California, at least, is de facto moving in this direction with complicated regulatory proceedings to encourage certain investments for certain cases to achieve the state's broader desired goals and objectives. So one could argue that this is already happening.

New York, by contrast, appears to want to walk away from micromanaging DERs by drawing lines in who can do what and to whom, while letting the various stakeholders figure out the details further described in the book's Introduction.

Here are a few examples of how this can be implemented:

- The installation of PV systems and energy storage on individual customer premises, commercial buildings, or as community solar projects can be handled as capacity expansion. Particularly when the externalities of environment and economic growth are considered. Schemes, such as net energy metering (NEM), would be adjudicated based on the value of the energy contributed. However, the capacity contribution would be considered as in the rate base and maintenance as an allowed operating expense.
- Subsidizing or funding the installation of hyperefficient and use technologies can be financed by electric utilities and considered as an investment to defer the need for generation or distribution capacity expansion.
- Funding the installation of electric vehicle–charging stations can be justified based on the societal benefits to all citizens.
- Funding research, development, and demonstration of DERs, demand-side management, and related technologies would be considered as allowed operating expenses and considered in the rate recovery process.

The danger may be that the IOUs may abuse their monopoly power if they are allowed to get into the DER business without adequate regulatory oversight. Moreover, the IOUs may also be tempted "gold plate" and "exaggerate" DER investments if proposals, such as those by Mike Florio previously mentioned, are in place. These problems are well known to regulators. The bottom line is that nothing is perfect and some form of regulatory oversight will most likely be required, regardless of which approach is implemented.

8 IS TRANSACTIVE ENERGY THE NEW MODEL?

TE is a relatively new concept, increasingly popular in utility media stories, roughly described as a market of multiple dimensions wherein consumers, utilities, and providers of all types can transact with anyone in the energy marketplace. It is hard to separate discussions about TE from those about the Smart Grid, the Internet of Things-Electric (IOT-E). One popular definition is provided by the GridWise Architecture Council as: "A set of economic and control mechanisms that allows the dynamic balance of supply and demand across the entire electrical infrastructure using value as a key operational parameter" (GridWise Architecture Council, 2014). Advocates of TE have effectively pointed out that in order for TE to work Distribution System Operators (DSOs) will be needed. As observed by Masiello and Aguero: "The complexity, fun and necessity for a distribution system operator (DSO) arise from the need to coordinate DER operations to preserve and improve, not jeopardize, grid reliability" (Masiello and Romero Aguero, 2016). However, creating effective DSOs is likely more difficult than simply assuring that communications and control are overlaid on the power-delivery infrastructure. TE by itself does not mean a new business model, but will require a great deal of evolving regulatory oversight.[2]

Some of the more critical questions regarding TE have been raised by the CA PUC's Policy and Planning Division, a few of which are listed below(paraphrased) here (Atamturk and Zafar, 2014):

- Is TE empowering consumers or complicating their lives?
- What pricing and services will TE develop?
- What is the cost and benefit of TE to consumers?
- Who should oversee the service providers?
- Can the TE systems and technologies be trusted?

In the spirit of an American folk wisdom saying: "If it ain't broke, don't fix it.", the author examined some of the needs that the power system of the future will provide, as well as its opportunities, and compared them to the tentative promises of TE. Table 11.1 summarizes this comparison.

Understandably many of the attributes of tomorrow's power system can be met by the existing regulatory model. However, what remains unclear is what the cost and benefits of each model will be and how difficult the transitions may become.

9 CONCLUSIONS

Debating concepts like these will help develop viable options in determining the landscape for tomorrow's power system. However, without orchestrating a careful transition from the existing financial arrangements that utilities enjoy, unleashing TE can place a great deal of risk on the ability to continue to support the existing power-delivery infrastructure.

Many of the proposed changes, some of which are described here and in number of other chapters in this volume, could conceivably be adopted and accomplished with the existing regulatory framework, enabling the existing electric distribution utility to become an "Uber Utility" of the future.

An Uber Utility would act as a one-stop shop for consumers, delivering a host of electric energy services they need. It would provide electric energy service along with service and installation of all related electric generation, storage, and energy-efficient devices. Such a utility would be allowed to include all of the assets that could provide overall support to the grid in their rate base. In return, it would retain an exclusive service territory in exchange for providing universal access to all at tariffs, which assure full cost recovery. As noted by Sioshansi in his chapter, the existing grid is a valuable asset already paid for by all. It will most likely remain indispensible to most consumers and most likely even more valuable to prosumers and prosumagers, who will continue to

---

2. For a more complete discussion on transactive energy (TE), the reader should visit various material produced by Ed Cazalet of TeMix, Inc.: www.temix.net

TABLE 11.1 Paticipation in the market will promote efficiency

Tomorrow's needs	Rate of return regulation	TE
Enable active participation by consumers	It is unclear that the technology needed to truly allow customers to actively participate in the buying and selling of electricity exist now or will evolve without research funding encouraged by IOU investment in R&D. Once systems mature, the regulated entities will be able to implement the new systems more effectively	IT firms are anxious to move into this space, but their focus on profit may leave some important options aside, such as hyperefficient appliances and devices
Facilitate participation in alternative service offerings and demand response programs	Regulators will need to encourage utilities to offer more options; they are often reluctant to do so when left to their own devices	ISOs have successfully offered Demand Response programs in active wholesale markets. Mechanisms must be developed to allow more consumers to participate in a variety of offerings
Accommodate all generation and storage options	Rate of return regulation is the most effective means to force utilities to make these accommodations and to provide the needed cost recovery	In TE, these resources must compete with conventional sources, and without laws or regulations will receive limited penetration
Enable new products and services	Regulators must remove all barriers and become more tolerant of investments that may potentially fail and not penalize utilities for taking risks in trying new products and services	New product and service offerings are subject to the whims of investors. Fewer innovations may succeed in TE, but those that make it to the demonstration phase are more likely to be sustainable
Provide power quality for the digital economy	Price-differentiated offerings of varying levels of reliability can be easily supported in the Rate of Return Business model. Regulators need to encourage this kind of innovation	The open market will come forward with Power Quality solutions, as they do now. In TE, not as many consumers may be able to afford these options

TABLE 11.1 Paticipation in the market will promote efficiency (*cont.*)

Tomorrow's needs	Rate of return regulation	TE
Optimize asset utilization and operate efficiently	In a regulatory model, regulators can set up ways to measure the efficiency of utilities and reward them financially	The incentives to be efficient are successful participation in the market itself. This is highly dependent on what part of the power system(s) the market participant is focused on. With <3000 distribution utilities in the United States, the opportunities are limited as to where vendors can apply new products and services
Improve system reliability	This is the holy grail of power systems and is the subject of extensive oversight in the regulated model	Both purchase power and transmission services arrangements will need to be crafted to offer incentives and penalties for TE to assure that reliability targets are met. Also it is not clear who sets those targets
Reduce system losses	Requires active regulatory intervention, as the price of energy is sufficiently low, so as to mask losses	Difficult to see how TE can assure losses are reduced unless the opportunities are substantial
Operate resiliently against attack and natural disaster	Requires regulation at the Federal and State levels	Requires regulation at Federal and State level.
Implement DSM	An ongoing part of the regulatory process now that can easily be part of tomorrow's power system	Will be left to the development of products and services and included as cost effectiveness is clear
Assure sustainability of the electricity system	Regulation at the State and Federal levels will be required to assure this is done most effectively	Regulation at the State and Federal levels will be required to assure this is done most effectively
Facilitate consumer access to information	Easily part of a regulatory model	Will be accommodated as part of open market promotional programs. May not be as effective in TE
Maintain high levels of safety	The hallmark of electrical workers. Likely to be continually reinforced in a regulatory model	Applies to all levels of electrical work by state and federal law. Likely not to be compromised in TE

DSM, Demand-side management; IOU, investor-owned utilities; ISO.

rely on it for reliability, voltage and frequency support, balancing, storage, and backup. For all these reasons, it is incumbent on the regulators to ensure that the integrated grid of the future is adequately funded and maintained by the widest population of possible consumers, prosumeers, and prosumagers, all of whom will continue to benefit from its many valuable services.

REFERENCES

Annual Energy Outlook, 2012. With projections to 2035. US Energy Information Agency. Available from: www.eia.gov/forecasts/aeo

Atamturk N., Zafar M., 2014. Transactive energy: a surreal vision or a necessary and feasible solution to grid problems? California Public Utilities Commission Policy & Planning Division.

Brattle Group, Freeman, Sullivan & Company, Global Energy Partners, LLC, Federal Energy Regulatory Commission (FERC), 2009. A National Assessment of Demand Response Potential (NADRP), Federal Energy Regulatory Commission Staff Report. Washington, DC.

California Current, 2016. Newsletter. Available from: www.cacurrent.com

Consortium for Energy Efficiency (CEE), 2016. Efficiency Expenditures reach $8.7 billion. Available from: www.cee1.org/content/efficiency-expenditures-reach-87billion

Demand-Side Management, 1984. Overview of Key Issues, EPRI EA/EM-3597.

EEnergy Informer, 2016. Newsletter.

Electric Power Research Institute, 2013. Program on Technology Innovation: tracking the demand from grid services. EPRI, Palo Alto, CA (Report 3002001497).

Energy Independence and Security Act, 2007. Available from: http://www.eia.doe.gov

Gellings, C.W., Gudger K., 1997. The emerging open market customer: market-smart consumers, new suppliers, and new products will combine to shape the "Fourth Energy Wave." Fourth Annual Utility Strategic Marketing Conference, April 1997.

GridWise Architecture Council, 2014. GridWise TE Framework. Available from: http://www.gridwiseac.org

GTM Research and the Solar Energy Industries Association (SEIA), 2016. Q4 2016 U.S. Solar Market Insight.

Jamison, M.A., 2005. Rate of Return: Regulation. Public Utility Research Center, University of Florida, FL.

Masiello, R., Romero Aguero, J., 2016. Sharing the ride of power. IEEE Power and Energy Magazine.

National Energy Technology Laboratory, 2007. The Modern Grid Initiative Version 2.0 (conducted by the for the US Department of Energy). Available from: http://www.netl.doe.gov/moderngrid

Chapter 12

Toward Dynamic Network Tariffs: A Proposal for Spain

Sergio Haro, Vanessa Aragonés, Manuel Martínez, Eduardo Moreda, Andrés Morata, Estefanía Arbós and Julián Barquín
Endesa, Regulatory Affairs, Enel Group, Madrid, Spain

1 INTRODUCTION

Since its inception more than one century ago, the electricity supply industry has undergone quite dramatic changes. New generation technologies have been developed, higher and higher voltage transmission lines built, and regional power systems have grown and merged to become national and, of late, transnational ones.

However, there are aspects in which the industry has been markedly conservative, namely, how consumption is measured and how consumers are charged. For one century, consumption meters have overwhelmingly been disk induction ones. They are cheap, reliable, and robust. Customers charging have consequently been based on the energy metered by the device at regular intervals, usually 1 or 2 months. Other simple load characteristics, such as the contracted power, have been used as well.

Historically, such techniques worked well, and it is important to understand why. Lack of sophisticated technology (smart meters) explains why flat, volumetric tariffs have been the norm, even if it has been always known that supply costs were different for different times of the day and of the year. On the other hand, it has also been rightly assumed that generally customers response was quite limited. Moreover, customers belonging to specific categories (e.g., residential urban consumers) displayed quite similar consumption profiles. As a consequence, volumetric charging did not induce inefficient behavior, and it allowed reasonable cost allocation. Management cost was also low. In any case, some customers that might exhibit significant flexibility or peculiar load profiles (e.g., some large industries) might also enjoy more complex meters and more sophisticated tariffs.

Technological advancements have put an end to this world. Distributed generation, storage, and automatic demand response means that customers response

Innovation and Disruption at the Grid's Edge. http://dx.doi.org/10.1016/B978-0-12-811758-3.00012-7

might be significant, and that they can shape their consumption profile, possibly becoming very different of the typical in their consumers' category (Perez-Arriaga and Bharatkumar, 2014). These new technologies offer opportunities to the power sector, as consumers' investments may advantageously substitute network or generation investments. This is particularly relevant as massive energy restructuring efforts are being made to decarbonize the economy.

Increasing tension between these new technologies and traditional tariffs is already evident, as described in a number of chapters in this volume. Quite often it arises because of the mismatch of volumetric tariffs and the network capacity-driven costs. Network costs do depend on the energy transportation capacity (kW) they have. The costs do not depend very much on the energy (kWh) that the network actually carries. If all, or at least most of the customers have similar consumption patterns, consumed energy can be a good proxy of the capacity demands put on the network (double the energy, double the required capacity). On the other hand, if consumption profiles are very different, consumed energy is no longer a good proxy. For instance, a constant load of 3 kW during all the day long amounts to an energy consumption of 3 * 24 = 72 kWh/day. The network capacity it requires is the maximum demand, namely 3 kW. Another load of 6 kW from 4:00 to 10:00 p.m. and nothing during the remaining hours requires 6 kW of network capacity but only consumes 6 * 6 = 36 kWh/day. Under volumetric tariffs the latter would pay half as much for network access as the former, being the actual network use twice as higher.[1]

In some jurisdictions, network costs have been met by charging the customer maximum capacity. This has been easy, as consumers are usually endowed not only with an energy meter, but also with a switch that opens the feeding circuit whenever the energy flow (power, kW) surpasses the switch rating. In the example above, the first customer would ask for a switch rated slightly above 3 kW, whereas the second one will need one above 6 kW. By charging customer maximum capacity, as provided by the switch-rated power, both customers would pay their fair share of network costs.

The problem with this approach has to do, again, with different load profiles. For instance, compare the previous customer demanding 6 kWh from 4:00 to 10:00 p.m. with another one that also demands 6 kWh for just 6 h, the difference being that the demand is now shifted from 1:00 to 7:00 a.m. Switch-rated power should be the same one, and both customers would pay the same amount under a maximum capacity tariff (incidentally, also in a volumetric or mixed tariff). But the new customer only imposes her load during a period when the network is largely unused. The load shifted to the small hours will not require new network investments, neither explains former network investments aimed

---

1. The example above is, of course, based on very much simplified profiles. However it is a fact that hugely different consumption profiles were a relatively minor concern in the past, but they are increasingly common today. For instance, compare the load profile of a prosumer with a PV panel and a battery with that of another consumer without.

at meeting peak load. Therefore it is neither efficient nor fair to charge the valley hours consumer so much as the peak hours one.

The moral is that time is of the essence. In order that consumers behave in a socially efficient manner, they must receive time varying signals reflecting actual network use. As explained below, failing to do so not only wastes a wonderful opportunity but it can paradoxically lead to an energy transition more expensive than the one if distributed technologies were not available.

Many, if not most, of this chapter's themes and conclusions are by no means unique (see, in particular, MIT, 2016). The authors' position has grown out from their work dealing with Spanish regulations. Therefore, the chapter focuses very much on the Spanish system. However, the matters it deals with are of interest for systems other than Spain. One reason is that Spain (a sunny and windy nation) is already acutely facing problems that will shortly become widespread in Europe and the world. Serious regulatory mistakes have been made, but there are also some success stories, as explained in chapter by E. Álvarez Pelegry in this volume. The study of both failures and successes might be useful in other jurisdictions.

The balance of chapter is organized as follows. In Section 2 the structure and amount of end prices for residential consumers is described. Section 3 describes in detail the hourly energy tariff already enacted in Spain, the so-called PVPC tariff. Section 4 proposes a general methodology to extend the hourly PVPC energy tariff to the regulated domain, yielding an hourly network tariff. Next, Section 5 is a case study based on actual Spanish data to get a feeling of the outcome of the proposed methodology, although it is not intended as an actual proposal to be implemented. Section 6 discusses the advantages of the new tariff from the point of view of the energy transformation, followed by the chapter's conclusion.

2   HOW MUCH AND WHAT FOR RESIDENTIAL CONSUMERS PAY?

Spanish residential electricity bills include both market and regulated terms. The market term comprises the cost of the energy purchased by the suppliers in the wholesale markets and sold to the final consumers, as well as other system services acquired in the wholesale markets (balancing and ancillary services). Energy at the connection point should be hourly metered. In case that no hourly metering is available, profiling is used.[2] This is a transient solution, as smart metering deployment should be finished by 2019. The supplier has the duty to purchase the metered energy, as corrected by loss factors computed by the System Operator, in the wholesale markets. The contractual conditions between suppliers and customers for the energy and energy services provided can be freely established. However the regulation establishes a fall-back, "PVPC

---

2. Profiling means that customers are charged according to the average consumption profile of consumers without smart meters, as derived from aggregated demand metered by the System Operator.

tariff," that residential consumers are entitled to request. The PVPC tariff[3] is a pass-through of hourly energy prices plus regulated terms, as described below.

The regulated term is also known as the access tariff. It includes network tariffs, intended to cover the network, a plethora of smaller regulated charges (mainly generation capacity payments, the cost of the Regulatory Agency and the System Operator, and others), and policy charges (mainly, but not only, renewable support charges and tariff deficit annuities[4]). The tariff structure is binomial: there are an energy term (€/kWh) and a capacity term (€/kW) with values conditional to the voltage connection level. Inconsistently, these terms are not hourly ones. The energy to be used is the aggregated monthly consumption. The power to be used is the contracted power that the consumer cannot exceed. There is the possibility of time-of-use tariffs up to three periods, conditional to the energy consumption and other characteristics.

The National Regulatory Authority[5] computes the network tariff. It is mostly paid through a capacity €/kW term, being the rational that network costs are mainly driven by the peak load to be served, and not by aggregated energy consumption. On the other hand the policy charges are established directly by the Spanish government. Most of them are charged as energy €/kWh payments, although the capacity €/kW term is also significant.

Fig. 12.1 shows the electricity bill for an average consumer who has requested the PVPC tariff. The size of the bars is proportional to the monthly euro payments. The market terms are hollow and the regulated terms dashed. Horizontally stripped bars are used for the network tariff (mainly network costs) and vertically stripped bars for policy charges (mainly subsidies). The diagonally stripped bar includes a number of additional regulated supply costs. In the sequel they will be ignored.

3 HOURLY PAYMENTS: THE FALL-BACK TARIFF (PVPC)

Residential consumers can freely contract with their suppliers the terms under which the energy and ancillary services are acquired. However they are entitled to the specific regulated PVPC tariff that acts as a reference. Supply under the

---

3. PVPC stands for "Precio Voluntario para el Pequeño Consumidor," Spanish for "Willing Price for the Small Consumer."
4. Policy charges discussion is not the focus of this chapter. However, to provide some background, it can be useful to point out their main components. First, there are support payments to renewable energy (mainly wind and PV generation), as well as to CHP facilities (usually gas-fired facilities providing industrial steam and electricity). Second, in the past, customers payments did not meet total regulated payments. As a consequence, system debt was created. Currently, the debt principal and interests are paid by the consumers as "tariff deficit annuities." Third, Canary and Balearic islands' electricity supply is more expensive than in mainland Spain. However, regulation requires that all consumers pay the same regardless of their geographical location. As a consequence, island consumers are subsidized by mainland ones. For a more detailed discussion, see Barquín (2014).
5. Comisión Nacional de los Mercados y la Competencia (CNMC) (https://www.cnmc.es/).

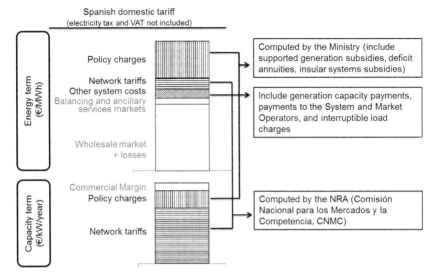

FIGURE 12.1 **Residential electricity bill breakdown.** *(Source: Ministry of Industry, own elaboration.)*

PVPC tariff must be provided by a small number of "reference retailers" (*comercializadores de referencia*).[6]

Since January 2014 the reference suppliers pass-through the hourly energy price as computed by the Spanish TSO, Red Eléctrica de España.[7] To the time-changing energy price, a regulated commercial margin, network tariffs, and policy charges are added up. The resulting price is the PVPC. The PVPC is published the day before, after the closing of the day-ahead market, but before that intraday prices and ancillary services prices are known. Therefore, these last components must be estimated by the TSO. The different prices are weighted according to the value of the energy or services traded in different markets. As most of the liquidity is in the day-ahead market, deviations from the "legal"

---

6. Spanish electricity system is a deregulated one. Generation and retailing (often called supply) are liberalized activities. Most wholesale electricity trading is carried out in the day-ahead market, although there is significant trading in markets closer to real time (intraday and reserve markets). The day-ahead and intraday markets are cleared by the Market Operator (OMIE) whereas reserve markets and the balancing mechanism is managed by the System Operator (Red Eléctrica de España) that also owns the transportation high voltage network. Distribution companies are regulated companies. If in the same holding as generation or retail companies, accounting and legal unbundling is required. Sectorial regulation is under the scope of the Anti-trust and Markets Regulation Commission (CNMC). The Ministry of Industry keeps an important regulatory role, in particular in the policy charges area. See CNMC (2015).

7. Prior to January 2014 the reference suppliers acquired the energy on quarterly procurement auctions (subastas CESUR) similar in organization and spirit to the regulated energy procurement auctions held on a number of PJM systems.

estimated PVPC from the "optimal" real PVPC are quite small. Fig. 12.2, a snapshot taken from the TSO website, shows the PVPC on a specific day, namely February 10th, 2016.

The curve at the left is the one of the standard PVPC 2.0A tariff, whereas the one at the right corresponds to a two-period time-of-use PVPC 2.0DHA tariff. Note that the energy and ancillary services prices are the same ones for both tariffs, being the difference the way that the regulated charges are added: constant along the day or in two periods. For each hour a number of market prices must be added: the day-ahead price, up to six intraday market prices, and several

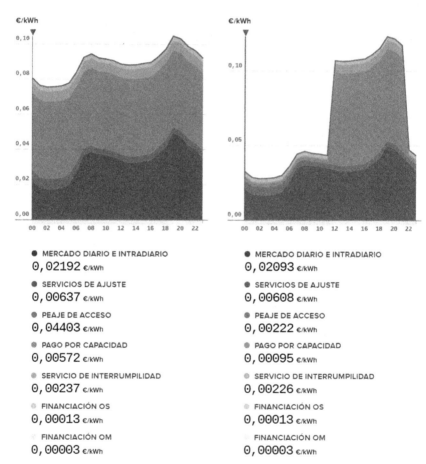

€/kWh	€/kWh
● MERCADO DIARIO E INTRADIARIO 0,02192 €/kWh	● MERCADO DIARIO E INTRADIARIO 0,02093 €/kWh
● SERVICIOS DE AJUSTE 0,00637 €/kWh	● SERVICIOS DE AJUSTE 0,00608 €/kWh
● PEAJE DE ACCESO 0,04403 €/kWh	● PEAJE DE ACCESO 0,00222 €/kWh
● PAGO POR CAPACIDAD 0,00572 €/kWh	● PAGO POR CAPACIDAD 0,00095 €/kWh
● SERVICIO DE INTERRUMPILIDAD 0,00237 €/kWh	● SERVICIO DE INTERRUMPILIDAD 0,00226 €/kWh
● FINANCIACIÓN OS 0,00013 €/kWh	● FINANCIACIÓN OS 0,00013 €/kWh
● FINANCIACIÓN OM 0,00003 €/kWh	● FINANCIACIÓN OM 0,00003 €/kWh

FIGURE 12.2 PVPC (fall-back tariff), February 10th, 2016. Translation: *Mercado diario e intradiario*, daily and intraday markets; *Servicios de ajuste*, ancillary services; *Peaje de acceso*, access tariff; *Servicio de interrumpibilidad*, payments to interruptible loads; *Financiación OS*, payments to the SO; *Financiación OM*, payments to the MO. Numbers are average prices. Note the different vertical scales. *(Source: Red Eléctrica de España (Spanish TSO) esios system, https://www. esios.ree.es/es/pvpc?date=10-02-2016.)*

ancillary prices. Therefore, the monthly price results from an average of thousands of prices. It should be stressed that even if some information technologies concerns were aroused in the past, this kind of billing is perfectly feasible and actually carried out on a regular basis.

Obviously this scheme is dependent on the feasibility of hourly metering. Smart meters rollout in Spain is well advanced and should be completed by the end of 2018.[8] Official statistics state that 14.49 million out of 28.09 (51%) had already been deployed by December 2015 (CNMC, 2016).

4 TOWARD AN EFFICIENT ACCESS TARIFF

The PVPC market energy term conveys hourly wholesale prices that reflect time-varying generation costs. Therefore this term incentivizes efficient behavior of the system users; not so the regulated terms. There are issues both with policy as well as with network charges.

Regarding the policy charges, there are several matters to consider (Newbery, 2015; Batlle, 2011). First it must be decided which part, if any, of these policy charges should be paid by the electricity consumer. For instance, presently about half of the insular systems subsidies are paid out from the State budget as they are considered to be State territorial policy. It is arguable if other subsidies, especially those to RES and CHP, should be paid in the same way as they arise from industrial or climate policies. On the other hand, the proceedings from the auctioning of carbon emission allowances are kept in the State budget. There is a case to share policy charges with other energy carriers, such as transportation or heating and cooling, if only to avoid economic distortions and implicit incentives to less efficient and dirtier alternatives (e.g., diesel vs. electric cars, or fuel-oil boilers vs. heat pumps heating).

In any case it seems very likely that a great deal of policy charges will continue finding their place in the electricity bill. These costs are unrelated to consumption patterns. A fixed monthly customer charge (€/month) seems the better way to reflect the underlying economics. The recourse to this kind of charges to pay for small supply costs (e.g., billing or metering) is usual in many US jurisdictions. However they are unknown in the Spanish context and unusual in the European one. Therefore, a power charge (€/kW) might be a preferable second-best, although it may induce inefficient storage investments.

On the other hand, network tariffs are the main regulated procurement costs. They are intended to pay for capital-intensive assets: variable cost is indeed comparatively very low. As a consequence, economic theory establishes that optimal signals are provided by low prices when network capacity is not binding and higher prices when it is.

---

8. Smart meters roll-out for all consumers is mandated by Royal Decree RD 1634/2006 dated December 29th, 2006, as developed by the Ministry Order ITC/3860/2007 dated December 28th, 2007.

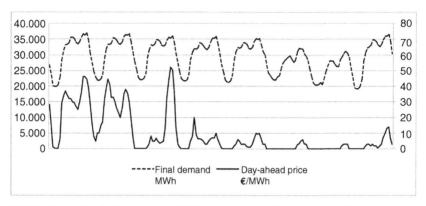

FIGURE 12.3   Hourly final demand (MWh, upper line, left hand scale) and day-ahead market prices (€/MWh, bottom line, right hand scale) from February 3rd to 10th, 2014. *(Source: Red Eléctrica de España, own elaboration.)*

Energy prices are not a good proxy of network use. Moreover, they will be increasingly bad as intermittent RES penetration does increase. Total demand can be considered as a much better proxy for the use of the network, as flows are roughly proportional to it. Fig. 12.3 below shows demand and day-ahead market prices during the week from February 3rd to 10th, 2014. It is noteworthy that during this period peak load was almost attained and, consequently, network use was arguably close to or at its peak value. On the other hand market prices were very low, as wind generation was extremely high.

All of these, points out that network fee methodologies need to be updated to faithfully reflect actual network use in a world with massive intermittent renewable and distributed resources deployment. Some of the most relevant issues are addressed in "Bringing DERS into the mainstream …" by J. Baak and "Access rights and consumer protections …" by F. Orton et al., both in this volume.

In any case, it can be argued that the marginal cost of the network is a good starting point to discuss what the optimal price signal should be. It seems sensible to focus in the long-term marginal cost. Under idealized circumstances in equilibrium short- and long-term marginal costs should be equal. However, that requires, among other things, taking fully into account reliability constraints, which is already a daunting task. Moreover, room must be made for assets indivisibility and other issues.

However, there are two serious drawbacks with marginal cost use, even if long-term:

● First, long-range marginal costs are notoriously difficult to compute. Their estimation requires very complex and not very transparent network expansion models (Mateo Domingo et al., 2011; Jamasb and Pollitt, 2008), difficult to check. Among other effects, that makes the tariffs derived from these models vulnerable to judicial or regulatory challenges.

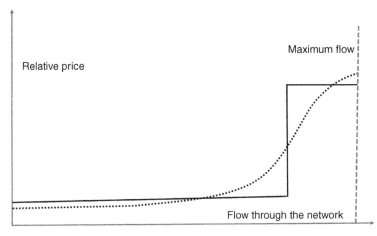

FIGURE 12.4 Shape of an ideal network tariff curve based on long-term marginal costs *(dotted)* and a stepwise approximation *(solid)*.

- Second, and arguably more importantly, long-range marginal cost tariffs do not guarantee cost recovery. Total proceedings may be above or below acknowledged costs, even if only because of unavoidable errors in the planning procedures. In any case it is a rising concern as investments in new technologies of uncertain costs and impacts increase.

However, long-term marginal cost figures carry out two important messages: they sharply increase when the network capacity is approached, and the main variable that explains them in each moment is the power flowing through the network.

The proposed approach is, in a nutshell, that the Regulator will provide, after consultation, a formula or curve to the System Operator to compute hourly network charges from the flow through the network (Fig. 12.4). The curve shape will approach as best as possible long-term marginal costs, as it guaranties cost recovery.

In particular, the proposed network tariff will sharply increase when network capacity is approached. The network tariff is time-changing, but not in a preset manner (e.g., higher on all working days from 5:00 to 10:00 p.m.). It rather fluctuates, as the wholesale market energy prices, according to the system operating conditions.

A specific consumer possibly makes use of several networks. For instance, a residential customer makes use of her low-voltage local distribution network, the medium-voltage regional network, and the transmission network. It is proposed to apply the approach to each network separately, and then add up the tariffs for all the networks the consumer makes use of.

The curve relates energy flow, that is power (kW), to a price. Ideally, every moment, every second, consumed power should be metered and charged.

This is clearly not feasible, neither needed nor useful. It is not feasible because of obvious equipment limitations. It is not needed because network flows do not change dramatically from second to second. And it is not useful because consumers response from second to second is not to be expected. Rather, a convenient time granularity should be used. In most places, including Spain, current market arrangements strongly suggest hourly aggregation. In this case, network tariffs do refer to average energy flow or power during each different hour. Numerically this happens to be the same figure as the energy consumed during that hour (if average power from 1:00 to 2:00 p.m. was 2 kW, this implies that energy consumption from 1:00 to 2:00 p.m. has been 2 kWh). This is why network tariffs are quoted below as €/kWh and can be aggregated to hourly energy prices. However it should not be forgotten that conceptually network charges are not an energy charge.

Rather than providing a complex curve derived from long-term marginal estimations, it might be more practical, and possibly no more inexact, to use a simplified curve, for example, a stepwise function. This has been actually the case in the case study described in the next section. In any case, as the curve reflects long-term costs, it includes the impact of future distributed resources if rightly forecasted. Of course, forecasts are rarely accurate. This is a reason why the Regulator must update periodically the curve (e.g., every 3 years). The approach can be extended to deal with disruptive levels of distributed generation and complex network topologies, and is explained below.

5 A SPANISH CASE STUDY

The goal of the case study is to provide a feeling of the outcome of the methodology sketched above in a real case, that of the Spanish system.[9] The starting point is the acknowledged network costs as published by the Ministry of Industry.[10] Network acknowledged costs are provided by voltage level. This information as well as yearly total energy flows are shown in Table 12.1.

Ideally, networks should be geographically differentiated. For instance, network use patterns in tourist areas, such as the Costa del Sol is very different from those in industrial areas, such as the Barcelona metropolitan zone.

---

9. See Haro (2016) for a more extended description of the methodology and case study results.

10. *Memoria de la propuesta de Orden por la que se establece la retribución de las empresas de distribución de energía eléctrica para el año 2016. Secretaría de Estado de Energía, Ministerio de Industria, Energía y Turismo.* Costs have been adjusted not to include Canary and Balearic islands' network costs, under a different regulation (265 M€). On the other hand, to be consistent with current Spanish regulation, Demand-Side Management and Capacity Payments costs have been included in NT4 costs (1062 M€). It is questionable if these last costs should be considered jointly with network costs. The costs provided in the Memoria have been allocated to the different voltage levels according to the CNMC methodology as described in *Circular 3/2014, de 2 de julio, de la Comisión Nacional de los Mercados y la Competencia, por la que se establece la metodología para el cálculo de los peajes de transporte y distribución de electricidad.*

TABLE 12.1 Acknowledged Network Costs and Flows by Voltage Level

Voltage level	Millions of €	Power flow (TWh)
NT4 $V > 145$ kV	2499	233.58
NT3 72.5 kV $< V <$ 145 kV	454	206.51
NT2 36 kV $< V <$ 72.5 kV	477	190.88
NT1 1 kV $< V <$ 36 kV	2243	175.57
NT0 $V < 1$ kV	1769	100.52
Total	7342	

Source: Spanish Ministry of Industry (see footnote 10).

However such data were unavailable from the Regulator, so the aggregated data by voltage level were used instead. Even so, at the end of the day, there is a much improved price signal as shown below. Power flow through a network comes from the final net consumption (consumption minus self-generation) of the customers connected at the network plus the net demand of users connected to lower voltage networks, as corrected by suitable loss factors.

There are also official data on electricity consumptions. Monotonic curves of power flowing through low (NT0) and high (NT4) voltage level networks are represented in Fig. 12.5. Note that the high-voltage curve is flatter than the low-voltage one. Intermediate voltage levels (not shown) take also intermediate positions.

A two stepwise approximation to the long-range marginal cost is used (Fig. 12.4). Derivation of this curve is a very complex task that would require developing new tools and methodologies. However it was felt that setting a demand threshold of 80% of the peak load could result in a sensible approximation. The threshold is represented in Fig. 12.5. In the high-voltage network, flow during 2823 h was above the 80% demand threshold. In the more peaky low-voltage one only 487 h were above.

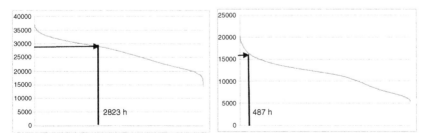

FIGURE 12.5 Monotonic network flow curves (high voltage on left, low voltage on right). *(Source: https://www.esios.ree.es/es/analisis/1201to/1026 (REE esios web).)*

To fix the relative height of the two periods, it was decided that 90% of the network costs should be recovered during the peak hours, as this is roughly the proportion of expenditures required to serve peak demand over total expenditures. So, for the low-voltage network, 90% of €1769 million (Table 12.1) must be recovered during the 487 h of peak network flow. Total flow during these peak hours happens to be 8.53 TWh, about 8% of total low-voltage demand. As a consequence, to recover all the network costs, a price of 1.76 €/MWh must be charged during valley hours, and a much higher price of 186.75 €/MWh during peak hours. For the flatter high-voltage network the figures are 1.07 and 24.57 €/MWh. Note that a flatter curve translates in a smaller peak/valley price ratio.

Each network user must pay for all the networks he makes use of. The charges should take into account T&D losses. So, a consumer connected to the high voltage only pays 1.07 €/MWh in valley hours and 24.57 €/MWh during peak hours. On the other hand, a consumer connected to the low-voltage network NT0 must also pay for the NT1–NT4 voltage networks, as consumed power flows also through this networks. The average loss factor from high to voltage level is 12.07%. So, a NT0 customer will be charged 1.07 * (1 + 0.1207) = 1.20 €/MWh for the high-voltage network during valley hours and 24.57 * (1 + 0.1207) = 27.54 €/MWh during peak hours. Similar figures should be derived for intermediate voltage levels. These computations should be performed for each hour, taking into account that a specific hour might be a peak hour for the high-voltage level but a valley hour for the low-voltage network or the other way around.

Fig. 12.6 shows the network tariff—duration monotonic curve for a user connected to the low-voltage network. Note the high network tariff of

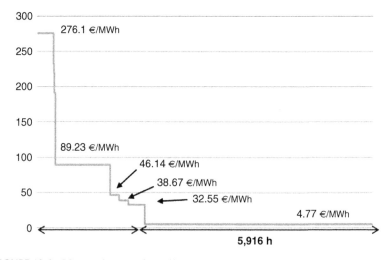

FIGURE 12.6   Monotonic network tariff curve.

| First Quarter | Second Quarter | Third Quarter | Fourth Quarter |

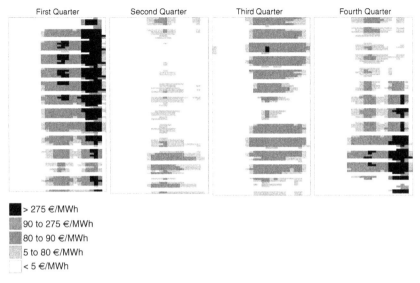

> 275 €/MWh
90 to 275 €/MWh
80 to 90 €/MWh
5 to 80 €/MWh
< 5 €/MWh

FIGURE 12.7   Network tariffs map. Horizontal, hours; vertical, days.

276.1 €/MWh that results from simultaneous peak charges in all network levels. Incidentally, it also illustrates the high correlation among peak hours of all networks. Note also the very low valley network tariff (4.77 €/MWh) that prevails during 5916 h.

As said above, the network tariffs are hourly ones. Fig. 12.7 shows a heat map of the low-voltage tariffs along a year. Note the high network tariffs during winter weekdays evening hours, and the low values prevalent during the small hours and the more template seasons.

There are no great technical barriers to implement these tariffs. The regulator might periodically (e.g., every year) publish the different network tariff curves computed according to a publicly known methodology. The curves relate flow through a network with the network tariff. At the time of publishing the PVPC tariff, the TSO not only knows the day-ahead market price and has accurate forecasts on the intraday and ancillary services prices, but also has accurate forecasts of the network flows. Consequently the TSO would be able to publish a new "PVPC tariff" by using hourly network tariffs. It is important to point out that no additional investments or equipment beyond those required by today's PVPC tariff are required; it is just a matter of regulatory reform.

Both the Regulator and the TSO forecasts are likely to be mistaken at some degree, and as consequence there will be a mismatch between the income of the network tariffs and the acknowledged network costs. However this problem also arises presently; there is no reason to assume that it is going to be more serious, and in any case can be addressed by the same mechanism (namely, by rolling out deficits or surpluses to the acknowledged costs of the following period).

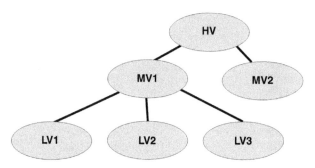

FIGURE 12.8 Different networks of a power system.

6 IS IT WORTH? NETWORK TARIFFS AND DECARBONIZATION

Note that there is nothing special about the "cascaded" structure considered in the example (a low-voltage network feeding from a medium-voltage one and so on). For instance, in the system in Fig. 12.8, a low-voltage LV1 consumer would pay the network tariffs computed for networks LV1, MV1, and HV, that is, the networks she makes use of. Network tariffs must naturally take into account proper loss factors. Solutions can be devised even for meshed, nontree networks. It is only required to define a proper metric of network use, such as the average participation method. However this sophistication does not seem needed.

The tariff network above described solves a number of relevant problems and can actually efficiently promote low-carbon generation penetration as well as other technologies. The remaining of this section deals with some of the issues.

- Self-consumption

  Badly designed self-consumption regulation can induce cross-subsidization because of two reasons:

  1. It may be a way to avoid policy charges. By just charging on the basis of net consumed energy, policy charges are avoided on gross consumption. As policy charges do not reflect supply costs but they rather are quasitaxes to be levied from the system users, policy charges are shifted from prosumers to the remaining consumers.
  2. It may provide a way to effectively sell energy at a price higher than its value. By exchanging possibly low-cost energy (both in terms of energy proper as well as network use) generated during sunny hours for high-priced energy in evening hours prosumers may obtain higher benefits than the value they provide.

  In some jurisdictions, as in the United States, the second effect seems to be the dominant one as explained in chapters by Sioshansi, Jones et al., and

Baak. In the high policy charges Spanish system the first effect is the most important.

Being more specific, in Spain, network costs are mostly paid through capacity (kW) charges.[11] Therefore investment in residential PV facilities does not affect very much the network payments, as prosumers' peak load is not much affected (it usually takes place after sunset). Besides, the prosumer pays the net hourly consumed energy at wholesale prices (see PVPC description above). In other words, from the point of view of energy and network charges, the customer can be thought as a sort of "black box." Only net hourly energy consumption (consumption minus self-generation) and peak demand are relevant here. No significant cross-subsidization is likely to arise.

However, a serious policy charges problem do remain. They are mostly volumetric (monthly kWh). For traditional consumers, the charges can be thought of as a quasitax on volumetric consumption. To ensure that prosumers also fully pay this quasitax, net energy consumption must be separated in consumption proper and self-generation. Otherwise self-generation benefits from a fiscal opportunity benefit (namely, quasitax avoidance because of substitution of taxed energy supplied by the system with self-generation of quasitax-free energy). This requires, in turn, metering of load proper or of self-generation (load can be estimated as net energy consumption plus self-generation). Therefore, setting meters inside the customers' premises is required. This is problematic from a legal or commercial viewpoint. In any case small residential prosumers are not currently required to fully pay the load policy charges.

Most of the Spanish PV developments are industrial-scale ones. This seems intuitively right, as industrial scale PV is typically significantly cheaper than small residential one. It also suggests that in some jurisdictions where residential PV is fast growing, the reason may lie in volumetric network or policy charges, that artificially enhance distributed PV attractiveness and have the potential to create significant cross-subsidization.

In any case, the use of the hourly energy and network tariff provides the right economic signals. This is very important if massive distributed PV deployment is expected. On one hand end prices should incentivize preferential deployment in sunny places. The energy term in the end prices address this issue. On the other hand, to excessively load networks in sunny places should be avoided, to defer or to avert expensive reinforcements. The network charges are to take care of that. Obviously the whole network must be divided into suitable regional ones: a single national distribution network price is clearly unable to provide optimal signals. As stated above, the proposed methodology can easily accommodate this

---

11. This part draws heavily on (Aragonés et al., 2016).

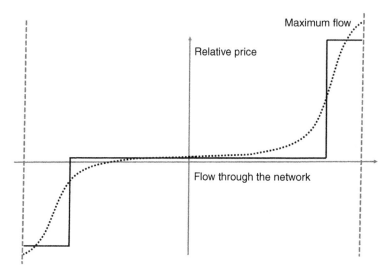

FIGURE 12.9 Shape of the long-term marginal network cost *(dotted)* and a stepwise approximation *(solid)*.

requirement. Final outcome should be a cheaper and more efficient PV generation deployment.

Ultimately, it might be the case that for certain areas and periods distributed generation outstrips consumption. These areas would become net power exporters. The proposed methodology can easily accommodate these developments. For instance, let us consider the admittedly extreme long-range marginal cost curve depicted in Fig. 12.9, analogous to Fig. 12.4.

Positive flow denotes net consumption and negative flow denotes net generation. Marginal costs are negative when the region is exporting power, as additional generation (that is a decrease in the flow through the network) entails additional cost. As a consequence, network tariffs should be negative in this regime. Revenues will be positive, as they are the product of negative prices times consumption, that is also negative (more generation than load). Among other effects, these tariffs should incentivize local demand, aligning it with local distributed generation.

- Storage and joint consumption

Massive intermittent RES deployment will be greatly facilitated by storage deployment as further described in a few chapters in this volume. As in other technologies, there is a choice to be made between centralized and distributed storage. Centralized storage benefits from scale economies and possibly better control. Decentralized storage can reduce network investment costs and losses, although it can also increase them if deployed in the wrong places and operates under wrong price signals. Possibly more than any other technology, storage requires time-varying prices to properly operate (i.e., to store energy during low-price periods and to release it during high-price ones). The network component of the hourly price embodies the implied network savings, and it is therefore the main signal that bias investment toward centralized or decentralized solutions.

In addition to the network tariffs, policy charges are also quite relevant when discussing storage, as suggested above. If policy charges are paid through energy (€/MWh) there is an undue incentive toward self-generation, as self-generation might save a prosumer part of these charges without contributing any social good (other consumer will pick up the charge). If policy charges are paid through capacity (€/MW) there is a parallel undue incentive to install batteries or other storage facilities to decrease contracted power beyond the point justified by network savings.

There is an additional problem related to capacity tariffs. Maximum power demanded by two or more consumers jointly will be typically less than the sum of the maximum power demanded by each one, as simultaneous maximum consumption is unlikely. Therefore, if policy charges are paid through capacity prices (€/MW) there is an incentive to jointly contract the supply. This is plainly inefficient, as revenues are reduced without a change in consumption patterns. Customer charges (€/month) to pay for policy charges, on the other hand, can be set in such a way to avoid these aggregation issues (e.g., by making them proportional to the served properties' values).

- Demand-side response and electric vehicles

Intermittent RES deployment implies that there will be periods when energy and network costs are likely to be very low. In these periods electricity consumption is economically sound even if electricity conversion efficiency is low. For instance, in a cold windy night, network use might be low and marginal cost of electricity very low as well. Therefore electric heating by low-cost heating rods is to be preferred to natural gas heating, especially if wind energy spillage is likely. Just this kind of switching may save significant amounts of carbon (Heinen et al., 2016). Presently there is no private incentive to procure this kind of equipment, as constant and relatively high network tariffs do not render electricity competitive along the day. Of course more efficient electric equipment, such as heating pumps might be considered and be found competitive, especially if policy charges are reduced.

Electric vehicles offer a way toward transportation decarbonization, so far as the implied power system is not coal dominated. They also bring important benefits related to local pollution in cities (specially NOX, SO_2, and particulate matter emissions). From the power system viewpoint, the most significant integration costs are not the energy ones but the distribution network ones. This is because electricity cars' energy consumption is much lower than that of petrol cars, but their widespread use can significantly increase flows in the owners' distribution networks (see "Powering the driverless electric car of the future" by J. Webb and C. Wilson in this volume).

In that sense, electric vehicles' smart charging is just another demand-side management measure (Eid et al., 2016). In order not to overload distribution networks, cars' battery charging during peak residential consumption must be avoided. It can be seen from Fig. 12.7 that network tariffs are very low during night hours, so a natural incentive to defer loading up to the small hours is already present. Whether the incentive is enough to further invest in vehicle to grid equipment can be also assessed from the hourly changing energy prices and network tariffs.

7   CONCLUSIONS

Smart meters rollout is already a reality in Spain. Presently they are used to confront the customers with the hourly changing energy prices of the wholesale market. Moreover information and billing systems are able to handle hourly changing network prices as well. Hourly network tariffs barriers are, in short, no longer technical ones.

This chapter proposes a specific methodology to compute hourly network tariffs. It is further argued that by approving such tariffs, efficient self-generation, storage, and electric cars deployment are incentivized.

There are other issues that are possibly relevant. First, the proposed approach allows the Regulator to focus in his traditional role of assessing the cost of the regulated assets and make sure that the distributors recover their cost and a fair return. In other words, it frees him from the duty to establish the social value of facilities indoors the consumers' premises and to compare with the value of regulated or even market assets. Computation of network costs and tariffs does not and should not require such information.

Policy charges, either subsidies to be granted to specific technologies, territorial compensation measures, or whatever, should be established by a different process, possibly also by a different agency. They should be clearly charged as such, either to the electricity or energy consumers or to the taxpayers. Charging should interfere as little as possible with efficient energy and network use signals.

Fair cost allocation is also an important matter of increasing social saliency. As discussed above, inefficient network tariffs can create cross-subsidies from traditional consumers to prosumers, who often are more affluent customers who live in properties spacious enough to set PV facilities and have the financial resources of doing it. Similar issues might arise as electric cars, batteries, and other distributed equipment are deployed. Efficient network tariffs are usually much fairer than the alternatives, as cross-subsidization is avoided. They also generally empower all consumers, who can see as changes in their behavior translate into billing savings.

REFERENCES

Aragonés, V., Barquín, J., Alba, J., 2016. The new Spanish self—consumption regulation. Energy Proc. 106, 245–257.

Barquín, J., 2014. The new Spanish electricity market design, ENERDAY 2014, 9th Conference on Energy Economics and Technology—A European Energy Market?, Dresden, Germany, April 11th, 2014.

Batlle, C., 2011. A method for allocating renewable energy source subsidies among final energy consumers. Energy Policy 39 (5), 2586–2995.

CNMC, 2015. Spanish Energy Regulator's National Report to the European Commission, 2015. CNMC, July 23rd, 2015.

CNMC, 2016. Estadística sobre la efectiva integración de los contadores con telemedida y telegestión de consumidores eléctricos con potencia contratada inferior a 15 kW (equipos de medida tipo 5) a finales del segundo semestre de 2015 (Statistics on smart meters effective integration for electricity consumers whose contracted power is less than 15 kW (meters type 5) at the end of 2015 second semester). CNMC, September 22th, 2016 (in Spanish).

Eid, C., Koliou, E., Valles, M., Reneses, J., Hakvoort S R., 2016. Time-based pricing and electricity demand response: existing barriers and next steps. Utilities Policy 40, 15–25.

Haro, S., 2016. Design of a tariff scheme based on cost causality. MSc Thesis, Comillas Pontifical University, June 2016.

Heinen, S., Burke, D., O'Malley, M., 2016. Electricity, gas, heat integration via residential hybrid heating technologies—an investment model assessment. Energy 109, 906–919.

Jamasb, T., Pollitt, M., 2008. Reference models and incentive regulation of electricity distribution networks: an evaluation of Sweden's Network Performance Assessment Model (NPAM). Energy Policy 36 (5), 1788–1801.

Mateo Domingo, C., Gómez San Román, T., Sánchez-Miralles, A., Peco González, J.P., Candela Martínez, A., 2011. A reference network model for large-scale distribution planning with automatic street map generation. IEEE Trans. Power Gen. 26 (1), 190–197.

MIT, 2016. The Utility of the Future. Available from: Energy.mit.edu/uof. MIT.

Newbery, D., 2015. Reforming UK energy policy to live within its means. EPRG Working Paper 1516, September 2015.

Perez-Arriaga, I., Bharatkumar, A., 2014. A Framework for Redesigning Distribution Network Use of System Charges Under High Penetration of Distributed Energy Resources: New Principles for New Problems. CEEPR WP 2014-006, October 2014. Available from: https://energy.mit.edu/publications/

Chapter 13

Internet of Things and the Economics of Microgrids

Günter Knieps
University of Freiburg, Freiburg im Breisgau, Germany

1 INTRODUCTION

The reform of electricity markets is gaining momentum worldwide, shifting increasing attention to innovations at the grid's edge within low-voltage electricity networks and challenging the traditional value chain from generation via high- and medium-voltage to the low-voltage household networks. Innovations from the bottom are strongly driven by embedded small-scale generation facilities (e.g., rooftop solar PVs), innovations in energy storage technologies (batteries and electric vehicles), and flexible demand response (e.g., by smart metering and remote control), enabling variable and flexible renewable producer and consumer (prosumage) activities.

In this context, microgrids are platforms for integrating locally and enabling real time–based generation, storage, and consumption of renewable energy. As such they may eventually serve as virtual power plants, aggregating the low-voltage electricity of several home networks and dispatching it to the medium-voltage distribution grid or actively trading through peer-to-peer open-trading platforms as further described in chapters by Biggar & Dimasi, Orton et al., Steiniger, Johnston, and others in this volume.

Although innovations regarding renewable electricity generation and battery technologies for storage outside and inside of electric vehicles are an important driver for the evolution of microgrids, the enormous innovation potentials of Information and Communication Technologies (ICT) are complementary and symbiotic elements in the evolution of smart grids in general, and microgrids in particular. Many chapters in this volume indicate the relevance of innovations in metering, sensors, real-time interactive machine-to-machine communication, and remote control. An illustrative example is AGL Energy, which is developing the world's largest virtual power plant involving 1000 homes and businesses in South Australia using solar PV and battery storage systems, and networked through a cloud connected control system as described in chapter by Orton et al.

Innovation and Disruption at the Grid's Edge. http://dx.doi.org/10.1016/B978-0-12-811758-3.00013-9

In the meantime, a large effort of international standardization organizations is driving innovations with regard to the virtual side of microgrids, with particular focus on home networks in the context of smart city initiatives (ITU-T, 2015a,b). In this context ICT innovations become increasingly relevant for the future evolution of microgrids and their interaction with virtual power plants and distribution/transmission networks.

The major goal of this chapter is to analyze the potentials of ICT for the organization of future microgrids, taking into account the multiple interactions with high- or medium-voltage electricity networks. Section 2 analyzes the role of ICT within microgrids, in particular the evolution of standards for data packet transmission (IP protocol), standards for sensor networks, and actuator standards for IP-based home networks (especially focusing on the evolution of G.hn and ZigBee standards). In Section 3, the outside interaction of microgrids is analyzed followed by the chapter's conclusions.

From an ICT perspective of virtual microgrids, real-time communication within and between microgrids, as well as between microgrids and distribution networks is required. Corresponding requirements regarding data transmission challenge the architecture of the traditional best effort ICT networks. Instead, the potentials and capabilities of Next Generation Networks (NGNs) (ITU-T, 2004, p. 2) to implement multiple broadband-based transport technologies become crucial. The future development and success of the Internet of Things hinges critically on meeting a wide array of heterogeneous Quality of Service (QoS) requirements, which cannot be met within the traditional best effort TCP/IP Internet due to the lack of QoS guarantees regarding latency, jitter, and packet loss.

2 ICT INNOVATIONS AND STANDARDS AS DRIVERS FOR MICROGRIDS

As is the case for intelligent transport systems and smart water and waste management systems, the challenge for smart grids is to cross-fertilize the traditional network industries with ICT. Innovations in ICT have large potentials for network industries. For the electricity sector ICT gains relevance not only within high- or medium-voltage smart electricity networks, but also within low-voltage microgrids. In particular, innovative ICT devices (e.g., sensors, tags, actuators, or meters) bridge the gap between the real world of devices and the digital world of ICT-based information. Thus, real-time adaptive prosumer interaction within microgrids requires high deterministic traffic qualities guaranteeing a very low latency (Fang et al., 2012, p. 18; ITU-T, 2015a, pp. 50 ff.; ITU-T, 2015b).

Large efforts are underway worldwide to develop the standards that are involved in the evolution of smart grids in general, and microgrids and related home networks in particular. The International Telecommunication Union (ITU) is the United Nation's specialized agency in the area of ICT, and Telecommunication Standardization Sector of ITU (ITU-T), in particular, provides recommendations for standard setting frameworks for control and management services of a microenergy grid, including advanced metering infrastructure service, demand–response management service, customer energy management service, and distributed energy grid management service (ITU-T, 2015b). Particular effort has been undertaken within the areas of home networks (ITU-T, 2015a, pp. 73–77; ITU-T, 2016a,b). Other standard setting organizations involved in the standardization not only of physical grids, but also of ICT are the International Organization for Standardization (ISO), the International Electrotechnical Commission (IEC) (ISO/IEC, 2012), and National Institute of Standards and Technology (NIST) (NIST, 2014). IPs for smart grids with particular focus on security issues have been developed by the Internet Engineering Task Force (IETF) (Baker and Meyer, 2011).

2.1   Characterization of Microgrids

Microgrids are located at the grid's edge and consist of two complementary parts: a low-voltage electricity network (physical microgrid), and a virtual network consisting of a complementary set of ICT components for two-way communications (virtual microgrid). A schematic illustration of physical microgrids in contrast to virtual microgrids is provided in Figs. 13.1 and 13.2. Whereas within a physical microgrid participating prosumers are balanced by an aggregator via a low-voltage electricity network, the complementary virtual microgrid consists of real time information flows between prosumage units and the aggregator.

The former can be connected to the latter via the Internet of Things, consisting of appliances embedded with sensors, electronic chips, and connectivity to communications networks. Devices serving as a bridge between physical electricity networks and the digital world (e.g., digital meters or sensors) are considered part of the digital world rather than part of the physical electricity

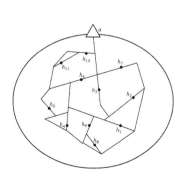

Key:

$\bullet$ = h_i; net load (sum of storage, generation, and consumption) of the prosumage unit i ($i = 1,...,n$)

a = aggregator of microgrid

$-$ = physical electricity flows

FIGURE 13.1   **Physical microgrid (low-voltage network).**

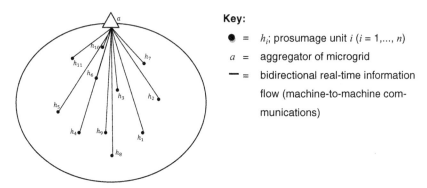

FIGURE 13.2 Virtual microgrid [Information and Communication Technologies (ICT) network].

network, although they also possess the characteristics of a physical entity (ITU-T, 2015a, p. 55).

The physical microgrid consists of a low-voltage electricity network, connecting the participating prosumers and providing access to physical electricity networks, either of other microgrids or of distribution networks, via the microgrid node as illustrated in Fig. 13.3.

On the one hand, the aggregator is assumed to balance production and consumption of low-voltage electricity within the microgrid. On the other hand, the aggregator ensures the import or export of electricity from the wholesale market of the medium-voltage distribution grid, based on injection or extraction fees charged at the microgrid node (Knieps, 2016a). Participation within a microgrid is voluntary; other prosumers in the local neighborhood may either belong to another microgrid or be served by a local/regional utility. The term (physical) node is used throughout this chapter to describe points of electricity injection or extraction within the distribution and transmission networks where location matters. As within (low-voltage) microgrids the location of consumption, storage or production of electricity does not matter, the term prosumage

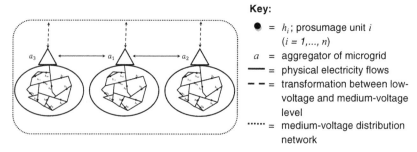

FIGURE 13.3 Networking physical microgrids.

gains relevance. Within a prosumage unit, several loads of different home appliances are relevant and, together with the small-scale storage and generation facilities as an aggregate, constitute the load of a prosumage unit.

It is important to differentiate between the ICT requirements inside a virtual microgrid and the outside requirements resulting from the connection either to other microgrids or to distribution networks. Standardization efforts within virtual microgrids are focused on different subparts, such as building-to-grid, home-to-grid, industry-to-grid, vehicle-to-grid, distributed renewables, generators, and storage (Fang et al., 2012; NIST, 2014, p. 43).

2.2   Microgrids and Virtual Networks

The organization and incentives within microgrids cannot be understood without analyzing the outside interaction of microgrids. The double fallacy of isolated microgrids should be avoided. First, from the physical electricity perspective import or export of electricity from other (low-voltage) microgrids or from (medium-voltage) distribution grids is without alternative (Knieps, 2016a). Second, from a complementary ICT perspective, real-time communication between home networks and aggregators, as well as between neighboring aggregators or between aggregators and wholesale distribution markets to order import or export via the microgrid distribution node becomes topical. After all, the ICT of microgrids requires agile QoS differentiations capable of providing high and deterministic QoS guarantees.

In the meantime, the challenge for ICT networks to take into account the requirements of the Internet of Things has gained increasing attention. "[T]he advancement of mass storage units, high speed computing devices, and ultra broadband transport technologies … enables many emerging devices such as sensors, tiny devices, vehicles, etc. The resultant new shape of ICT architecture and huge number of new services cannot be well supported with current network technologies." (ISO/IEC, 2012, p. vi). The term Future Networks has been coined, drawing attention to the overall challenges to ICT developments: "network of the future which is made on clean-slate design approach as well as incremental design approach; it should provide futuristic capabilities and services beyond the limitations of the current network, including the Internet" (ISO/IEC, 2012, p. 1).

The evolution of ICT physical network architectures is characterized by a change from specialized network infrastructures toward shared multipurpose IP-based communications infrastructure. In this context the concept of network virtualization has evolved: "Network virtualization is a method that allows multiple virtual networks, called logically isolated network partitions (LINPs) to coexist in a single physical network" (ITU-T, 2012, p. 2).

ICT logistics of microgrids possess the character of virtual networks reflecting the QoS requirements of data packets within microgrids and their interconnection to the multipurpose All-IP Internet. From the perspective of virtual networks, Recommendation ITU-T (2016a) on home networks can be considered as an important innovation complementary to All-IP communication infrastructures of access and core networks with seamless data packet transmission, irrespective of the physical communication network infrastructure (Cisco, 2014).

Focusing on virtual microgrids, the requirements of complementary application services and virtual networks comprise the identification-based connectivity between a "thing" and the Internet of Things, application support, security and privacy protection, interoperability, reliability, high availability, and adaptability. ICT-based networking of virtual microgrids also entails a large potential for enabling virtual power plants capable of dispatching significant amounts of energy to the wholesale markets as described in chapter by Steiniger.

Depending on the real-time price signals, prosumage behavior is incentivized. Given the sun and wind conditions for generation and the preferences for timing flexibility for consumption, rational prosumage decisions can take place to sell electricity to the virtual power plant. However, price signals should reflect the opportunity costs of marginal prosumage behavior. Separate metering of all embedded generation with different charges for generation and consumption leads to ample disincentives, which can only be avoided by net metering treating consumption and generation within a microgrid equally as proposed in chapter by Biggar & Dimasi. More generally it can be shown that the opportunity costs of generation and consumption (loads) within a microgrid are completely triggered by its outside options either to import or export from the wholesale market of the distribution network (via the microgrid node) or to trade directly with other microgrids in the neighborhood (Knieps, 2016a, pp. 276 ff.).

2.3 Internet Protocol–Based Virtual Microgrids

Traditionally the IP was developed for data packet transmission between a sender and a receiver to deliver a package of bits (an internet datagram) by means of addressing and fragmentation to enable communication between two hosts (Postel, 1981). In contrast, the Internet of Things uses the IP to provide communication between many objects. The aim is for objects to be addressed and controlled via the Internet, not by specific communication protocols as in the past with radio frequency identification (RFID) networks, but having an IP address and using the IP (ITU-T, 2015a, p. 57). The limitations of conventional IP architecture using IP addresses as node identifiers (IDs), as well as node locators for mobility and multihoming, has led to the development of a network layer–based protocol (the Locator/ID Separation Protocol) that enables the separation of IP addresses into two separate numbering spaces for network topology independent IDs and node locators. IPv6 networking components include IP addressing, QoS, routing, and network management and security (Baker and Meyer, 2011; Cisco, 2010, 2014; Farinacci et al., 2013; ITU-T, 2009).

Adoption of IP provides end-to-end bidirectional communication capabilities between any devices in the network. New application fields entail demand/response, distributed energy resource integration, and electric vehicle charging. Significant challenges for data packet transmission networks arise to deal with the strongly increasing amount of IP addresses, rapidly growing data traffic, as well as mobility and QoS requirements due to complex bidirectional communications with increasing traffic volumes and low latency requirements.

The physical microgrid network can be connected to the virtual microgrid via the Internet of Things based on All-IP communication infrastructures, enabling the seamless application of different communication infrastructures, such as fixed telecom access networks, cable access networks, and mobile access networks (Knieps and Zenhäusern, 2015, pp. 339 ff.). Although sensor networks and other ICT components within microgrids could be based on non-IP standards, the advantages of IP protocol evolution toward IPv6 from the perspective of the Internet of Things is pointed out. The microgrids of the future are embedded within overall Internet-of-Things architectures of smart sustainable city infrastructures, and thus definitively based on the IP protocol (ITU-T, 2015a).

Hardware and software requirements of the application services are beyond the scope of the All-IP communication infrastructures and traffic management of NGNs and related QoS differentiations. Nevertheless, application protocols have also been developed by the IETF, for example, for billing and identification purposes (Knieps, 2015, p. 743). Additional efforts have been provided by ITU-T in general and also in relation to several cases focusing on the role of virtual networks in general and of home networks in particular.

2.4   The Evolution of IP-Based Standards for Virtual Microgrids

Networking the home appliances within prosumer units is also part of the virtual networks of microgrids, which may result in virtual power plants able to dispatch power to the distribution grid. As the physical electricity networks of microgrids are interconnected with medium-voltage distribution networks and thereby indirectly also connected to high-voltage networks, the communication requirements of microgrids are not isolated from the complementary smart grid ICT requirements. A topical example is the linkage of solar PV and battery storage systems via a cloud-connected control system, similar to the one described in chapter by Orton et al.

Prosumer units act independently (noncooperatively), basing their consumption and generation decisions on scarcity signals (price signals) provided by the aggregator. The basic idea is that price signals from wholesale markets provide incentives for the prosumers to adjust generation, storage, or consumption using highly automated machine-to-machine communications, as illustrated in chapter by Steiniger. The aggregator collects the resulting consumption and generation quantities and orders the remaining quantities from the distribution grid operator. All decisions are on real-time quantities; location within a microgrid where consumption or generation takes place is irrelevant, only the total amount of net import or net export at each given moment in time is relevant. However, the location of the microgrid node within a distribution network is relevant because injection or extraction charges are to be based on the location-dependent opportunity costs of distribution network usage (Knieps, 2016a, pp. 275 ff.).

2.4.1 IP-Based Sensor Networks: ZigBee IP

During the last decade, large innovation potentials have led to significant evolution in sensor and actuator technologies. Fundamental innovations range from the development of one-way thermostats or energy control (e.g., paging broadband technology) to two-way communication channels. As a wireless platform, ZigBee was developed in the 1990s and gained standard level in 2003 (IEEE 802.15.4). It has a focus on applications that do not need high bandwidth, but do need low latency, such as sensors and control devices (which also require very low energy consumption for long battery duration) (ITU-T, 2015a, p. 45). IPv6 was considered by the IETF as the perfect solution for transmitting IP data packets over wireless platforms (IEEE 802-15.4 networks) using IPv6 over Low Power Wireless Personal Area Networks (6 LoWWPANs) header compression and easily allowing auto configuration (Kushalnagar et al., 2007).

In the meantime the ZigBee IP standard version IEEE 2030.5 (ZigBee Smart Energy ProtocolV2.0) was designed to enable the use of multiple link layer technologies (e.g., WiFi, Ethernet, and Home Plug); it also serves for applications requiring higher bandwidth. Due to the original choice of IP, seamless integration of sensor networks with different link layer technologies is possible. The goal of this new standard version is not only to inform and to act via thermostat, but also to enable plug-in electricity vehicle charging and other energy service interfaces based on networked information systems. Several advantages arise due to the possibilities of a seamless combination of various link layer technologies by using the IP and already existing IP-based heterogeneous communication infrastructures (Simpson, 2015). For example, smart phones with WiFi can be applied seamlessly for electricity metering using ZigBee, as well as for plug-in electricity vehicles by using Home Plug. The ZigBee Smart Energy Protocol V2.0 specifications define an IP for monitoring, controlling and for real-time automatic delivery and consumption of energy (application to water consumption is also possible, in contrast to water sourcing).

ZigBee IP is a mobile communication standard that can be applied (often in combination with other ICT technologies) to different fields of applications, not only smart grids or microgrids, but also autonomous driving; other areas of smart cities, such as water consumption and garbage collection; and other intelligent and smart applications, including e-health, bus on demand, car sharing, etc. (ITU-T, 2015a, pp. 7 ff.).

2.4.2   IP-Based Wired Home Networks: Evolution of G.hn

Focusing on the home network aspects of energy, ITU-T Study Group 15 developed home networking specifications for smart electrical grid products coined G.hn. "The main objective of this project is to define home network devices with low complexity for home automation, home control, electric vehicles and smart electrical grid applications" (ITU-T, 2015a, p. 74). The focus is, among other things, on microgrids, which can function connected to the distribution network or isolated from it, on enabling the network to accommodate users with new needs, enhancing efficiency in real-time grid operation with particular focus on active control, and demand response of distributed generation and distribution grids management, improving planning of future network investment, as well as improving market functioning by time-based, dynamic energy and demand response and load control programs, ensuring network security, system control, and quality of supply (ITU-T, 2015a, pp. 73-77; ITU-T, 2016a,b).

The focus of home networks is on high-speed cable-based communication networks supporting the functioning of smart electricity microgrids. "This Recommendation specifies the system architecture and functionality for all components of the physical (PHY) layer of home network transceivers designed for the transmission of data over premises' wiring, including inside telephone wiring, coaxial cable, power-line wiring, plastic optical fibers, and any combinations of these" (ITU-T, 2016a, p. 1). "With G.9960's ability to use any wire in the home as a possible Smart Grid connection, every device in the home can have its energy consumption monitored and managed, as well as interconnecting any wired device into a smart network where data accessibility is as valuable as energy efficiency" (ITU-T, 2010, p. 3).

The basic role of transceivers is to enable two-way communication embedded in a Home Area Network, consisting of a communications network in the premises, or operating as part of a smart grid utility access network outside the home, as the "last leg" link in a smart grid access network attached to the home. Smart grid networking technology may entail advanced metering services, communication between electric vehicles and their charging stations, or communication between smart appliances, such as heaters, air conditioners, and other appliances (ITU-T, 2010; Cisco, 2014).

It can be concluded that recent efforts worldwide can be observed regarding the standardization of ICT components at the grid's edge within cable-based and wireless networks, enabling real time and adaptive prosumage behavior with automated machine-to-machine communications. As communication networks for virtual microgrids are provided within broadband multipurpose access networks, QoS requirements are not only limited to the low latency requirements of data packet transmission for smart grids, but also comprise the QoS requirements of other Internet-of-Things applications, for example, very low latency communication for networked vehicle services (ITU-T, 2011b, 2015a).

2.5   Data Protection and Cybersecurity

The innovations of smart bidirectional meters and other sensor-driven data communication and processing technologies are providing a powerful new resource for "Big Data" mapping consumer preferences and behavioral patterns, such as those described in chapter by Woodhouse & Bradbury. Although there is no doubt that prosumage activities at the grid's edge may strongly benefit when aiming to harvest the fruits of renewable energy policies, e-privacy, and cyber security concerns gain increasing relevance, in particular at the grid's edge. Implementing ICT-driven applications within smart grids using sensor networks, interaction between home networks via platforms and cloud computing to enable decentralized distributed computing systems, transactive energy, and peer-to-peer trading raise important cyber security and data protection issues.

The European Commission (2014) considers data protection and security as an important challenge for smart metering, and more general smart grid systems. The aim is to guarantee end-to-end security and to refer to the interplay of general rules for data protection and sector-specific rules. The question arises, as to how guarantees of end-to-end security can be achieved via the principle of layer-based security for ICT networks beyond home networks. For example, data security can be promoted by prohibitions to exchange data between prosumer units and prohibitions for aggregators to send data of individual units to other units or to distribution operators, with only aggregated data allowed to be transmitted. Security standards within home networks are developed by ITU-T (2016b, pp. 348–376) based on advanced encryption, authentication, and confidentiality standards developed by NIST. The focus is on information metering and measurement, as well as information transmission, and the goal to avoid vulnerabilities enabling attackers to obtain user privacy, gain access to control software, or alter load condition to destabilize the grid in unpredictable ways (Fang et al. 2012, p. 26).

Network layer security (IP security/IPsec) is developed by the IETF in Kent and Seo (2005) and Baker and Meyer (2011). The focus is on the IP network layer, offering a set of security services for traffic at the IP layer in IPv4, as well as IPv6, including access control confidentiality (via encryption), data origin authentication, etc. These security services are provided at the IP layer, guaranteeing protection in a standardized fashion for all protocols that may be carried over IP. This architecture differs between a more time-consuming authentication and key exchange protocol step on the one hand, and the actual data traffic protection on the other. In addition to network layer security, further security measures can be installed on upper layers of the IP stack, including application layer–dependent security mechanisms, such as digital signatures (Baker and Meyer, 2011, pp. 15 f.).

Security within data link layer specifications for wireline-based home networking transceivers (ITU-T 2016b, pp. 348–376) can be implemented by means of encryption and authentication and key management procedures. Advanced encryption standards and encryption algorithms referring to standard specifications recommendations of the NIST are developed to guarantee the required security inside a network domain and the security of a network containing more than one domain.

In a topical report published in November 2016 the Broadband Internet Technical Advisory Group (BITAG, 2016) issues the following recommendations (BITAG, 2016, pp. iv–vii):

- Security for Internet-of-Things devices should be improved, following security and cryptography best practices.
- Internet of Things devices that have implications for user safety must be able to continue functioning if Internet connectivity is disrupted or cloud back-end fails.
- Rights to remotely decrease Internet-of-Things device functionality by a third party should be disclosed.
- Cybersecurity programs and a transparent vulnerability reporting process should be established.

3 MICROGRIDS AND THEIR RELATION TO NEXT GENERATION NETWORKS

The future Internet of Thing has many different applications with heterogeneous data traffic QoS requirements. Security problems that can be traced back to insufficient traffic quality within the Internet should be avoided by implementing the possibilities of NGNs with deterministic QoS guarantees.

3.1 Outside Communications of Microgrids

The growing importance of real-time sensitive ICT applications is no longer driven exclusively by real-time sensitive telecom applications, but increasingly from Internet of Things–driven applications and subsequent machine-to-machine communications, such as connected cars and other intelligent traffic systems, or smart grid and microgrid applications. The communication subsystem of smart grids in general and of microgrids in particular must support high QoS requirements, such as very low latency. This is because the critical data (e.g., the grid status information) must be delivered promptly. Electricity allocation procedures become real-time instead of day ahead. Traditional network policies were dealing with a single snapshot of the network state, ignoring time-sensitive load changes. The evolution of smart networks strongly increases the necessity to implement mechanisms for automatically responding to a wide range of events that may occur. Within smart networks, the network state changes continually, providing the challenge for network operators to implement flexible, responsive network policies.

Powerline Access services are communication services connecting households or businesses to the aggregator or the utility. "With the arrival of G.9960 technologies, power line communications (PLC), which has traditionally been a "best effort, moderate speed" communications option, is now a very high speed, high quality communications path" (ITU-T, 2010, p. 6). However, the communication protocol between the home networks and the utility or the aggregators (e.g., between the meter and the billing entity) is considered outside the scope of G.9960 and could more adequately be located at network layer protocols, including IPv6 (ITU-T, 2010, p. 7).

The Automated Metering Infrastructure (AMI) is based on two-way communications systems requiring real-time data delivery and a high degree of cybersecurity for a large number of consumers (NIST, 2009). The communication protocol between meter and utility or billing entity is outside the scope of home networks G.9960 initiatives, but is compatible with IPv6 protocol of data packet transmission (ITU-T, 2010, p. 7). As G.hn supports the IP, it can be readily interconnected with IP-based access and core networks (Cisco, 2010, 2014).

3.2   Virtual Microgrids and Next Generation Networks

The concept of virtual networks of microgrids reveals the QoS requirements of ICT within microgrids and their interaction with the smart grid ICT environment of distribution and transmission networks. ICT infrastructure cannot only be realized seamlessly within IP-based virtual microgrids, but also requires active traffic management for data packet transmission within the All-IP Internet. QoS of transmission is considered important for ICT within home networks, as well as for services communicating between different home networks, virtual power plants, and microgrid nodes of distribution networks. End-to-end QoS guarantees for connections outside of a microgrid are also considered relevant (ITU-T, 2016a, pp. 17 ff.). Broadband access can provide an energy services channel and a public broadcast channel, as well as other channels, such as a broadband channel for Internet connections. As a consequence, innovations emerging in conjunction with the development of communication networks within microgrids, and in particular home networks, cannot be considered isolated from the wide array of QoS innovations regarding data packet transmission in the All-IP Internet. Together, these innovations fundamentally challenge the traditional best effort Internet without traffic guarantees.

The evolution from best effort TCP/IP Internet toward NGNs with active traffic management is required for future virtual networks, not only for microgrid applications, but also for the increasing variety of other smart networks, such as networked cars, smart road infrastructures, etc. (Knieps, 2015, pp. 743 f.). The basic definition of NGNs is as follows: "A packet-based network able to provide telecommunication services and able to make use of multiple broadband, QoS-enabled transport technologies and in which service-related functions are independent from underlying transport-related technologies. It enables unfettered access for users to networks and to competing service providers and/ or services of their choice. It supports generalized mobility which will allow consistent and ubiquitous provision of services to users" (ITU-T, 2004, p. 2). The question arises as to how the required transition from best effort TCP/IP protocol toward active traffic management can be achieved to fulfill the QoS requirements of virtual networks and in particular of virtual microgrids.

3.3   Incentive-Compatible QoS Differentiation Within NGNs

The economic incentives for QoS differentiations based on the introduction of multiple traffic classes within NGNs can only be analyzed if the entrepreneurial QoS potentials and related traffic architectures are taken into account. Traffic QoS requirements can not only be considered from the perspective of Internet of Things only, but must also take into account all other application services provided within a NGN as illustrated in Fig. 13.4.

Considering virtual microgrids and their relation to NGNs, complex capacity allocation problems arise regarding QoS guarantees of bandwidth capacities that are not yet identified within the following definition: "network virtualization should provide the capability of regulating the upper limit bandwidth usage by each LINP in order to maintain the overall throughput and performance" (ITU-T, 2012, p. 2).

Although both levels are vertically complementary they constitute two independent decision units: traffic network providers (QoS for All-IP network) and virtual service network providers. One basic principle is that property rights and decision competency for the traffic network providers

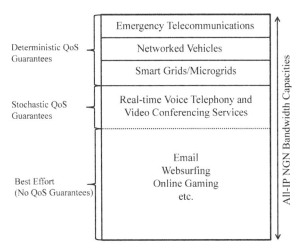

FIGURE 13.4   Virtual microgrids and Next Generation Networks (NGNs).

of NGNs are different from those for the virtual service network providers. The advantage of network virtualization has been described as follows: "The virtual resource management is recommended to allow LINP operators to configure LINPs quickly without imposing complexities of physical network operation including management of network addresses and domain names" (ITU-T, 2014, p. 3).

Virtual networks are based on QoS differentiated traffic capacities. They constitute an essential input for the provision of downstream applications in the Internet of Things and cannot be bypassed by separated traffic management within virtual networks. NGNs may entail a large variety of network services implemented via the concept of virtualization of different application service networks that are all competing for the same physical infrastructure capacities. A large variety of different QoS requirements emerges due to Internet of Things, as well as the various Internet and telecommunications services. The question arises as to how to measure the opportunity costs of different QoS guarantees. Traffic architecture should enable the variety of application services within NGNs in such a way that the question of how to optimally solve the network capacity allocation problem can be addressed. This permits as a corollary to approach the question of how to allocate capacities for cloud computing, cooperation, and coordination between different virtual networks, and the division of labor between intravirtual network capacity allocation and QoS traffic capacity imported from the All-IP network. Under the assumption that the QoS class implementation of NGNs is chosen in such a way that traffic logistics are sufficient to enable the variety of application services, the capacity allocation problem can be analyzed.

Irrespective of how the microgrid ICT traffic management is organized, from the perspective of the traffic network providers which manage All-IP infrastructure capacities, only the bandwidth usage of the virtual service network is relevant. As it turns out, heterogeneous virtual service networks rely on traffic capacities that give differentiated QoS levels. Thus, not only are incentives created for the organizers of virtual networks to economize bandwidth consumption by optimizing intraservice network usage, but also the viability of virtual networks can only be guaranteed if the traffic network providers that operate the NGNs provide required QoS levels.

During the last decades there have been intensive efforts within standardization committees, in particular the ITU-T and IETF, to develop guidelines for IP QoS traffic classes (Babiarz et al., 2006; Ash et al., 2010, pp. 4 f.; ITU-T, 2011a, Table 2, pp. 12 f.; ITU-T, 2013). In particular, a monotone decreasing ordering of traffic classes regarding delay, jitter, and packet loss requirements has been developed. To encompass the variety of heterogeneous traffic requirements based on the All-IP broadband infrastructures, the concept of Generalized DiffServ architecture is applied, which enables QoS differentiation of different traffic classes with deterministic and stochastic traffic quality guarantees (Knieps, 2015, pp. 739 f.).

Describing the special case of stochastic QoS guarantees based on prioritization between data packets, a pricing scheme based on interclass externality pricing was developed by Knieps (2011). Knieps and Stocker (2016) present an extended model describing the case of deterministic traffic classes (without explicitly including a hierarchy) and lower traffic classes with stochastic QoS guarantees, and introduce an incentive-compatible pricing scheme based on interclass externalities between deterministic and stochastic traffic classes. The more general case of a hierarchy of deterministic traffic classes is modeled in Knieps (2016b). Traffic classes requiring deterministic guarantees for higher QoS levels consume relatively more bandwidth. If QoS parameters are more strictly defined (e.g., the highest traffic class guaranteeing very low latency and packet drop probability) bandwidth capacity requirements are higher than those required for providing less stringent levels of QoS.

This permits the implementation of economic incentives for the underlying capacity allocation problem in NGNs, resulting in economic priority ordering between different QoS traffic classes. Differentiated pricing for a hierarchy of traffic classes with deterministic, as well as stochastic QoS guarantees, is based on the opportunity costs of network usage for providing the relevant QoS requirements in different virtual networks. Thus, active entrepreneurial traffic management of NGNs providing QoS guaranteed ICT traffic services is essential for meeting the QoS requirements of smart grids and virtual microgrids.

4 CONCLUSIONS

In contrast to the former hierarchical organization of electricity systems, important changes are driven by the increasing role of distributed generation, with prosumers taking on an active role at the edge of low-voltage microgrids with short-term generation of renewables and short-term response of demand, taking into account the rising potentials of storage (topics covered in other chapters of the book). An evolution is occurring, from top–down passive consumers toward active prosumage behavior and the development of related adaptive communication network policies reacting to various types of events and enabling real-time geopositioning, and content relevancy for many users via cloud computing and dynamic data exchange.

Based on the recent developments of active traffic management within the All-IP Internet (Knieps and Bauer, 2016), ICT logistics of microgrids are characterized as virtual networks, pointing out the QoS requirements of ICT within microgrids and their interconnection to the multipurpose All-IP Internet providing heterogeneous QoS traffic classes. The question is analyzed how the required change from best effort TCP/IP protocol toward active traffic management can be achieved to fulfill the QoS requirements of virtual networks, and in particular of virtual microgrids.

During the last decade data protection and cybersecurity have gained top agenda status worldwide, not only for smart electricity networks, but also for other Internet-of-Things applications, for example, autonomous driving or intelligent traffic systems, and more general concerns regarding consumer protection and security within the app economy (OECD, 2013). As it turns out, security and privacy are challenging issues for the future Internet of Things and related application fields, such as smart grids, microgrids, and home networks (BITAG, 2016). It seems to be crucial to differentiate between security issues due to unauthenticated communications, unencrypted communications or the lack of automatic, secure software updates versus insufficient data transmission qualities within the traditional best effort TCP/IP Internet without data traffic quality guarantees.

As it turns out, standardization activities for enabling compatible microgrids and related home networks have accelerated considerably over the past years. Corresponding innovations are closely related to the broader ICT innovations enabling real-time and adaptive network services, which are commonly subsumed under the heading of the Internet of Things. Thus it can be expected that local initiatives to implement microgrids and related home networks will gain increasing momentum and will further spur the evolution of the All-IP ecosystem.

ACKNOWLEDGMENTS

Helpful comments by Volker Stocker are gratefully acknowledged. Special thanks are due to the editor of this volume for his ongoing encouragement and suggestions.

REFERENCES

Ash, G., Morton, A., Dolly, M., Tarapore, P., Dvorak, C., El Mghazli, Y., 2010. Y.1541-QOSM: Model for Networks Using Y.1541 Quality-of-Service Classes, RFC 5976.

Babiarz, J., Chan, K., Baker, F., 2006. Configuration Guidelines for DiffServ Service Classes, RFC 4594.

Baker, F., Meyer, D., 2011, Internet protocols for the smart grid, Internet Engineering Task Force, RFC 6272.

BITAG (Broadband Internet Technical Advisory Group), 2016. Internet of Things (IoT) Security and Privacy Recommendations: A Uniform Agreement Report. Available from: www.bitag.org/report-internet-of-things-security-privacy-recommendations.php

Cisco, 2010. Why IP is the right foundation for the smart grid, White Paper.

Cisco, 2014. A standardized and Flexible IPv6 Architecture for field area networks.

European Commission, 2014. Commission Recommendation of 10 October 2014 on the Data Protection Impact Assessment Template for Smart Grid and Smart Metering Systems, OJ, L 300/63.

Fang, X., Misra, S., Xue, G., Yang, D., 2012. Smart grid—the new and improved power grid: a survey. IEEE Commun. Surv. Tut. 14 (4), 944–980.

Farinacci, D., Fuller, V., Meyer, D., Lewis, D., 2013. The Locator/ID Separation Protocol (LISP), RFC 6830.

ISO/IEC, 2012. Information Technology—Future Network—Problem Statement and Requirements—Part 1 Overall Aspects (Technical Report), TR 29181-1. Available from: http://standards.iso.org/ittf/publiclyavailablestandards/index.html

ITU-T, 2004. Next Generation Networks—Frameworks and Functional Architecture Models: General Overview of NGN, ITU-T Recommendation Y.2001 (12/2004).

ITU-T, 2009. General Requirements for ID/Locator Separation in NGN, Recommendation ITU-T Y. 2015 (01/2009).

ITU-T, 2010. Applications of ITU-T G. 9960, ITU-T G. 9961 transceivers for smart grid applications: advanced metering infrastructure, energy management in the home and electric vehicles, Technical Paper (06/2010).

ITU-T, 2011a. Network Performance Objectives for IP-Based Services, Recommendation ITU-T Y. 1541 (12/2011).

ITU-T, 2011b. Framework of networked Vehicle Services and Applications Using NGN, Recommendation ITU-T Y. 2281 (01/2011).

ITU-T, 2012. Framework of Network Virtualization for Future Networks, Recommendation ITU-T Y. 3011 (01/2012).

ITU-T, 2013. Network Performance Objectives for IP-Based Services, Amendment 1 New Appendix XII—Considerations for Low Speed Access Networks, Recommendation ITU-T Y. 1541 (2011)—Amendment 1 (12/2013).

ITU-T, 2014. Requirements of Network Virtualization for Future Networks, Recommendation ITU-T Y. 3012 (04/2014).

ITU-T, 2015a. Overview of Smart Sustainable Cities Infrastructure, Focus Group Technical Report, ITU-T FG-SSC (05/2015).

ITU-T, 2015b. Framework of a Micro Energy Grid, Recommendation ITU-T Y. 2071 (09/2015).

ITU-T, 2016a. Unified High-Speed Wire-Line Based Home Networking Transceivers—System Architecture and Physical Layer Specification, Recommendation ITU-T G. 9960 (2015)—Amendment 2 (04/2016).

ITU-T, 2016b. Unified High- Speed Wire-Line Based Home Networking Transceivers—Data Link Layer Specification, Recommendation ITU-T G. 9961 (2015)—Amendment 2 (07/2016).

Kent, S., Seo, K., 2005. Security Architecture for the Internet Protocol, RFC 4301.

Knieps, G., 2011. Network neutrality and the evolution of the internet. Int. J. Mgmt. Netw. Econ. 2 (1), 24–38.

Knieps, G., 2015. Entrepreneurial traffic management and the Internet Engineering Task Force. J. Compet. Law Econ. 11 (3), 727–745.

Knieps, G., 2016a. The evolution of smart grids begs disaggregated nodal pricing. In: Sioshansi, F. (Ed.), Future of Utilities—Utilities of the Future: How Technological Innovations in Distributed Energy Resources will Reshape the Electric Power Sector. Academic Press/Elsevier, Amsterdam, pp. 267–280.

Knieps, G., 2016b. Internet of Things (IoT), future networks (FN), and the economics of virtual networks. Available from: https://ssrn.com/abstract=2756476; http://dx.doi.org/10.2139/ssrn.2756476

Knieps, G., Bauer, J.M., 2016. The industrial organization of the Internet. In: Bauer, J.M., Latzer, M. (Eds.), Handbook on the Economics of the Internet. Edward Elgar, Cheltenham, pp. 23–54.

Knieps, G., Stocker, V., 2016. Price and QoS differentiation in all-IP networks. Int. J. Mgmt. Netw. Econ. 3 (4), 317–335.

Knieps, G., Zenhäusern, P., 2015. Broadband network evolution and path dependency. Compet. Regul. Netw. Ind. 16 (4), 335–353.

Kushalnagar, N., Montenegro, G., Schumacher, C., 2007. IPv6 over Low-Power Wireless Personal Area Networks (6LoWPANs): Overview, Assumptions, Problem Statement, and Goals, RFC 4919.

NIST, 2009. The role of the Internet Protocol (IP). Advanced Metering Infrastructure/AMI Networks for the Smart Grid. 24 October 2009, National Institute of Standards and Technology U.S. Department of Commerce (PAP 01).

NIST, 2014. NIST Framework and Roadmap for Smart Grid Interoperability Standards, Release 3.0. NIST Special Publication 11083r3. NIST National Institute of Standards and Technology U.S. Department of Commerce.

OECD, 2013. The app economy, OECD Digital Economy Papers, No. 230. OECD Publishing. Available from: http://dx.doi.org/10.1787/5k3ttftlv95k-en

Postel, J., 1981. Internet Protocol, DARPA Internet Program Protocol Specification, RFC 791.

Simpson, R., 2015. IEEE Standards Association, IEEE 2030.5TM-2013 (Smart Energy Profile 2.0), An Overview for KSGA, GE Digital Energy. Available from: http://robbysimpson.com/prez-zos/IEEE_2030_5_Seoul_Simpson_20150424.pdf

Part III

Alternative Business Models

14. Access Rights and Consumer Protections in a Distributed Energy System   261
15. The Transformation of the German Electricity Sector and the Emergence of New Business Models in Distributed Energy Systems   287
16. Peer-to-Peer Energy Matching: Transparency, Choice, and Locational Grid Pricing   319

17. Virtual Power Plants: Bringing the Flexibility of Decentralized Loads and Generation to Power Markets   331
18. Integrated Community-Based Energy Systems: Aligning Technology, Incentives, and Regulations   363
19. Solar Grid Parity and its Impact on the Grid   389

Chapter 14

Access Rights and Consumer Protections in a Distributed Energy System

Fiona Orton*, Tim Nelson*, Michael Pierce** and Tony Chappel*

*AGL Energy, Sydney, NSW, Australia; **AGL Energy, Docklands, VIC, Australia

1 INTRODUCTION

The supply of electricity to residential and small business customers within Australia's National Electricity Market (NEM) is considered to be an "essential service," meaning that it underpins health, well-being, and a reasonable quality of life. Electricity retailers are, therefore, required to meet a range of government licensing conditions that deliver additional consumer protections for these customers, to guarantee access to high-quality electricity supply, regardless of the capacity to pay. These protections go well beyond Australia's consumer laws, which apply to the sale of *all* products and services.

As with much of the National Electricity Law (NEL) and supporting regulatory framework in Australia, these provisions were implemented before the widespread availability of "behind-the-meter" energy technologies, and with the assumption that electricity would be exclusively supplied to customers from the grid, under a linear supply model that had been in place for decades, where electricity was generated in large, centralized, and predominantly thermal power stations, and transported via transmission and distribution networks to end users. With the rapid expansion of home energy technologies, opportunities are emerging for end users to participate more directly in electricity markets, redefining the services that customers expect from energy providers. As distributed energy resources (DERs) become increasingly mainstream, market rules and regulations must be reviewed and updated to ensure that they reflect consumer choices about how essential services are accessed, and create a level playing field on which innovative and conventional business models can compete at the grid's edge.

Many new energy products, such as rooftop solar, energy storage, electric vehicles (EVs), digital metering and energy efficiency, along with the digitally enabled energy management and trading platforms that make use of these technologies, have not historically been covered by electricity supply regulations. As a result, markets are not "technology neutral," with different rules often applying to the pricing, connection, and sale of demand- and supply-side appliances, based on technology type rather than fundamental properties or system impacts- positive or negative. Australian State and Federal Energy Ministers—collectively known as the COAG Energy Council (2016)—have noted this inconsistency, stating that "different customer protections apply to products and services for electricity supply behind the meter depending on the business model employed." Policymakers and regulators are considering how market rules and structures should evolve to better incorporate emerging technologies, and to facilitate a more distributed, bidirectional, and customer-centric energy system[1]—a challenge faced globally, as described by energy regulators from the USA, Australia, and Europe in this volume. Nevertheless, some observers argue that reform is failing to keep pace with technology advancement and that legacy regulations are complex and inherently biased toward traditional supply models, resulting in market distortions, inefficiencies, and restricting the range of products and services available to customers (Parkinson, 2016).

While "cost-reflective pricing" will be a key determinant of the uptake rates of new technologies, the regulatory framework is equally important. This chapter explores how consumer protections and grid access rights may need to be redefined in energy markets featuring widespread DERs, using Australia's NEM as a case study. Section 2 provides a summary of recent market developments in the NEM for context. Section 3 outlines the current uptake and future potential of DERs in the NEM. In Section 4, the impact of these products on household energy demand is examined, highlighting the heterogeneity of customer needs that may emerge. Section 5 discusses how access rights and consumer protections are currently applied and where reform may be required, followed by the chapter's conclusions.

2 CONSUMER MARKET DEVELOPMENTS IN THE NEM

Between 2008 and 2015, household electricity prices rose substantially, more than doubling in some NEM jurisdictions (Simshauser and Nelson, 2013), primarily due to a very significant increase in capital expenditure on distribution networks to meet tighter reliability standards and to service projected growth in peak demand, which in many cases did not eventuate. Concern about rapidly rising bills prompted customers to cut their consumption and install solar PV as a substitute for grid supply, and between 2009 and 2014, the average household demand in the NEM dropped by 18% (ESAA, 2015).

The prominence of flat "average rate" tariff structures encouraged customers to reduce consumption at the times most convenient to them, rather than at times

---

1. For example, The Power of Choice review (AEMC, 2012), which included a series of proposed reforms to increase demand-side participation.

that would produce system-wide benefits. As a result, peak demand continued to grow or remained high with the use of home air-conditioning units on hot summer days,[2] while underlying demand declined, leading to a significant reduction in network capacity utilization. As energy volumes fell, unit prices were adjusted upward to recover network revenues, leading to further conservation or substitution; a phenomenon known as an "energy market death spiral." For many customers, bills have increased significantly in return for no perceptible improvement in service quality and despite efforts to reduce consumption.

Reflecting the results of market research, Energy Consumers Australia (ECA, 2016) argues that consumers are dissatisfied with the "value for money" they are receiving for electricity services from conventional suppliers, ranking the sector below a range of others, including banking, water, telecommunications, and insurance. Energy Consumers Australia (ECA) states that customers generally neither believe that the electricity market is working in their best interests, nor that it will improve in future. Most consumers are unlikely to recommend their current electricity retailer to a friend or colleague, with "Net Promoter Scores" for the industry substantially negative. As a result, customers are actively seeking out alternative technologies and supply arrangements that reduce their reliance on supplies from conventional utilities, allowing them to reduce utility bills and take greater control over their spending, including solar PV, battery storage, and energy efficiency. New retail energy products are also emerging to provide customers with different ways to control usage and spending. For example, Origin Energy, one of the largest integrated energy companies in Australia, has introduced the "Predictable Plan," where customers pay the same fixed bill for 12 months, regardless of consumption; and new entrant retailers, such as Mojo Power and Powershop, now offer energy "subscriptions" and online prepurchasing of low-price "Powerpacks," which are marketed to customers as an alternative to conventional retail energy plans.

The introduction of more cost-reflective tariff structures for energy networks—particularly "demand tariffs"—has been considered for several years as an option to reduce *future* price rises by promoting more efficient use of the existing infrastructure, and reducing cross-subsidies between customers, as discussed in the Foreword by Conboy. Demand tariffs involve a proportion of network charges being linked to the customer's maximum demand from the grid during predefined windows, such as weekday afternoons and evenings. In doing so, they better reflect the high capital costs associated with grid expansions to meet demand peaks that occur for only a few hours each year, and provide incentives for customers to shift demand from these times. While the broad introduction of demand tariffs would arguably provide fairer pricing and encourage the efficient deployment of energy technologies, the implementation of pricing reform is slow. In Victoria, the only NEM jurisdiction with a mass rollout of

---

2. Between 2000 and 2010, the penetration of home air conditioning more than doubled, from 35% to 73%, which is close to market saturation (Productivity Commission, 2013).

digital meters, demand tariffs are to be implemented on an opt-in rather than opt-out basis, and of the customers who could benefit, most would make only modest savings: typically less than approximately A$100 per year. Networks are therefore predicting low uptake rates—less than 2.5% of customers over the 2017–20 period (AER, 2016a). Adoption is likely to be even slower in other jurisdictions without widespread advanced metering infrastructure.

The public—and therefore policymakers—appear to be skeptical of pricing changes that increase complexity, and which may leave some customers worse off, despite the opportunity for lower costs and greater efficiencies in the long term. The Consumer Action Law Centre (CALC, 2016) states that for innovation and competition to thrive in Australian energy markets, consumers need to be willing to participate, and must "trust that the market will deliver the outcomes they expect in terms of service, quality and price." Given the prevailing low levels of consumer trust in energy market participants and governance institutions, convincing consumers of the benefits of tariff reform remains a significant challenge.

3 OUTLOOK FOR DISTRIBUTED TECHNOLOGIES IN THE NEM

In this section, the current adoption and future potential of a range of distributed energy technologies are discussed, including solar PV, home energy storage, EVs, and virtual power plants (VPPs).

3.1 Solar PV

Australia has some of the highest rates of solar PV installation globally; by June 2016 almost 5 GW of small-scale solar PV had been installed across 1.55 million rooftops (AEC, 2016). In South Australia and Queensland, penetration of household solar exceeds 25%, with the national average around 17%. Prior to 2012, solar installations were largely driven by government renewable energy policies, which provided significant capital subsidies and premium feed-in tariffs (FiT). Over time, generous subsidy and FiT programs were scaled back, removed, or closed to new entrants. Nevertheless, as technology costs have fallen, new installations have remained popular, with self-generated solar power now representing a cost-effective substitute for grid supplies in several states with little or no subsidy.[3]

The installed capacity of small-scale solar PV is expected to grow steadily for the foreseeable future, particularly with new financing and business models emerging that provide solar with low or no upfront costs. Origin Energy has estimated that Australia could have as many as 5.3 million rooftops that are suitable

---

3. AEC (2016) found that the payback period for a 5-kW household solar PV system was between 5 and 10 years in Adelaide, Brisbane, Darwin, Sydney, and Perth. Mountain and Harris in this volume discuss the payback for solar PV investments in Victoria.

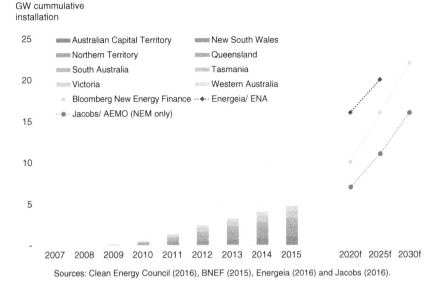

Sources: Clean Energy Council (2016), BNEF (2015), Energeia (2016) and Jacobs (2016).

FIGURE 14.1   Australian small-scale solar PV installation and forecasts to 2030.[4] *AEMO*, Australian Energy Market Operator; *ENA*, Energy Networks Australia; *NEM*, National Electricity Market.

for solar, but are currently without installations (Keane, 2015), and Jacobs (2016) forecasts that installation rates will remain high until the early 2030s when market saturation will be reached in some regions. Morgan Stanley (2016) observes: "Given the unlimited supply, large market size, and the compelling economics, the only remaining constraint is rooftop space." Fig. 14.1 presents historical installations and a range of forecasts to 2030, all of which predict that the installed capacity will more than triple over this period. Bloomberg New Energy Finance (BNEF, 2015) predict that by 2040, around half of all residential buildings will have rooftop solar and 18% of all generation will occur behind the meter.

3.2   Energy Storage

While the current capacity of battery storage installed behind the meter in Australia is very low, installations are expected to grow rapidly as system costs fall. Lithium-ion energy storage is not generally considered to be economically viable yet, with payback periods for most customers above the 10-year warranty period of commercially available units. Nevertheless there is an intense public interest in the technology, with 46% of survey respondents answering that they would "definitely" or "maybe" consider installing solar and battery

---

4. The Jacobs forecasts include PV installations in NEM states, excluding Western Australia and the Northern Territory.

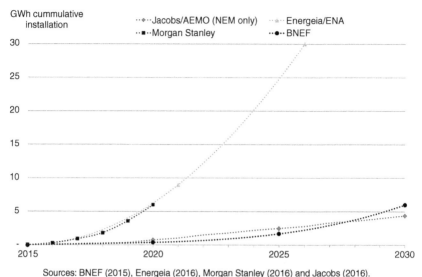

Sources: BNEF (2015), Energeia (2016), Morgan Stanley (2016) and Jacobs (2016).

FIGURE 14.2 Australian behind-the-meter storage forecasts to 2030.

storage (Morgan Stanley, 2016). For those customers who were "not interested" or "unsure," the top reasons provided were high costs and belief that the technology would improve, or costs would go down, in future, indicating that the market is likely to grow. Indeed some analysts suggest that installing solar with storage is already cheaper for average residential customers than grid supply alone (Mountain, 2016).

There are several battery storage product offerings already available to Australian households, including from the NEM's three large, vertically integrated electricity providers: AGL Energy, Energy Australia, and Origin Energy; several smaller energy retailers; and independent solar installers. Morgan Stanley (2016) predicts that the average installed cost of a 7-kWh residential battery system could drop by around 40% in 2 years, from almost A$10,000 in 2016 to around $5,500 in 2018, mostly due to falling battery pack costs.[5] This indicates a potential market size across the NEM of 1–2.2 million homes by 2020—between 11% and 25% of households—including half a million "retrofits" of battery storage units to existing solar installations. A range of forecasts are shown in Fig. 14.2. While these analysts all agree that battery technology will play a significant role in energy markets, their predictions about when installations will accelerate are somewhat divergent.

The main reason that customers give for being interested in home energy storage is the potential to reduce grid power use and thereby cut energy bills. For many households, solar generation and household demand are not closely

---

5. For example, the Tesla Powerwall 2 was launched in October 2016 and at 14 kWh, it is double the size of the Powerwall 1, offered for the same price.

correlated, and battery storage provides the opportunity to store surplus daytime generation for use in the evening when demand typically peaks. Dissatisfaction with traditional energy providers and a strong desire to control energy spending may encourage some households to install battery storage, even if the economics are not considered to be conventionally viable.[6] If technology costs drop to the point that a home energy storage system can be installed for less than A$5000, mainstream uptake may expand quickly, becoming a "credit card" purchase for many households, rather than an economic decision based on cost benefit. Solar installations in Australia accelerated rapidly once the typical out-of-pocket expense for households fell below this threshold, and the payback period of the investment was less than 3 years, and similar trajectories have been forecast for battery storage if these conditions are met. Households may also seek to install energy storage systems to avoid "bill shock" when their premium solar FiTs expire, suddenly exposing them to much higher energy bills. For example, the New South Wales Solar Bonus Scheme[7] closed in December 2016 for over 146,000 customers, many of whom have paid low or no electricity bills for several years. The ability to maintain power supply during blackouts may also motivate some customers to install home storage.[8]

The structure of energy tariffs will also influence storage installation. In 2014, the Australian Energy Markets Commission (AEMC) made rules requiring networks to provide more "cost-reflective" pricing, and as shown in Table 14.1, most will begin offering demand tariffs for residential customers from 2017 on an opt-in basis. Both demand and Time-of-Use tariffs can improve the economics of battery storage, by providing incentives for pricing arbitrage; charging the battery when prices are low for use when prices are high, or using stored energy to moderate or eliminate demand from the grid during peak times. New storage products are already being developed to help customers optimize usage under new tariff arrangements, such as from Sunverge Energy (2016), which recently added demand charge management to its platform, which it estimates could reduce power bills for Australian customers by up to 50%.

Despite forecasts of rapid deployment for behind-the-meter energy storage, coexistence with the grid, rather than full substitution is the most likely outcome for most customers (Nelson and McNeil, 2016). Wood and Blowers (2015) found that for typical customers to achieve 99.9% reliability with an "off-grid" system—equivalent to around 9 h of outage per year—5 kW of solar PV and 85 kWh of storage would be needed, at an estimated cost of A$72,000; both the large area of roof space required and the system expense would be prohibitive for most households. Hence most customers are likely to remain connected to

---

6. Typically when the payback period is less than 7 years.
7. The solar bonus scheme provided a gross feed-in tariff (FiT) of A$0.60 per kWh, enabling a payback period of 2.1 years and a net profit per household of up to A$10,000.
8. For example, on September 28, 2016 there was a state-wide "System Black" outage in South Australia that lasted for several hours and which has been the subject of extensive media coverage.

TABLE 14.1 Cost-Reflective Network Tariffs From 2017

Network business	Proposed tariff
Ergon	Opt-in Seasonal Time-of-Use Energy and Seasonal Time-of-Use Demand tariffs
Energex	Opt-in Demand tariff includes Hot Water Tariff for Demand customers
Ausgrid	Opt-in Time-of-Use tariffs. From July 1, 2018, all new customers assigned to Time-of-Use tariffs with opportunity to opt-out to a transitional tariff. All existing customers with digital metering to be assigned to this transitional tariff on July 1, 2018
Essential	Time-of-Use tariff default for new customers, new solar PV installations and metering upgrades. Opt-in Demand-based tariffs also available
Endeavour	Opt-in Time-of-Use tariffs. All new customers with interval meters assigned to Time-of-Use tariffs from July 1, 2018 on an opt-out basis
ActewAGL	Time-of-Use tariff default for all new residential and small business customers. Small business customers can opt-in to Demand tariffs. Possible gradual introduction from December 1, 2017 of residential demand tariff
Citipower. Powercor, United Energy, Jemena, AusNet	Opt-in residential Demand Tariffs (not available in AusNet service area until 2018). Opt-in Demand tariffs for all small business customers consuming <60 MWh per annum. United and Jemena: Demand tariffs mandatory for small businesses consuming >60 MWh per annum. Powercor, CitiPower, and AusNet: transitional Demand tariff mandatory for small business consuming >60 MWh per annum. Cost reflectivity of the transitional tariff will increase between 2017 and 2022
South Australia Power Networks	Opt-in cost-reflective residential Demand tariff. Opt-in "fully" cost-reflective demand tariff for small business customers. Mandatory assignment to transitional Demand tariff (50% cost reflective Demand) for new three-phase customers and progressive increases in cost-reflectivity until 2022

Source: Distribution network businesses tariff statements.

the grid for the foreseeable future; although for households with solar and storage, this may no longer be a major source of energy supply.

3.3 Electric Vehicles

Batteries currently comprise around a third of EV costs, so as lithium-ion battery costs fall, EVs may become cost competitive with conventional vehicles on a life-cycle basis. The pairing of driverless technology with EVs could also accelerate uptake; as explored by Webb & Wilson in Chapter 6. While there are currently very few EVs in Australia, Energeia (2015) estimates that the economically efficient adoption rate is 4 million by 2035, representing around 22% of the light passenger fleet, with interim targets of around 900,000 in 2025 and 2 million in 2030. However, EV sales appear likely to fall short of these figures due to market barriers, including the high purchase price of EVs; low public familiarity; and unpriced externalities, such as greenhouse gas emissions, lack of public charging infrastructure, and technical limitations, including range.

Few EV models are currently available to Australian consumers, and costs are high and adoption rates low. Australia has not introduced national policies to support EV uptake, which have proved effective in other global markets, such as vehicle emission standards, capital subsidies, exemptions from charges and taxes, preferential access to transit lanes and parking spaces, or government subsidized charging infrastructure. EV manufacturers have, therefore, been reluctant to invest in the small Australian market, and many popular models internationally are not available in Australia, including the Nissan LEAF Gen 2, Chevrolet Volt and Bolt, and Renault ZOE (ClimateWorks, 2016). While vehicle availability and customer choice remain low, mass adoption of EVs in Australia appears unlikely, and may lag behind other global markets.

EVs are likely to have a fairly minor impact on electricity demand in the NEM. Projections prepared for the Australian Energy Market Operator (AEMO) show that even if uptake is high, 2035 electricity demand would only increase by 4% (Jacobs, 2016). However, for individual households, EV ownership can increase electricity use by 30%–50%. It is therefore unsurprising that energy retailers are beginning to develop new products tailored to the needs of early EV adopters. In June 2016, AGL Energy announced that it would offer an "all you can eat" EV plan, where EV customers receive unlimited vehicle charging, with carbon offsetting, for A$1 per day. Click Energy has also launched a new energy plan for EV drivers, offering lower energy rates for the household's total energy purchase, not just vehicle charging.

As mobile energy storage devices, EVs also have the potential to discharge energy for use during peak times, or to provide grid-support services. To date these kinds of "vehicle-to-home" or "vehicle-to-grid" applications have generally not been supported by vehicle manufacturers. For example, Tesla has developed a range of both stationary energy storage products and EVs rather than combining their functionality into a single product. However, if new markets emerge for these services that could provide significant value to EV owners, vehicle specifications may follow.

3.4 Virtual Power Plants

Distributed energy technologies have the potential to provide services to local networks and energy markets, particularly if they can be aggregated and dispatched in unison. As a larger share of power generation moves behind the meter, and wholesale energy markets are increasingly supplied by intermittent renewables, "orchestrated DERs"—where a third-party aggregator has partial or full control over energy storage or "smart" appliances—may play a growing role in balancing instantaneous supply and demand, and providing "firm" capacity needed to maintain security. For networks, DER can be deployed to reduce or shift demand from the grid at peak times, as well as to manage voltage fluctuations, avoiding or deferring capital expenditure on infrastructure augmentation, and improving the quality of power supply. The specific capabilities of DERs to provide services along the supply chain are only beginning to be demonstrated, and the markets that would allow these value pools to be accessed are largely undeveloped. The way in which services are monetized, and value is shared among customers, aggregators, retailers, and networks will ultimately influence the economics and uptake of the associated technologies. Steiniger in this volume provides further discussion of "VPPs" and their application in global markets.

DERs may offer a range of services simultaneously, allowing value created in wholesale markets, networks, and within the home to be "stacked" to maximize customer benefit. Under such arrangements, Morgan Stanley (2016) estimates that home energy storage could become cost effective in the NEM by approximately 2018, with the value of network services comprising around half of the potential revenue pool; the remainder being from solar self-consumption, tariff arbitrage, and a relatively small wholesale component accessed via VPP aggregation.

Energy Networks Australia (ENA, 2016), an industry group representing network companies, states that the consumer value of DERs could be "unlocked" through the implementation of demand tariffs for all customers on an opt-out basis, and by giving customers with DERs the option to opt-in to providing network services, incentivized by dynamic and location-specific pricing, which would target regions where infrastructure is constrained. ENA estimates that if networks buy grid services from DER customers, this "orchestration" could replace the need for $1.4 billion in conventional network investment by 2026, and $16.2 billion by 2050. A number of distribution networks across the NEM are conducting small trials of distributed battery storage to demonstrate the technology's capabilities. In the longer term, networks will be required to strictly ring-fence their nonregulated energy services from regulated "poles-and-wires" work. This will be important for facilitating consumer choice and innovation (AER, 2016b).

At an aggregated level, the output from a large number of distributed systems can become a VPP, able to dispatch power to the grid to be traded at wholesale pool prices. VPPs could provide hedges for retailers, to limit their exposure to volatile wholesale pricing, either through traditional financial instruments, such as "cap" contracts, or by reducing exposure to high prices by directing homes to use stored energy from batteries rather than grid supplies. Reposit Power has begun offering a "GridCredits" product, provided in partnership with selected renewable energy retailers, which links customer batteries to wholesale energy markets and enables the export of power when prices spike, for a premium payment of A$1.00 per kWh. Batteries may also be suitable to provide a range of ancillary services, which have, to date, been mostly provided by thermal generators with "spinning reserve," such as fast-response frequency control.

In August 2016, AGL Energy announced that it was developing the world's largest solar VPP demonstration plant, involving 1000 homes and businesses in South Australia equipped with solar PV and battery storage systems, linked through a cloud-connected control system (AGL, 2016). Once in place, the VPP will provide the equivalent of 5 MW of peaking capacity. In return for a heavily discounted home energy storage system—A$3499, including hardware, software, and installation—customers enter into an agreement that allows AGL to occasionally direct the batteries to discharge to the home or to the grid, at times of high demand or grid instability. For the majority of the time, the batteries will be used to help customers self-consume their stored solar power. Projects, like this VPP, will help to demonstrate how relationships between networks, retailers, consumers, and the market operator can create and access new value streams.

Within the NEM, the South Australian market has, by far, the highest concentration of intermittent wind and solar generation, and with the recent closure or mothballing of aging thermal plant, there are concerns about system stability and a potential shortfall of "firm" capacity available to meet peak demand (Nelson and Orton, 2016). As a result, average wholesale energy prices are higher and more volatile than in other jurisdictions. VPPs may represent a new way of managing energy peaks and grid stability, to support energy systems with high penetrations of intermittent renewable generation. If successful, this would provide an alternative to building new thermal generation, or increasing interconnection with other regions, which in a disruptive market environment, are investments that could become stranded before the end of their design lives. DER assets, with multiple potential revenue opportunities, could prove to be a more flexible and adaptable solution.

Alternative VPP models are also being trailed in the NEM, albeit at smaller scale. Ergon Energy is installing grid-connected battery and solar systems at 33 customers' premises, but is retaining ownership and charging a fixed service fee in return for cooptimizing several value streams. Cloud-controlled demand response can also behave as a VPP by reducing energy demand during peak periods in an orchestrated way. During the summer of 2015–16, AGL Energy and United Energy undertook a trial involving the installation of "smart" air conditioners for 68 customers. During certain hot weather events, the devices were sent commands to slightly increase the set point temperature, with the aim of reducing peak demand by 25 kW.

The high level of activity in this space involving technology trials by both competitive and regulated industry participants would suggest that Australian utilities see a significant potential for VPPs to deliver customer and shareholder value in a distributed energy system. Recent findings from the AEMC (2016) suggest that behind-the-meter battery storage should be defined as "contestable"—and therefore delivered by entities appropriately ring-fenced from regulated participants. This should facilitate the development of competitive markets for services along the supply chain, and the emergence of new service providers—aggregators—to bundle these into compelling consumer products.

4   GROWING CUSTOMER HETEROGENEITY: IMPACTS OF TECHNOLOGY ADOPTION ON HOUSEHOLD DEMAND

As households adopt behind-the-meter energy technologies to supply and manage their usage, the changes to their demand for supplies from the grid are profound. In this section, the "load profiles" of residential customers with and without DER technologies, including solar PV, energy storage, and EVs are discussed, considering both typical and peak demand days.[9]

4.1   Typical Demand Days

The averaged "load profile" of a sample of residential electricity customers in South Australia is presented in Fig. 14.3, for a representative mild summer day: December 13, 2015. As is typical for Australian households, daily demand peaks significantly in the early evening.

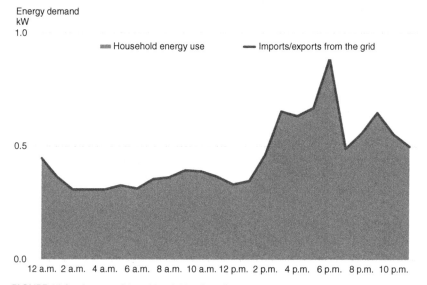

FIGURE 14.3   Average SA residential load profile on a mild summer day.

9. Analysis in Section 4 is based on the averaged load profile of a sample of South Australian residential customers with digital meters, and normalized PV generation based on a sample of monitored systems in South Australia.

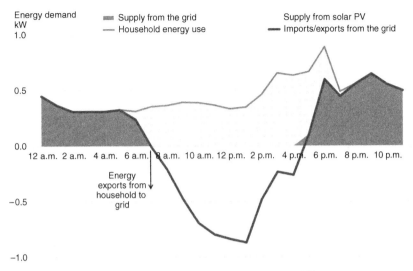

FIGURE 14.4 Average residential load profile with a 3-kW solar PV system.

Fig. 14.4 illustrates how the averaged load profile would change if a 3-kW solar PV system were installed. During daylight hours, the system generates more electricity than the household consumes, with the surplus exported to the grid. While the electricity consumed by the home remains the same, the solar system now supplies most of the home's daytime energy needs, reducing energy demand from the grid by 48%. The volume, time, and shape of grid demand are significantly different from the previous example.

When a home energy storage system—with 6-kWh useable capacity[10]—is installed in conjunction with the solar system, dependence on the grid is further reduced. On a typical summer day, the household's solar generation could supply 88% of its own energy needs, either consumed directly or stored for later use, with the home drawing from the grid for only a few hours in the early morning once the battery is discharged, as shown in Fig. 14.5. For the remainder of the day, the home neither needs to import or export power to the grid.

In this simple configuration, surplus solar generation is used to charge the battery, and power stored in the battery is subsequently discharged to the home in preference to drawing supply from the grid. Once the battery is fully charged, any additional solar generation is exported, and once the battery is fully discharged, supply from the grid resumes. In this scenario, the household self-consumes as much of their solar generation as possible. The battery does not discharge directly to the grid, and nor has its use been optimized for specific tariff structures or to provide grid support services –however reducing evening demand peaks would likely occur in all cases.

---

10. Equivalent to a 7–8 kWh system with a depth of discharge around 80%.

(A)

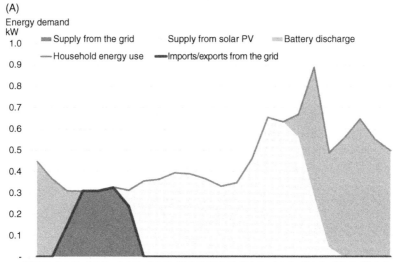

(B)

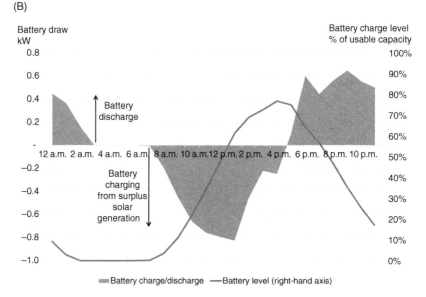

FIGURE 14.5 (A) Average residential load profile with 3-kW solar PV and 6-kWh home energy storage systems; (B) Energy storage charge profile.

4.2 Peak Demand Days

The averaged load profile of the same sample of customers is presented in Fig. 14.6 for a hot summer day,[11] when the South Australian system-wide demand peaks tend to occur, and when the grid can become strained. On this day, household electricity use was 98% higher than on the mild day, with higher demand throughout the day as a result of greater air-conditioning use. The evening demand peak was 70% higher than in Fig. 14.3.

Considering the household that has installed a 3-kW solar PV system, on the hot summer day the home, both, uses more electricity and generates more solar power than on the mild day, as shown in Fig. 14.7. The use of solar reduces peak energy demand from the grid by 20% relative to the nonsolar household; however, during the middle of the day when solar production exceeds the household's energy use, surplus energy must be exported to the grid. In fact, the household's peak use of grid capacity is now for *exporting* solar generation, rather than *importing* power supply; the maximum capacity required by the home has only fallen by 1% compared to Fig. 14.6. In the Foreword, Conboy discusses that in some Australian neighborhoods, 50% of households have rooftop solar systems. As solar penetration continues to grow, solar exports, rather than energy imports may increasingly define the network capacity required in these areas.

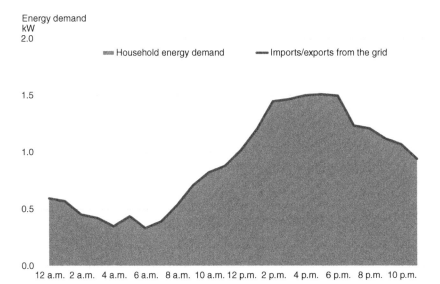

FIGURE 14.6   Average SA residential load profile on a hot summer day.

---

11. December 16, 2015.

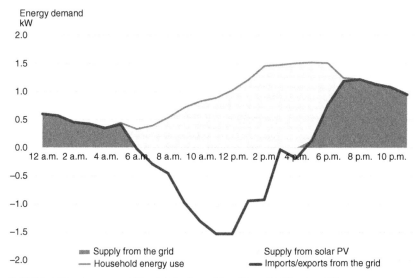

FIGURE 14.7 Average residential load profile with a 3-kW solar PV system on a hot summer day.

When home energy storage is added to solar in the same configuration as previously discussed, on the hot summer day the household is able to self-supply 84% of its energy requirements, and its peak demand is 52% lower, as shown in Fig. 14.8. Importantly, during the hours when the grid is most strained on peak demand days—typically 3–8 p.m. in South Australia—the home is largely self-sufficient, and draws no energy from the grid at all.

The addition of EV charging can also significantly change the shape of a household's energy-demand profile. If charged during the overnight off-peak period, the EV increases energy use on the hot summer day by 28%—or 56% on a typical day—and almost doubles the household's peak grid demand, as illustrated in Fig. 14.9. However, this peak occurs well outside of the problematic system-wide peak demand period, when there is likely to be surplus available capacity.

4.3   Divergent Customer Requirements

As households and businesses adopt different energy technologies, the services they require from traditional utilities will diverge. A summary of the analysis in this section is provided in Table 14.2. As discussed in Section 3, current projections from a range of analysts suggest that by the early-to-mid 2020s, around 40% of NEM households may have installed behind-the-meter technologies and will, therefore, have nonstandard connection requirements. Customers with solar and storage may draw less energy supply from the grid, but it will continue to provide an important source of backup capacity in the event of poor weather

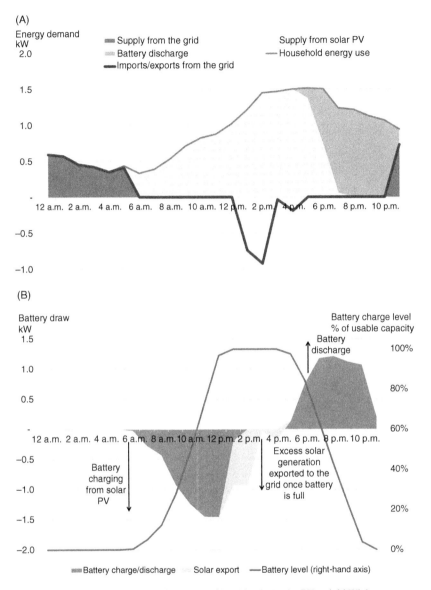

FIGURE 14.8   (A) Average residential load profile with 3-kW solar PV and 6-kWh home energy storage systems on a hot summer day; (B) Energy storage charge profile.

or equipment failure. EV customers may seek product offers with low off-peak rates that correspond with their charging needs.

New service agreements may need to be developed to reflect changing customer expectations, such as the ability to both import and export energy and to be compensated for the value created for networks and wholesale markets.

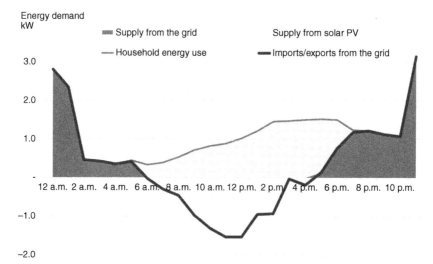

FIGURE 14.9 Average residential load profile with solar PV and off-peak EV charging on a hot summer day.[12]

TABLE 14.2 Customer Requirements From the Grid Depend on Technology Choices

	Typical demand day		Peak demand day	
	Daily grid consumption (kWh)	Share of home energy use from grid (%)	Maximum demand from grid (kW)	Maximum export (kW)
Grid only	11.0	100	1.5	—
Solar PV	5.7	52	1.2	1.5
Solar PV and Energy storage	1.3	12	0.7	0.9
EV	17.2	100	3.1	—
Solar PV and EV (overnight charging)	11.9	69	3.1	1.5

EV, Electric vehicle.

12. EV charging load is based on average driving for a vehicle that travels 15,000 km per annum and slow charged from a 15-A plug. Off-peak charging is consistent with the findings of Cook et al. (2014) that EV owners in San Diego responded well to market pricing signals, with the vast majority charging overnight and in the early morning.

While the application of more cost-reflective demand tariff structures could provide additional flexibility, with costs faced by customers more representative of the services they actually use, new pricing models may also evolve. For example, customers with low grid use could be offered a new grid "subscription" product, such as a fixed fee for a defined level of grid capacity access. The importance of utility pricing structures for DERs in the Australian market is further described in chapters by Biggar & Dimasi, MacGill & Smith, and Mountain & Harris.

5 EVOLUTION OF CONSUMER RIGHTS AND PROTECTIONS

In this section, the application of consumer protections and access rights for electricity supplies from distributed and grid sources are discussed.

5.1 Consumer Protections in a Distributed Energy Marketplace

Since the 1990s the responsibility for supplying electricity has shifted from governments to privately owned entities, following the structural separation of monopoly state-owned electricity commissions, the introduction of retail market contestability, and privatization of energy retailing. Whereas governments have a responsibility to pursue social welfare objectives, regulatory consumer protection frameworks were introduced in each jurisdiction to ensure that access to this essential service was maintained as ownership transitioned to profit-maximizing commercial enterprises.

Between 2012 and 2015 the National Energy Consumer Framework (NECF) was adopted in all NEM states except Victoria, to harmonize obligations and retail requirements. To be licensed, electricity retailers must comply with NECF requirements or equivalent regulations, including maintaining supply for customers who are having difficulty paying their bills. Customers are guaranteed access to an offer of supply for electricity and gas, minimum contractual terms, access to ombudsman and dispute resolution services, and energy-specific marketing rules that ensure customers provide "explicit and informed consent" before entering into a contract. Retailers must provide flexible payment options, hardship programs, and energy efficiency advice, and strictly enforced rules govern how and when energy services can be disconnected. Additional support measures are targeted to vulnerable customer groups, such as government-funded energy rebates for low-income households, concession cardholders, and those who require electricity to power life-support equipment.

While these requirements typically apply to the "sale of energy," they do not easily accommodate the proliferation of behind-the-meter technologies and evolving business models. The sale of energy *devices*, such as solar PV, energy storage, and energy-management systems, is not covered by energy consumer protection frameworks, rather the Australian Consumer Law (ACL) provides customers with the generic protections, which apply to the sale of all goods and services nationally.[13] The Australian Energy Regulator (AER) and Victorian Government have also developed approaches to exempt some nontraditional energy sellers from NECF provisions that are not applicable or overly onerous relative to the type and level of supply. Examples include the on-selling of power to a small number of customers within a microgrid, such as apartment complexes, retirement villages, or shopping centers, and the sale of energy from solar PV via "power purchase agreements" (PPA) to customers that are connected to the national electricity grid. A hierarchy of consumer protections has therefore been established: supply from, and connection to, the national electricity grid are "primary," whereas supplies from other sources and technologies are considered as "supplementary" and subject to lesser consumer protections.

As a result, the business model by which a technology is sold to a customer is a key determinant of the level of protections afforded for the supply, including eligibility for energy rebates and concessions, rather than the degree to which the customer depends on the source to meet their energy needs. Table 14.3 presents a summary of how this framework currently applies to different solar energy products currently available to customers in the NEM.

As energy-supply arrangements for customers become more complex, and with grid-supplied energy likely to represent a shrinking share of total electricity use for many households, this framework may not be sustainable; recall from Section 4 that customers with solar PV and battery storage systems may source as little as 12% of their daily energy supply from the grid. Some customers may seek multiple trading relationships with retailers, for example, to charge an EV separately from household supply, raising questions of primacy, while others may wish to leave the grid altogether. Under the current pricing structures, customers that are exclusively supplied from the grid, including low-income households with limited access to DERs, would bear a disproportionate share of the costs of providing a consumer protections safety net for *all* customers. Opportunities for regulatory arbitrage also arise when suppliers are required to comply with different standards for the provision of products that are effectively in competition with one another. Distributed technologies are often marketed as favorable compared to the cost of grid supply, but consumers may not realize that a reduced obligation to provide consumer protections forms part of the cost differential.

5.2 Technology Connections and Access Rights

A key assumption underpinning the technology uptake projections presented in Section 3 is that customers will be *allowed* to install their preferred

---

13. The Australian Consumer Law (ACL) provides all Australian consumers with protections relating to product safety, sales practices, consumer guarantees, unfair contract terms, and unfair business practices. Energy-specific consumer protections apply in addition to those provided under the ACL.

TABLE 14.3 Consumer Protections for the Use of Solar Energy

Product type	Description	Protection framework[a]
Solar PV	Consumer-owned system, purchased outright	ACL
Solar PV financing plan	Consumer-owned system, purchased with or without an upfront deposit, and with monthly repayment obligations	ACL
Solar PPA	Third party–owned system installed at customer premises. Under the PPA, the customer buys solar power from the system provider for an agreed period of time, for example, 10 years	ACL and retail license exemptions framework
Solar leasing	Third party–owned system installed at customer premises. The customer pays a monthly fee to lease the system for an agreed period of time, for example, 10 years	ACL
GreenPower	Purchase of remotely generated renewable energy from a licensed energy retailer via the energy grid	ACL and NECF or equivalent jurisdictional framework

ACL, Australian Consumer Law; NECF, National Energy Consumer Framework; PPA, power purchase agreements.
[a]*If products involve loans or product leasing, the National Consumer Credit Protection Act also applies.*

technologies and connect them to the local electricity network. As discussed above, the NEL and supporting rules guarantee customers access to an offer of grid-connected electricity supply under reasonable terms, but customers must *apply* for a new connection with their local distribution network service provider (DNSP) if they wish to install embedded generation, storage, or digital metering.[14] The process and costs involved with grid connections depend on a number of factors, including the system size, whether it is to be used to export electricity to the grid and whether the DNSP is satisfied that its infrastructure can accommodate the connection of the device without augmentation or other electrical works (AEMC, 2016).

While efforts have been made to streamline these connection processes for retail customers—such as requiring DNSPs to provide a connection offer for all "basic" installations and to publish relevant information on their websites—application processes are not standardized and each DNSP has established different technical requirements, with jurisdictional safety regulations in some cases overlaying additional complexity. The level of technical information

---

14. As per Chapter 5A of the National Electricity Rules.

required for applications, associated costs, and processing times vary widely. In some distribution network areas, solar PV systems up to 10 kW are considered to be basic, while in others the threshold can be as low as 3 kW. Some DNSPs in the NEM regularly process simple applications within 24 h, while others take the maximum 10 business days allowed by the National Electricity Rules, and a couple of DNSPs have introduced fees of up to A$200 to process new connection applications.[15] Upon application, some consumers are being informed that they cannot install solar systems of a certain size, or cannot install systems at all, in some cases because the network is already saturated. With continued technology uptake, this experience may become more common, particularly given that networks in the NEM are not required to publicly disclose the level of available capacity at the local feeder or transformer level. Long and complex application processes, additional costs, connection refusals, or other limitations on the use of DERs—such as system-size restrictions or output curtailment—may pose barriers to optimal technology installation.

Under current rules, connecting new energy-consuming devices, such as air-conditioning units or swimming pool pumps, does not typically require network approval, despite the potential for high coincident demand to strain infrastructure and necessitate additional augmentation. This may be changing, with some DNSPs proposing that home EV charging be placed on a "controlled load" circuit, where customers receive lower electricity rates, but are restricted as to the times they could charge their vehicle. However, this may not provide customers with the flexibility they desire. While the introduction of dynamic and cost-reflective pricing can incentivize customers to optimize their use of both demand- and supply-side technologies, a technology-neutral framework for grid connections should also be considered.

As consumer preferences change, regulatory frameworks may need to evolve to define a "right" for customers to install behind-the-meter technologies, particularly where primarily intended for self-consumption. The community may not consider connections processes to be equitable or reasonable if they result in some customers being granted permission to install devices that reduce their energy costs, while neighboring properties are denied the same opportunity. Likewise, it is unclear why safety and technical requirements for simple connections should vary significantly between jurisdictions, presenting an opportunity for standardization. With distributed technologies offering a partial or full substitute for grid supplies, policymakers could also consider whether it is appropriate to establish an independent arbiter to assess connection applications to ensure that information asymmetry is not being used to prevent legitimate deployment of new technology. New tools that provide transparent and granular information concerning the grid's capacity to host DERs, similar to those being

---

15. For example, AusNet Services in Victoria, and Ausgrid in New South Wales have introduced application fees, although most distribution network service providers (DNSPs) do not charge for this service.

developed in California as described by Picker in this volume, could also be valuable in the NEM.

6 CONCLUSIONS

This chapter has explored how distributed technologies may shape the way that households produce and consume energy into the future. According to the forecasts discussed in Section 3, by the early-to-mid 2020s, around 40% of households in the NEM may install one or more DER technologies. The growing heterogeneity of grid requirements suggests that there is an urgent need for network product and pricing reform, so that customers are able to access and pay for only the services they need. However, the introduction of cost-reflective tariffs across the NEM has slowed, with few customers expected to opt-in to voluntary demand tariffs before 2020, which may exacerbate existing cross-subsidies between customers and prevent the efficient deployment of DER.

According to CALC (2016), low levels of industry trust and complex offerings may lead to customers engaging with energy markets in economically irrational and unpredictable ways:

> The challenge Australia's energy market now faces is that effective competition, innovation and market efficiency require informed customer participation, but evidence shows that consumers don't trust, and are not engaged in the energy market. Moreover, people don't make the decisions expected of them, almost always preferring the status quo and feeling that choices in the energy market are too confusing, too much 'hassle' or not genuine as the products are all the same.

Indeed, the primary reason that households have sought to install new energy technologies is to reduce energy bills by limiting their reliance upon conventional utilities. Households that have made substantial investments in energy technologies, such as rooftop solar, may quite understandably object to pricing reform that requires them to pay more, and such an outcome may further erode industry trust. It is also questionable whether the development of an even more complex suite of network pricing and access arrangements requiring sophisticated user engagement could be successful in this environment. Low levels of consumer trust in incumbent industry participants may also limit their ability to attract customers in emerging DER markets, as energy market value increasingly shifts "behind the meter."

Importantly, reform of the regulatory frameworks governing consumer protections and grid access can build consumer trust and confidence in market participation. As the adoption of behind-the-meter technologies continues—particularly battery storage—the existing regulatory frameworks that treat DER as a *supplementary* supply source may need to be replaced by a technology-neutral approach that better reflects the experience of households that wish to source the majority of their energy use from DERs. An ability to install DERs may become as essential as access to grid supplies. Many households would already consider that it is their "right" to install technologies to manage or reduce energy costs, in much the same way that they can switch retailers in competitive markets to access better offers.

Policymakers should consider how appropriate consumer protections can be extended to all energy products, given that the level to which consumers consider a supply source to be "essential" is likely to be linked to the extent it is used to provide energy to their household. Consumer protections for all energy products should include three key elements: first, ensuring that customers receive appropriate information about energy products, including associated benefits and risks, to enable confident decision making; second, establishing frameworks for managing payment difficulties, hardship, and service disconnections to ensure customers maintain access to an essential service; and third, ensuring that government support is delivered in a technology-neutral manner, including energy concessions and access to ombudsman and dispute-resolution services.

REFERENCES

AGL Energy (AGL), 2016. AGL launches world's largest solar virtual power plant battery demonstration to benefit customers. Available from: http://www.agl.com.au/about-agl/media-centre/article-list/2016/august/agl-launches-world-largest-solar-virtual-power-plant

Australian Energy Council (AEC), 2016. Solar Report June 2016. AEC publication, Melbourne.

Australian Energy Markets Commission (AEMC), 2012. Power of Choice Review—Giving Consumers Options in the Way They Use Electricity Final Report. AEMC Publication, Sydney.

Australian Energy Markets Commission (AEMC), 2016. Integration of Storage: Regulatory Implications, Draft Report. AEMC Publication, Sydney.

Australian Energy Regulator (AER), 2016a. Tariff Structure Statement Proposals Final Decision—Victorian Electricity Distribution Network Service Providers: CitiPower, Powercor, AusNet Services, Jemena Electricity Networks and United Energy. AER Publication, Melbourne.

Australian Energy Regulator (AER), 2016b. Draft: Ring Fencing Guideline Explanatory Statement. AER Publication, Melbourne.

Bloomberg New Energy Finance (BNEF), 2015. New Energy Outlook 2015. Presentation by Kobad Bhavnagri, Sydney, July 9, 2015.

Clean Energy Council, CEC, 2016. Clean Energy Australia Report 2015. CEC Publication, Melbourne.

ClimateWorks Australia, 2016. The Path Forward for Electric Vehicles in Australia—Stakeholder Recommendations. ClimateWorks Publication, Melbourne.

Consumer Action Law Centre (CALC), 2016. Power Transformed—Unlocking Effective Competition and Trust in the Transforming Energy Market. CALC Publication, Melbourne.

Cook, J., Churchwell, C., George, S., 2014. Final Evaluation for San Diego Gas and Electric's Plug-in Electric Vehicle TOU Pricing and Technology Study. Nexant, Inc.

Council of Australian Governments (COAG) Energy Council, 2016. Consumer protections for behind the meter electricity supply consultation on regulatory implications. Available from: http://coagenergycouncil.gov.au/sites/prod.energycouncil/files/publications/documents/Consumer%20Protections%20Consultation%20Paper%20-%20August%202016.pdf

Energeia, 2015. Review of Alternative Fuel Vehicle Policy Targets and Settings for Australia. Prepared for the Energy Supply Association of Australia.

Energeia, 2016. Network Transformation Roadmap: Work Package 5—Pricing and Behavioural Enablers, Network Pricing and Incentives Reform. Energeia Publication. Prepared for the Energy Networks Association and CSIRO for the Electricity Network Transformation Roadmap.

Energy Consumers Australia (ECA), 2016. Energy Consumer Sentiment Survey Findings. ECA Publication, Sydney.

Energy Networks Association (ENA), 2016. Unlocking Value for Customers—Enabling New Services, Better Incentives, Fairer Rewards. Electricity Network Transformation Roadmap Publication, a partnership between ENA and CSIRO.

Energy Supply Association of Australia (ESAA), 2015. Electricity Gas Australia. ESAA Publication, Melbourne.

Jacobs, 2016. Projections of Uptake of Small-Scale Systems for Australian Energy Market Operator, Final Report. Jacobs Group, Melbourne.

Keane, A., 2015. Origin Energy calculates Australia has $4.4 billion of wasted roof space. Available from: http://www.news.com.au/finance/money/costs/origin-energy-calculates-australia-has-44-billion-of-wasted-roof-space/news-story/dd18cb3637aeb04d818e503362a0012c

Morgan Stanley Research, 2016. Australia Utilities Asia Insight: Solar and Batteries. Morgan Stanley Australia Limited Publication, Melbourne.

Mountain, B., 2016. How battery storage can cut home electricity bills by one quarter. RenewEconomy. Available from: http://reneweconomy.com.au/how-battery-storage-can-cut-home-electricity-bills-by-one-quarter-73736/

Nelson, T., McNeil, J., 2016. The role of the utility and pricing in the transition. In: Sioshansi, F. P. (Ed.), Future of Utilities—Utilities of the Future. Academic Press, Amsterdam.

Nelson, T., Orton, F., 2016. Climate and electricity policy integration: is the South Australian electricity market the canary in the coalmine? Electricity J. 29 (4), 1–7.

Parkinson, G., 2016. Energy incumbents fight changes that could accelerate battery storage. RenewEconomy. Available from: http://reneweconomy.com.au/energy-incumbents-fight-changes-that-could-accelerate-battery-storage-73260/

Productivity Commission, 2013. Electricity Network Regulatory Frameworks. Productivity Commission Report No. 62, Canberra.

Simshauser, P., Nelson, T., 2013. The outlook for residential electricity prices in Australia's national electricity market in 2020. Electricity J 26 (4), 66–83.

Sunverge Energy, 2016. Sunverge Announces new demand charge management features. Available from: http://www.sunverge.com/sunverge-announces-new-demand-charge-management-features/

Wood, T., Blowers, D, 2015. Sundown, Sunrise—How Australia Can Finally Get Solar Power Right. Grattan Institute Publication, Melbourne.

Chapter 15

The Transformation of the German Electricity Sector and the Emergence of New Business Models in Distributed Energy Systems

Sabine Löbbe and André Hackbarth

Reutlingen University, Reutlingen, Germany

1 INTRODUCTION

Induced by a societal decision to phase out conventional energy production—the so-called *Energiewende* (energy transition)—the rise of distributed generation acts as a game changer within the German energy market. The share of electricity produced from renewable resources increased to 31.6% in 2015 (UBA, 2016) with a targeted share of renewable resources in the electricity mix of 55%–60% in 2035 (RAP, 2015), opening perspectives for new products and services. Moreover, the rapidly increasing degree of digitalization enables innovative and disruptive business models in niches at the grid's edge that might be the winners of the future. It also stimulates the market entry of newcomers and competitors from other sectors, such as IT or telecommunication, challenging the incumbent utilities. For example, virtual and decentral market places for energy are emerging; a trend that is likely to speed up considerably by blockchain technology, if the regulatory environment is adjusted accordingly. Consequently, the energy business is turned upside down, with customers now being at the wheel. For instance, more than one-third of the renewable production capacities are owned by private persons (Trendresearch, 2013). Therefore, the objective of this chapter is to examine private energy consumer and prosumer segments and their needs to derive business models for the various decentralized energy technologies and services. Subsequently, success factors for dealing with the changing market environment and consequences of the potentially disruptive developments for the market structure are evaluated.

Innovation and Disruption at the Grid's Edge. http://dx.doi.org/10.1016/B978-0-12-811758-3.00015-2

The remainder of this chapter is organized as follows: Section 2 describes the German market, including its regulatory framework. In Section 3, major game changers in the Business-to-Customer (B2C) segment are outlined. This covers new consumer preferences and their effects on the market potential of distributed energy and energy communities. On this basis, Section 4 demonstrates emerging business models for distributed energy systems in Germany. In Section 5, key success factors of business models, as well as an outlook of the probable future market structure are derived, followed by the chapter's conclusions.

2 THE GERMAN ENERGY MARKET IN TRANSITION

Induced by European law (Directive 96/92/EG), the liberalization of the German electricity market started in 1996. The adjoining process of amendments to the law and their implementation steadily increased competition. The process culminated in the amendment of the German Energy Industry Act in 2005 and its associated regulations, which required transmission and supply to be unbundled. This process was accompanied by introducing the legal framework for the supply and support of energy from alternative, decentral generating facilities in the early 2000s.

2.1 Market Structure

In the German electricity sector, generation, transmission, distribution, and retail activities are unbundled, resulting in 879 distribution system operators and around 1240 suppliers, so that the more than 47 million households and more than 3 million industrial and commercial customers can, on average, choose between around 100 different suppliers. The transmission grid at the highest-voltage level is divided into four autonomous zones currently operated by the transmission system operators: Tennet, Amprion, 50Hertz, and TransnetBW (Bundesnetzagentur/Bundeskartellamt, 2016).

The market share of the four largest utilities—E.ON, RWE, EnBW, and Vattenfall—in German electricity generation fell from 84% in 2008 to 76.2% in 2015. This change is taking place due to the increasing share of renewable resources being predominantly owned by new players. While the big four power companies own most conventional generation capacities, in 2012 they owned only about 5% of renewable resources (Bundesnetzagentur/Bundeskartellamt, 2016; RAP, 2015; TrendResearch, 2013). Furthermore, retail competition has been increasing in the last few years. In 2015 24.9% of household customers were supplied by a competitor, that is, not the default supplier, and the "big four" had a market share of 41% (Bundesnetzagentur/Bundeskartellamt, 2016).

To organize the corresponding processes between the market partners, a comprehensive energy data management to synchronize physical and financial flows is essential. Its backbone is a system of balancing groups, that is, the virtual energy-volume accounting containing any number of entry and exit points, managed by a balance responsible party. All producers and consumers are included into balancing groups to report and follow the balanced load and generation schedules. Unpredicted imbalances are settled by the transmission system operators. Finally, the distribution system operator concentrates consumption and production data and by default operates the meters (Section 2.3). The aggregated data is then transmitted to the respective market partners and the settlements between customers and suppliers, power companies and distribution system operators, or balancing group managers and transmission system operators are done (BMWi, 2015, 2016a).

2.2   Support Schemes for Renewable Energy

In 2000 the regulation and support schemes for renewable energy generation were set up with the Renewable Energy Sources Act (EEG). The EEG introduced fixed feed-in tariffs for electricity from renewable sources, which were much higher than those for conventionally produced electricity and were guaranteed for a period of 20 years. Since then, the grid operators are obligated to prioritize electricity from renewables. Electricity consumers pay the arising costs through a surcharge, the EEG apportionment, which steadily increased to 6.88 €-cents/kWh in 2017, while the fixed feed-in tariffs were steadily reduced (Fig. 15.1).

In 2014 the EEG was amended, and for the first time caps in MW for each renewable energy source were set. Furthermore, since that time subsidies for large photovoltaic (PV) and wind sites are granted through tendering instead of fixed feed-in tariffs. Moreover, a "direct marketing" scheme was introduced, replacing the feed-in tariff, which obliges owners of larger installations to sell the produced electricity themselves or through a third party. Subsequently, the difference between the feed-in tariff and the revenues earned on the wholesale electricity market are rewarded. Additionally, self-generated and self-consumed power from solar panels is exempt from the EEG reallocation charge (small PV < 10 kW) or granted a discount (Hake et al., 2015).

In 2017 a further amendment of the EEG law came into effect, which exempts electricity from battery storages from the EEG reallocation charge and allows green electricity to be marketed regionally via a certification system.

As a consequence of this massive subsidization of green and distributed electricity production and substantial exemptions for many large industrial customers the average price per kilowatt-hour for residential customers more than doubled in the past decade to 29.69 €-cents in 2016, making them the second highest in Europe (Eurostat, 2016a). The largest share of the energy bill consists of taxes and levies, with a total of 54% for residential customers (BDEW, 2016).

As a consequence of the support scheme, the share of electricity produced from renewable sources increased to about 32% in 2015 (BMWi, 2016b) with about 1.5 million private and corporate producers; more than 6 times as many

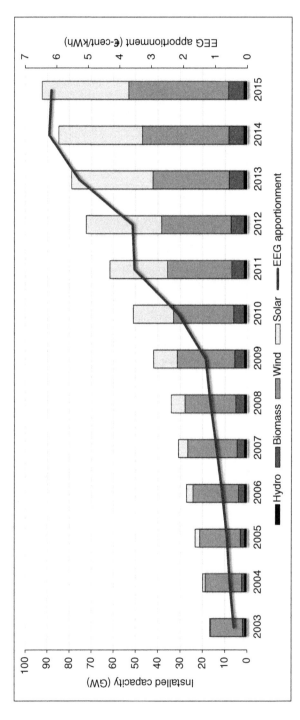

FIGURE 15.1   Development of the installed capacity of Renewable Energy Sources Act (EEG)–compatible plants (in GW) and the EEG apportionment (in €-cent/kWh). *(Sources: Authors based on data from Netztransparenz.de (2016) and Bundesnetzagentur (2016).)*

as in 2000, the starting year of the EEG (Fig. 15.1). They form an important potential for actors offering services to complement and optimize operation, management, and extension of their respective installations.

2.3   Smart Metering

For some of the emerging new business models in the energy market, especially the peer-to-peer energy networks and innovative contracting offers, the installation of smart meters is essential. In Germany, however, residential customers with consumption up to 100,000 kWh still are equipped with conventional electricity meters and billed based on a standard load profile. Thus, emanating from EU requirements, smart meters are to be introduced during a transition period of 15–20 years, equipped with bidirectional communication devices and automated and remote meter reading for ¼-h values. Based on the Act on the Digitisation of the Energy Transition, the smart meter rollout starts in 2017 for customers consuming more than 10,000 kWh and for subsidized renewable and cogeneration production larger than 7 kW. Those consuming more than 6000 kWh—about 10% of the market—are to be equipped with smart meters by 2020. For the remaining 90%, the technology is optional (BMWi, 2016c). This long transition phase does not attract business relying on smart meters and the real-time and high-resolution measurement of real electricity flows, such as the exchange of electricity in peer-to-peer energy networks. For the next years, two target groups will most probably be addressed with priority:

1. The still small group of prosumers as well as consumers with large consumption and/or production with obligatory smart meter installation.
2. All consumers demonstrating a willingness to pay for smart meters either as stand-alone option or as part of other energy products.

In contrast, most EU countries follow the European Commission's rollout target of 80% market penetration for smart meters by 2020, so that at the end of 2016 almost one-third of the 283 million electricity customers in the European Union had a smart meter (Ryberg, 2016).

3   THE B2C MARKET: POTENTIALS AND MAJOR GAME CHANGERS

The market potential for distributed electricity generation of private households is described in the following subsections. The analysis addresses, first, customer preferences as an important basis for the evaluation of future market potentials; second, the housing situation and the existing distributed energy solutions, as they form the fundament of predictions of future sales of PV and battery storage devices—the main technology installed in prosumer households—and, thus, also play a major role in various business models; and, finally, the emerging competition from energy communities, which are arousingmotivation for customers to switch their current suppliers and, hence, are transforming the economic landscape.

3.1 Target Groups and Customer Benefit

The knowledge about consumer preferences and requirements is crucial for the success of business models and the diffusion of innovations, such as smart meters or combined PV and storage systems. Therefore, a broad range of studies[1] on a wide variety of related topics can be found in the literature. Interestingly the results regarding the decision criteria and product attributes found to be most important to energy consumers show a great correspondence, albeit the diversity of the analyzed products and services or behaviors: consumer investment decisions, preferences for specific technologies (e.g., home storage systems and smart home technology), green electricity, or specific producers (e.g., local energy communities), as well as reasons for switching the supplier and for participating in community energy or prosumerism.

Thus, and as also reflected elsewhere in this volume, the seven major consumer needs or expected benefits concerning distributed energy can be summarized as follows:

- Economy: Energy cost savings or increases in payments for energy production, secure investment, acceptable payback period, and return on investment (i.e., for assets, such as PV and cogeneration).
- Autonomy: Self-sufficiency, independence from incumbents, possibility to (actively) participate in the energy transition.
- Community: Desire to share and to integrate into a community (democracy and codetermination).
- Ecology: Energy savings, emission mitigation, environment and resource protection (renewable energy), and possibility to promote certain energy sources.
- Regionality: Regional or local production and ownership structure of supplier (energy community, municipal utility, and power company).
- Comfort and safety: Accessible, trouble-free, and time-saving service or personal assistance (all-inclusive or care-free package), reliability and trustworthiness of the supplier (transparency), data security, and privacy.
- Technology: Individualized offers (mass customization), technical interest (do-it-yourself), or simplicity of technology (plug-and-play).

---

1. Among others: Kairies et al. (2016), Forsa (2015), Shelly (2014), Oerlemans et al. (2016), Gamel et al. (2016), Gossen et al. (2016), Radtke (2016), Dóci and Vasileiadou (2015), Tabi et al. (2014), Sagebiel et al. (2014), Herbes and Ramme (2014), Reichmuth et al. (2014), Rommel et al. (2016), Hamari et al. (2015), Bauwens (2016), Balck and Cracau (2015), and Kalkbrenner and Roosen (2016).

Moreover, customers can be found in specific milieus with similar lifestyles and value systems. One of the typologies is the so-called "Sinus milieus."[2] Within these milieus, five consumer segments strongly support the *Energiewende* (BMUB, 2015) and turn out to be promising target groups for products and services in decentralized energy:

- "Established conservatives" compose 10.2% of the German population. They are the classic establishment with its sense for responsibility and success ethic, aspirations of exclusiveness, and leadership. Their homeownership rate is 57%.
- "Liberal intellectuals" represent 7.3% of the German population. They are the fundamentally liberal, educational elite with postmaterial roots, and desire for self-determination. Their homeownership rate is 54%.
- "High achievers" constitute 7.4% of the German population. They are the efficiency-oriented top performers and stylistic avant-garde with global economic thinking and high IT skills. Their homeownership rate is 49%.
- "Movers and shakers" compose 7.2% of the German population. They are the unconventional creative avant-garde, hyperindividualistic, digitally connected, and mobile, looking for new frontiers. Their homeownership rate is 41%.
- "Social ecologists" represent 7% of the German population. They are globalization skeptics, idealistic, critical, and aware consumers with distinctive ecological and social conscience. Their homeownership rate is 40% (Sinus, 2015; GIK, 2015).

Accordingly, these consumer segments show the highest rates of solar installations for heating purposes already (except for the "movers and shakers") and are also those preferring regional electricity suppliers most (GIK, 2015).

In total, results show that the most important target group can be described as being environmentally aware, male homeowners with high income and high educational level, having an above-average technical interest and good knowledge or experience with renewable energies, and living in suburban and rural areas (Soskin and Squires, 2013; Oerlemans et al., 2016; Gamel et al., 2016; Kalkbrenner and Roosen, 2016).

Moreover, in the current situation—high liquidity in the financial market in conjunction with low interest rates—an important target group is attracted by alternative financial investments in assets and investment-driven products and services, such as decentralized energy systems, either directly or through crowdfunding. For instance, the typical home storage customer is in his beginning 60s, desiring to spend his money for a somewhat reasonable cause in his own premises.

---

2. The Sinus milieus group people above 14 years of age according to their fundamental value orientations, ways of living, and social background. The milieus can be contextualized according to social status, for example, education, income, and profession, and basic orientation (Sinus, 2015).

As also described by Johnston et al. in this volume, electricity consumers, especially the young, affluent, and environmentally committed, prefer not only renewables, but also renewables from local sources. In this context, research suggests that consumers are willing to additionally pay around 1 €-cent/kWh for green electricity without specific provenance, and about 3–4 €-cents/kWh for regionally or locally produced electricity, especially if it is generated by energy cooperatives or municipal utilities instead of power companies (Reichmuth et al., 2014; Sagebiel et al., 2014; Rommel et al., 2016).

3.2 Market Potential

Homeowners are the most promising target group for decentralized energy services. This incorporates combined PV and storage systems, but also microcogeneration and heat pumps. However, compared to other countries, Germany in 2015 was characterized by a comparably low ownership rate of 43% (Destatis, 2013)—in contrast to almost 70% in the EU (Eurostat, 2016b) and 63.5% in the United States in 2016 (USCB, 2016)—so that only about 30%, that is, about 12 million German households, own a single-family house.

The other 70% of the population are tenants and apartment owners and, thus, are restricted or not able to purchase a decentralized energy system due to their housing situation. To address this large potential, new products and services for housing companies and tenants are currently developed, such as energy delivery for tenants—energy produced on the premise, using a combination of PV, cogeneration units, and batteries.

In 2015 in total more than 1.5 million PV systems with a solar capacity of about 40 GW_p and about 34,000 battery storage units were installed in Germany (BSW, 2016). This represents only a small fraction of the total market potential for PV and batteries, which amounts to around 300 GW_p and 300 GWh, respectively, overall translating into a €400 billion business (Rothacher, 2016).

Given the ambitious political goals and rising customer demand, the market diffusion for PV alone or in combination with battery storage is expected to continue in the upcoming years. In 2015 about 40,000 PV installations (<10 kW_p) and 18,000 battery storage units were newly installed in German households, with retrofits accounting for only 10% of the latter. Today already, increasing battery adoption rates in the course of the installation of new PV drives the demand so that researchers expect a steady acceleration of battery sales figures up to around 50,000 units/year in 2020 (Ammon, 2015; Bräutigam, 2016). However, it is expected that from 2021 onward, when the fixed feed-in tariffs for solar electricity expire for the first PV installations, retrofits will start to increasingly push storage demand. Depending on the scenario assumptions (future development of battery adoption rates and retrofitting rates), in 2020 130,000–230,000 storage systems can be expected in total in Germany, as can also be seen in Fig. 15.2 (Hagedorn and Piepenbrink, 2016).

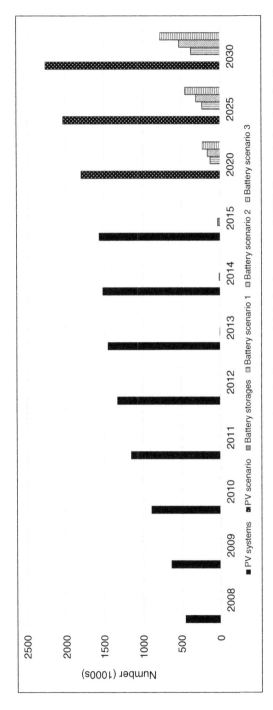

FIGURE 15.2   Number of photovoltaic (PV) systems and batteries and a forecast of their development until 2025. (*Source: Authors based on Bundesnetzagentur (2016), Hagedorn and Piepenbrink (2016), and Kairies et al. (2016).*)

Small-scale CHP systems are the second cornerstone of distributed energy generation in the household sector, as they do not suffer from volatile energy production patterns, such as PV. The market potential for minicogeneration units (<50 kW$_{el}$) and micro-CHP systems (<2 kW$_{el}$) is predicted to amount to 23,600–32,800 and 132,000 installed units in 2020, respectively, depending on the scenario, that is, the assumed rate of refurbishment and adoption rate of modern heating technologies (Adolf et al., 2013).

Heat pumps offer the possibility to absorb the volatile renewable electricity for supplying buildings with heat. Their market volume is projected to lie between 1.61–2.37 million installations in 2030 (BWP, 2016).

3.3 Competition

The market for decentralized energy is further influenced by green-energy suppliers, cooperatives, and crowdfunding platforms, as all compete for the same customer groups. For instance, the market for green electricity products is very competitive. In 2013 19.1% of German customers purchased green electricity (Bundesnetzagentur/Bundeskartellamt, 2016). The market leader, LichtBlick SE, has about 500,000 residential customers, followed by another roughly 10 competitors with more than 100,000 customers each (E&M, 2016). However, after years of growth, induced by the Fukushima disaster, the market currently languishes.

Instead of changing the power supplier or tariff, proponents of distributed and green energy have the possibility to directly invest in renewable energy generation and/or energy efficiency projects through cooperatives or crowdfunding. In 2015 around 860 cooperatives were operating in Germany (DGRV, 2016). However, following 15 years of growth, the number of cooperatives stagnates, mainly due to a change of the regulatory regime. In contrast, the crowdfunding investment volume in green-energy projects grew by 167% to €6.9 million in Germany in 2015 (Crowdfunding.de, 2016).

3.4 Game Changer: Energy Communities

Recently, energy communities based on peer-to-peer energy networks have started to emerge (see also Johnston et al. as well as Koirala & Hakvoort in this volume), trying to attract the consumer groups described in Section 3.1, desiring ecologically and/or authentic, verifiable, and trustworthy regionally generated products and suppliers. These customer preferences and the advancing digitalization foster the collaborative and intertwined production and consumption of (user generated) physical and nonphysical goods and services through transactions, such as renting, swapping, borrowing, trading, and financing (Hamari et al., 2015). Such networks first emerged in computer applications, where peer-to-peer architecture is defined as follows: "In a P2P architecture there is minimal (or no) reliance on always-on infrastructure servers. Instead the application exploits direct communication between pairs of intermittently connected hosts, called peers." (Kurose and Ross, 2010). Thus, the main characteristic of a community is the lack of a central instance that handles, organizes, or observes the interactions between peers, which, on the contrary, are directly connected with each other. Once a peer sends information to the network it is available to all connected peers, which also have access to the same set of functional abilities. Therefore, peer-to-peer energy networks can usually be found at the grid's edge, that is, the lowest-voltage level, close to the majority of consumers (Thompson, 2013).

The base case of a peer-to-peer energy network—transactions between individual users made by use of a provided online platform—is depicted in Fig. 15.3.

Within this framework, distributed *producers* represent the sources of energy and supply electricity to the network. Producers are characterized by the amount of energy they produce, the volatility and predictability of supply, the cost of production, and controllability. These parameters lead to operational options for trading via price signals and grid operation via switching commands, given precise and real-time meter data (BMWi, 2014). In a peer-to-peer environment, this information can be made available through the entire network.

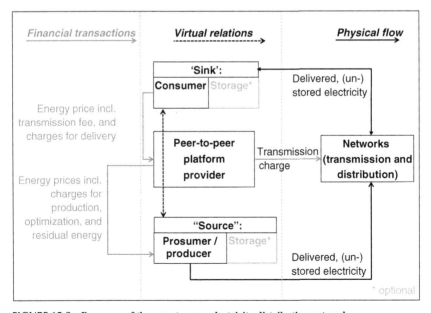

FIGURE 15.3   Base case of the peer-to-peer electricity distribution network.

On the other side, *consumers* are the sink of energy in the network and currently acquire electricity from the power company or platform provider, paying for the amount of energy used. Integrating these consumers in the information flow of the network also enables demand-side energy management measures, such as schedulable load shifts; price optimizations, for example, with real-time pricing; or increases in energy efficiency (see also chapters by Gellings and Haro et al. in this volume).

The so-called *prosumers*, who both produce—mainly based on PV systems—and consume electricity, are in the focus of peer-to-peer business models. Storage technologies are increasingly integrated, opening chances for further revenue and rent seeking by shifting the moment energy is released to or used from the grid. In this case, local micromanagement based on the aforementioned network information is essential, as it allows the prosumer to use the energy for themselves first and store the surplus to optimally release it to the grid—for example, at a profitable price.

Main market actors are the *service providers* of the peer-to-peer platforms, coordinating the distributed producers and consumers of electricity. Assuming the physical or grid layer as given, the service provider implements a virtual layer through his online community. This way, he enables the information flow between consumers and producers/prosumers. Moreover, he supplies services like market communication and balancing group management (Burger et al., 2016).

3.4.1 Limitations of Communities in the Energy Sector

Referring to the above-mentioned IT-centered definition of a peer-to-peer architecture, the delivery or swapping of electricity should be relinquished to consumers and producers entirely. However, today no such "pure" individualized peer-to-peer business models can be found in the German market, or elsewhere, as pointed out by Koirala & Hakvoort in this volume. This is due to the following reasons:

- Regulatory requirements cannot be fulfilled by individual consumers so far, especially regarding trading and billing, so that the support of service providers is mandatory.
- Residential customers are for the most part not yet equipped with smart meters.

Therefore, platform providers or their key partners currently resume these tasks for their customers (BMWi, 2014). They physically trade energy to optimize revenues and to guarantee security of supply. This means that currently producers and consumers are matched in a virtual layer, but the energy is sold to and bought from the service provider. However, the energy market offers other ways of differentiation, based on the aforementioned customer preferences: community-based models collect the generated electricity and redistribute it, either on a national level or within local communities. This is depicted in Fig. 15.4. In a "pure" peer-to-peer network producers would connect and sell

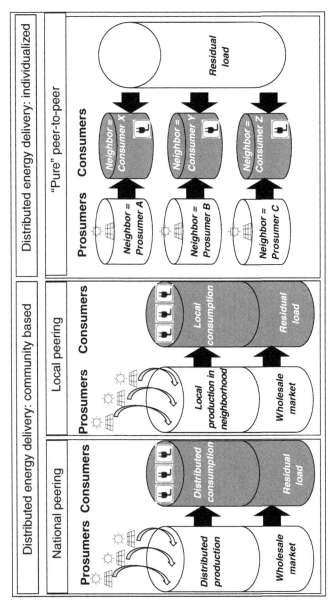

FIGURE 15.4   Options for energy communities.

directly to their consuming peers, a feature that could be enabled by the blockchain technology, as described in Section 4.4.

3.4.2 Lessons Learnt From Established Peer-to-Peer Business

As other industries possess more mature community-based services, it is worthwhile to derive lessons learnt from these. One of the best-known examples is Airbnb, a multisided online marketplace for homestays, founded in 2008 and at present with a market share of nearly 10%. Hosts accommodate guests at relatively low prices, so that in Europe Airbnb rentals are about 30% cheaper than hotels (Capital-Redaktion, 2016). Trust in the platform, hosts, and guests is crucial. Therefore, Airbnb provides the website, sets standards, and offers a host-protection insurance, a community and a 24/7 service. Marketing, search engine optimization, social media promotion, and integrated third-party services, such as a currency calculator or a charging infrastructure with Tesla, are key activities (Korosec, 2015). Key resources comprise a well-networked and structured digitalized community. The community cares for competition between members, as hosts and guests rate each other. Airbnb's revenues consist of service fees for bookings, which amounted to $900 million in 2015. The company expects to become profitable in the near future (Chafkin, 2016). Table 15.1 summarizes similarities and differences between Airbnb and peer-to-peer electricity networks.

Obviously, both industries differ physically regarding the role of the electric grid as physical and organizational intermediary. Moreover, the energy sector

TABLE 15.1 What Peer-to-Peer Energy Networks can Learn From Airbnb

	Airbnb	Energy peer-to-peer
Regulation	Important; country specific	Very important; country specific
Market potential	Global	Limited to size of the electric grid
Customer benefits	Hosts: Revenues and acquaintances Guests: low price, "feeling at home"	Producers: sharing, revenue maximization, and autonomy Consumers: green and regional product and community
Revenue streams	Service fee for bookings	Membership fee and energy rates
Cost structure	High share of fixed costs	High share of fixed costs, energy data management, and marketing cost

is influenced more strongly by regulation than the accommodation sector—especially concerning prerequisites that have to be met in the area of billing and market communication—and characterized by a high uncertainty regarding the future political framework. However, important lessons for the energy industry can be drawn from similarities of both sectors:

- Emotional benefits to attract and retain customers are important.
- Trust in and authenticity of the platform provider is key: website, energy-management software, apps, etc., need to be accessible, flawless, and secure; an excellent service, based on a well-networked and structured digitalized community, is essential.
- Gaining relevant market share and a consistent growth strategy is vital to amortize high fixed costs in IT, processes, and marketing.

4 EMERGING BUSINESS MODELS FOR DISTRIBUTED ENERGY SYSTEMS

4.1 Overview

New business models in distributed energy in Germany have begun to emerge from the following groups of companies:

- Independent start-ups, such as Buzzn, providing a peer-to-peer platform based on distributed resources, resulting in a high level of independence from incumbents.
- Utilities offering contracting and community products linking comfort and safety arguments with green and local production features, such as EnBW.
- Start-ups with financial support and corresponding control of utilities, such as Beegy, linking a high credibility as a relatively independent actor with access to resources and know-how from the established players.
- Battery suppliers, such as Sonnen, offering contracting products and energy communities that add value to their storage products, for example, the optimization of production, consumption, and storage or the sale of balancing services based on an aggregation of customer batteries.
- Service providers offering white-label products for peer-to-peer communities and energy management services to established players, for example, Lumenaza, a start-up with intense cooperation with EnBW.

As all offers are based on digitalized services with high upfront cost, all actors have a vital interest to gain market share. Utilities further participate to retain and gain customers, to optimize their portfolio and to learn from innovation. The business models focus on peer-to-peer communities and contracting packages for distributed energy, as depicted in Table 15.2 and laid out in the following subsections.

TABLE 15.2 Customer Benefit of Emerging Business Models for Distributed Energy Systems

Product/service	Energy delivery				Contracting/packaging	
	Peer-to-peer delivery		Regional peer-to-peer delivery		PV & storage, cogeneration	Local energy for tenants
Customer benefit	Producer	Consumer	Producer	Consumer		
Economy	◐	○	◐	○	◐	◐
Autonomy & community	●	●	●	●	●	●
Ecology	●	●	●	●	●	●
Regionality	○	○	●	●	○	●
Comfort & safety	○	○	○	○	●	◐
Technology	◐	◐	◐	◐	●	○
Examples described in this chapter	B2C: Buzzn, Lumenaza		B2C: White label for suppliers, cooperatives, etc.: Lumenaza		B2C: Sonnen	

Notes: Level of additional customer benefit compared to standard energy delivery. ● = High, ◐ = medium, ○ = null or very low.

4.2   Peer-to-Peer Energy Delivery

Peer-to-peer delivery is offered on national and regional scales. In total, market volume of these offers is still limited. In 2016 the number of contracts of all providers should be well below 10,000. Two out of a multifold of examples are explained further.

4.2.1   National Peer-to-Peer Energy Delivery

The number of players offering national community delivery is limited. One of the most prominent examples is Buzzn. Fig. 15.5 shows the distribution of

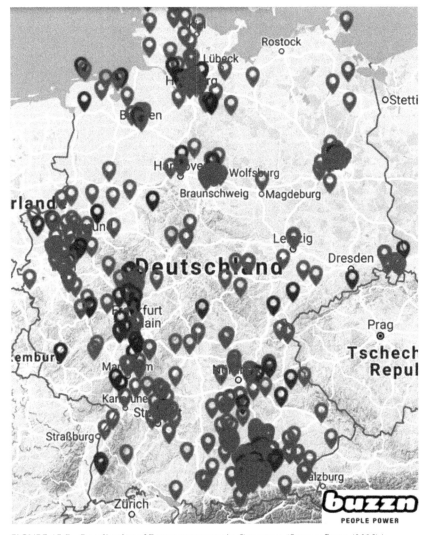

FIGURE 15.5   **Localization of Buzzn customers in Germany.** *(Source: Buzzn (2016).)*

participants in the "Buzzn Community" across Germany; being producers, consumers, or both.

Buzzn

Founded in 2009, Buzzn links prosumers and consumers across Germany and delivers energy based on intracommunity production. Value proposition encompasses community spirit and independence of incumbents. Consistency and credibility is the key, that is, consumers are offered green, distributed power of known origin. Producers receive a price premium compared to the standard feed-in tariff or can realize additional revenues for those PV systems that run out of this subsidy. Buzzn clearly focuses the customer benefits, while integrating the customers into the value creation process, be it as energy producers or as supporters of a clean and distributed energy production. Distribution channels encompass the website, the "Buzzn Community," social networks and online marketing instruments.

The company handles services for both producers and consumers, including the switching process, and runs the community on a technical and customer-management level. Regulatory and marketing know-how and resources, as well as a company culture based on core beliefs in line with the community approach are key. To guarantee independence and, hence, credibility, the company is auto-financed. Key partners include a service provider for energy-balancing management and trading and the affiliate "Buzzn System" delivering metering services.

Buzzn generates revenues from optimizing generated and consumed energy in the network, that is, the margin obtained in the wholesale market, and from delivery and installation of smart meters.

Buzzn's core interest is to maximize the number of participants to gain market share and to be attractive for peers. Additionally, Buzzn is starting to develop further products and services, such as local energy for tenants, called "Localpool."

4.2.2 Regional Peer-to-Peer Energy Delivery

As outlined in Section 3.1, customers tend to pay higher prices for regionally produced electricity. Many players, therefore, develop corresponding products, as demonstrated by the next example, which offers a two-stage distribution system.

Lumenaza

Founded in 2013, Lumenaza claims to be "the energy community." The company provides software as a service, including a peer-to-peer energy platform for producers, utilities, housing companies, landlords, and cooperatives. For example, Lumenaza supports the realization of a regional electricity product called *Fichtel-gebirgsstrom* of Stadtwerke Wunsiedel, based on regional prosumers. Lumenaza delivers the white-label product, including the template and content management for the website, takes over the generated ecoenergy and energy balancing, and sells the electricity in the wholesale market. On this basis, the utility, as customer of the white-label product, is in charge of local marketing and sales.

Lumenaza closely cooperates with EnBW, for example, in realization of its product EnBW Solar Plus, a contracting product including PV and storage.

4.3 Innovative Contracting

At first place, contracting is an established product in a slightly growing market. In 2011 around 45,000 on-site contracting services, including delivery, installation, and maintenance of heating devices and the delivery of energy over the entire lifetime were sold in Germany (Prognos, 2013). However, distributed energy leads to a relaunch and new product development. Today, packaged components also include renewable production, heat pumps, or storage devices. The offers are either based on existing installations or address consumers with old or without respective installation in place and include storage, energy management, or smart home devices. Some of the business models additionally incorporate the participation in a virtual power plant, where the load of the customers and the delivery of the producers are aggregated and optimized and the residual energy is bid into wholesale markets to gain revenues. Examples cover the business models of Sonnen, which is described below, and Next, which is focused on in the explanations by Steiniger in this volume.

Suppliers encompass about 500 utilities, energy service companies, or facility managers. For example, 30% of larger municipal utilities offer storage devices in combination with on-site renewable production. Smart home products, virtual power plants, and demand-side management solutions grow dynamically (Prognos, 2016). In the following, the business model of Sonnen is characterized.

Sonnen GmbH

Sonnen GmbH is the market leader in lithium batteries with a market share of 27% in Germany and 23% in Europe (EuPD Research, 2016). Sonnen's existing products are currently installed in over 13,000 households globally. The company started in Germany in 2011 and since then expanded to Switzerland, Austria, Italy, and the United States. In 2015 the company employed 200 people, generating €26 million revenues. The objective to sell 10,000 batteries in 2016 will most probably be outperformed (Handelsblatt, 2016). Further growth is accelerated through new products and the extension of the sales network, nationally and internationally (Hannen, 2016).

Customer segments include residential and commercial customers either with or without PV or cogeneration units. In terms of Sinus milieus, the most promising target groups are "high achievers" and "movers and shakers," due to their multioptional, efficiency-oriented thinking, and digitally connectedness, as well as "social-ecologists," because of their distinctive ecological and social attitudes.

Products are based on rather long-lasting and modularly extendable lithium iron phosphate batteries. The main battery stores 2–16 kWh and is sold with inverter, energy-management system, and smart home software to optimize self-supply. Combined with a PV installation, it is designed to replace up to 80% of the electricity from the grid. Membership in the peer-to-peer network "SonnenCommunity" includes the optimization of the energy production, storage, and delivery of the residual load. Thus, the value proposition encompasses the independence from incumbent suppliers and the sharing effects of the community.

The contractual relationship is a vital component for risk management. Customers buy the battery based on a 10-year contract, a warranty (e.g., for 10,000 load cycles), remote supervision, and an emergency hotline, while community membership is based on a 1- or 2-year contract. Those who buy a new Sonnen battery, additionally become member of the SonnenCommunity—for a membership fee—and allow Sonnen to access and steer the battery receive a discount. Prices for residual load deliveries and the feeding-in of stored electricity vary. In 2016 Sonnen started offering a flat rate guaranteeing zero additional costs for residual electricity under certain conditions.

The main distribution channels are its own website, expositions, energy consultants, and certified electricians. Key activities include production of battery systems, management of multiple revenue and cost factors, as well as marketing, smart metering, and billing processes. Given the growth ambitions, financing is crucial: in 2016 General Electric and several financial investors injected €76 million (Handelsblatt, 2016). In Germany, energy data management and the optimization of production and consumption are realized in cooperation with a utility. Moreover, Sonnen works together with sales partners and electricians handling installation and maintenance.

Revenues originate from the sale of battery systems, installation and maintenance fees, membership fees, and energy delivery. Thus, calculation is based on several profit margin–generating components, such as spot market prices and the provision of balancing services.

Targets for growth shape the cost structure, for example, upfront investments in the development of hardware, inverters, IT services, R&D, as well as in sales and marketing. For example, currently 45 developers work on future solutions. Sonnen's objective is to become a dominant player in the market for "new energy" in 10-years' time. Target markets include the Business-to-Business (B2B) segment, alternating and direct current systems, electric mobility, smart home solutions, and the integration of different components into comprehensive energy systems (Fuhs, 2016).

Sonnen, as well as its competitors, constantly adapt their business models to consumer demand and changes in the regulatory framework. Four examples are described here.

4.3.1   Storage Cloud

Some battery suppliers promise independency from utilities via a "storage cloud." Customers can store their self-produced, but unconsumed solar power virtually in a centralized battery and consume this electricity later on. The customer pays a monthly fee for utilizing these virtual cloud services.

4.3.2   PV Leasing and Contracting

House owners provide their roof surface to the contractor who installs and operates a PV unit. The house owner pays a fee for this service and generates electricity with this rented PV. The product makes use of the current regulatory environment, reduces the effort for the customer regarding installation and operation, and corresponds to customers' desire of being energy self-sufficient. These products show sales figures with increasing growth rates.

4.3.3   Flat Rates

Willingness to pay for flat rates tends to be higher compared to pay-per-use tariffs, so that expected margins are higher as well (Ascarza et al., 2012; Lambrecht and Skiera, 2006). Moreover, the share of fixed costs of electricity prices rises with increasing production from renewable energies. Consequently, costs are becoming more projectable, so that fixed prices can be guaranteed for longer time spans. This yields a considerable number of new products and business models, as demonstrated in the Sonnen example.

   These new offers relaunch contracting products and unlock new target groups. Concerning the Sinus milieus, the "established conservatives" and the "liberal intellectuals" seem promising due to their high homeownership rate and interest in ecofriendly, easy-to-use, and care-free solutions.

4.3.4   Local Green Energy for Tenants

To address the large potential of tenants and flat owners described in Section 3.2, new business models are being developed. House owners are enabled to offer energy generated on-site—for example, with PV or cogeneration units—to their tenants directly, without using the public grid. The product consists of this self-produced energy plus the delivery of the residual load. Services include project development and realization, operation, maintenance, and energy delivery. Competitors incorporate housing companies, utilities, and energy or metering service providers. The product appeared around 2014, and already in 2016, almost 40% of the larger municipal utilities included it into their portfolio (Prognos, 2016). In total in 2016, in around 80 projects local energy for tenants from PV and cogeneration units was delivered to about 15,000 apartments. The growth potential is high, as PV systems so far have been installed mainly on single-family houses and not on apartment houses. Regulation meanwhile supports the business model, resulting in a positive, but unstable outlook of 700–900 mainly PV-based projects until 2020 (TrendResearch, 2015).

   Profitability of such products depends on the subsidy schemes for renewable energy and cogeneration, turning it into a "rent capture" business model (Al-Saleh and Mahroum, 2015). Its value proposition for tenants contains a convenient access to locally produced and community-based energy and independence from utilities. For housing companies, the model generates additional revenue and supports an ecofriendly image. Contractors are able to gain loyal customers and support regional value creation.

To summarize, products for energy delivery and contracting are adapted to new customer demands and to regulatory framework. This is further supported by future technological opportunities, such as the blockchain technology.

4.4 Outlook: Business Models Based on Blockchain Technology

As also reflected in other chapters in this volume, future business models envisage the provision of a virtual marketplace where peers are able to trade their energy products, reaching from very individualized offers to regular supply agreements on a bilateral basis, based on digitalization, prosumage, and the appreciation of customers.

To develop such a "pure" peer-to-peer network, regulatory and technological changes are needed. One solution for the latter could be the blockchain technology, which underlies the better-known cryptocurrency Bitcoin (Nakamoto, 2008). Blockchain is a cryptographically secured distributed database system operated by a peer-to-peer network. Its main purpose is to keep track of transactions between the peers in the network, while guaranteeing the integrity of the single transactions and the whole system at the same time. The lack of a central authority that controls and verifies the transactions, such as a bank in the area of financial services, opens up manifold possibilities to accomplish these transactions.

Thus, unlike the current and common situation where an intermediary is needed, the verification process is provided by the blockchain and executed by the entire network (Fig. 15.6). To keep track of every single transaction they are time stamped and recorded. After a given time interval a block is created by proof-of-work—a computationally intense process called mining—guaranteeing that the created block cannot be manipulated without spending an even greater amount of computer power into rerunning the mining process. As a result of this proof-of-work, the validity of the blocks and transactions in the blocks can easily be determined. As soon as the block is verified,

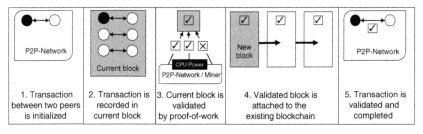

FIGURE 15.6 **Operating principle of the blockchain technology.** *P2P*, Peer-to-peer.

it is added to the blockchain, referencing to the preceding block in the chain. In this way, transactions can be tracked backward through the blockchain. With every new transaction performed and block added, the chain grows (Sieverding, 2016).

In peer-to-peer energy networks, the provider offers an online platform, while a decentralized network based on blockchain handles the transactions without an intermediary. So-called smart contracts, which are to be arranged between peers directly and verified by the blockchain, take over the role of the central authority. Producers are able to trade their energy themselves and prosumers can share their energy consumption and production with other prosumers (Lacey, 2016). Smart contracts could, for instance, include the energy quantity, date of delivery, and price. As a consequence, the system can gain in flexibility, pace, and accuracy because tasks are to be increasingly automated or even substituted by smart contracting. Additionally, costs could be reduced. Thus, implementing blockchain technology could be a key success factor of decentralized energy applications (Sieverding, 2016).

However, blockchain technology is in a very early development stage in the energy industry. In Germany, blockchain-based projects comprise, for example, the regional green certificate "GrünStromJeton" and RWE's electric mobility solution "Share & Charge" (Burger et al., 2016; Sieverding, 2016).

GrünStromJeton

A joint venture of a German start-up, the utility Stadtwerke Energieverbund (SEV), the metering supplier Discovergy, and the software supplier Sunride, which aims to develop a market for blockchain-based certificates, based on an open-source project called "GrünStromJeton." Energy customers are to receive a validated and secure certificate for the quality of their consumed energy instead of conventionally validated proofs of origin via verified generation. Based on smart meters, the consumption and the simultaneous local generation are measured and put into relation. If the customer consumed in times with high renewable electricity in the local network, he is credited with green electricity tokens. Otherwise, he receives gray electricity tokens. On this basis, regional blockchain-based proofs of origin on postcode level are generated and valued through decentralized applications (dAPPs). The customer is then able to trade his tokens, for example, to energy suppliers who can upgrade their conventional production portfolio, comparably to existing certification processes. The blockchain-based process is accessible for everyone possessing an internet connection. The processes promise to be economical and efficient because of low transaction costs, that is, the public Ethereum blockchain and a dAPP. The calculation processes of the GrünStromJeton are based on open data, which is available for most electricity grids in Germany. Transactions are designed to be transparent for all participants. So far, GrünStromJeton functions as a pilot project without sales.

Share & Charge

In 2016 Innogy Innovation Hub (subsidiary of RWE) started testing its community charging station product "Share & Charge" in cooperation with Slock.it. Drivers of electric cars get access to home-charging stations of peers. Prices are defined on an individual basis and a tool for calculation is provided. A mobile application helps drivers to localize and reserve free charging stations. Transactions are based on smart contracts, smart plugs, and blockchain technology, with payments executed automatically between member accounts (Eble, 2016).

5 THE TRANSFORMATION PROCESS

Based on the previous sections, success factors for the development of peer-to-peer business models can be derived and their impact on the development of market structures can be assessed.

5.1 Success Factors

The recent experience with peer-to-peer networks and innovative contracting offers in the energy industry is reflected in Fig. 15.7. The business models described in the previous section are all placed in a considerably changing market and regulatory environment, linked to the assumption of growing market potentials and the belief to create substantial added value for the customers.

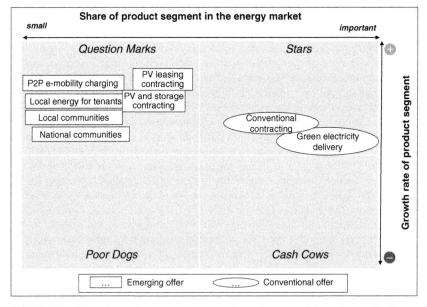

FIGURE 15.7   Transformation process gives rise to development of "question marks."

To develop a sustainable market position, the following success factors need to be considered, similar to those highlighted by Woodhouse & Bradbury in this volume.

5.1.1 Marketing

The more the energy business is decentralized, that is, divided into smaller units, the more the customer becomes part of the value chain as a producer or a storage provider, and, consequently, the more the customer becomes the focal point. Marketing has to be based on the needs and motivations of well-defined customer groups. Results from the literature (Gangale et al., 2013; Herbes and Ramme, 2014) and own research conducted in 2016 show that so far, value propositions of suppliers of energy products and services focus on rational arguments:

- information about technical features,
- price advantages,
- ecological arguments, and
- comfort and ease of use.

In this environment marked by an overflow of replaceable rational communication, offers with actual unique selling propositions and outstanding benefits, covering emotional consumer needs (such as independence and autonomy), contribution to social well-being, and to the energy transition (see Section 2.2 for details), are able to attract customers. The following considerations indicate directions for a customer-oriented marketing mix:

- For peer-to-peer electricity networks, a consistent "us-strategy" is key. This necessitates shaping the community based on the needs of the network members, integrating offline and online marketing channels. However, the Sinus milieus "high achievers" and "movers and shakers" might be more interested in "real" grassroot approaches.
- For contracting products, the "established conservatives" and the "liberal intellectuals" seem promising customers, due to their high homeownership rate and interest in ecofriendly, easy-to-use, and care-free solutions.

5.1.2 Digitalization, Customization, and Size

The dividing line between success and failure in the long run is size. Virtually all products and services based on distributed energy, from energy data management, data analytics to community management, must be based on digitalized processes. In the digitalized world, average costs drop with a growing number of customers. This enhances concentration in service delivery, as services can be copied and scaled up without additional cost, so that the costs per unit of a dominant market player drop faster, when sales grow, than those of his competitors. Therefore, maximizing the number of customers and chargeable services is key to gain noteworthy revenues. In this context, it is fundamental to balance the efficiency of processes and economies of scale on the one side and the individualization of the customer dialogue on the other side.

5.1.3   Entrepreneurship

Entrepreneurship means to take risks, for example, in financing a growth period. The acquisition of a relevant market share is the basis for longer-lasting success, even more so in markets driven by digitalization. One important instrument to handle this risk in the intertwined world of energy is risk sharing between customers and partners. All products and services for contracting and peer-to-peer-communities have to deal with this risk allocation. This covers the risk of future regulatory development, the development of spot and balancing market prices, or the management of warranties of assets, such as PV installations or storage devices, their availability, operation, and maintenance. Sharing the risk means to define equilibriums based on diverging contract durations with different partners, corresponding cost and price structures and components, all based on data analytics. As the Sonnen example and others demonstrate, German market actors are currently starting to gain experiences with these components on a broad scale.

5.1.4   Realization and Adaptation

Finally, it is good to have a good idea, but it is its realization that makes the difference. Accordingly, in an increasingly competitive environment, it is vital to be there at the right moment and to be fast. This encompasses strategic, structural, and process-oriented business development. Incumbent utilities are struggling to be ready with the right product at the right time in high quality, integrating new business via internal development, incubators, and open innovation.

5.2   Future Market Structure for Distributed Energy Systems

Future market structure is determined by centralization of services on the one hand and the motivation of different shareholders on the other hand.

As demonstrated in Section 4, business models so far are realized based on centralized energy delivery and backstage processes. Innovations, such as blockchain applications, can contribute to a more distributed delivery and control of processes.

Moreover, some innovation, especially in the peer-to-peer business, originates from a for-benefit motivation of the initiators, where profits potentially are subsumed to a social goal. Then, other business models are initiated in for-profit–orientated companies, where any social good is subsumed to the goal of shareholder profit (Kostakis et al., 2016). "The emerging and disruptive P2P economy is led by profit-driven corporations, such as Airbnb and Uber, but they owe their innovative business model to initial local civil organizations seeking to maximize the value of their resources" (Wainstein and Bumpus, 2016). As

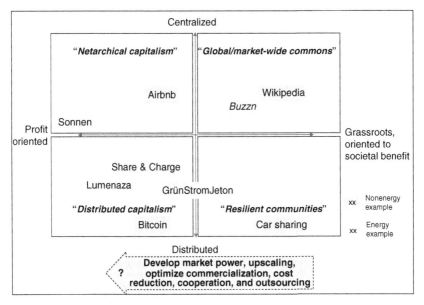

FIGURE 15.8   Typology of competitors in digitalized business models.

Fig. 15.8 demonstrates, most providers in the energy industry can be found to be profit-oriented companies, but companies like Buzzn started with motives of sharing and doing "something useful."

Based on the concept of Kostakis et al. (2016), "netarchical capitalism" and "distributed capitalism" differ in the distribution of control over the productive infrastructure, with both being profit-oriented. Companies uniting centralized governance with peer-to-peer infrastructures comprise, for instance, Airbnb or Sonnen. Distributed companies with peer-to-peer infrastructure and profit-orientation comprise Bitcoin, "Share & Charge", and Lumenaza. "Resilient communities" and the "global commons" are oriented toward societal, nonprofit benefit. Wikipedia and Buzzn are good examples for global and nation-wide initiatives, respectively. "Resilient communities" encompass movements, such as car sharing and renewable energy cooperatives.

As long as customers pay for a superior, authentic, and convincing offer, "resilient communities" and the "global commons" might stay in the market. However, with intensified competition efficiency and commercialization will be crucial for success. This does not necessarily imply moving toward profitability as a key objective in general. However, each and every single process needs to meet professionalized standards and benchmarks, so that the economic principle penetrates the business. The core interest is to develop a relevant market share for covering costs and generating value for stakeholders.

6 CONCLUSIONS

In a decentralized energy market, the view changes from a production–networks–trading–sales model toward a sources–storage–sinks model. Within this context, the energy business is becoming more intelligent and diverse and value generation partially migrates from central production to the customer's premises. The customer becomes the focal point. This development is heavily supported by German politics with an ever-changing regulation in the past 2 decades.

Thus, in constantly changing niches, with customers getting more knowledgeable and emotionally involved—seeking for participation, codetermination, and self-sufficiency—competitors nurture peer-to-peer business or innovative contracting services. While some solutions for homeowners have been in the market for a few years now and the main part of the newly launched products and services focus on single-home households, the segment of flat owners, as well as housing companies and their tenants offer an important potential—especially in Germany with its high rental rate—that currently is almost entirely untapped. Competitors comprise start-ups and incumbents, large and small utilities, energy service companies, storage provides, e-commerce, or IT companies. Success will be closely linked to the ability to develop and realize a clear strategy in turbulent times regarding technological development—especially sector coupling, smart grids, and blockchain technology—and political interventions, indispensable customer orientation, and a ceaseless quest for efficiency and effectiveness.

As also pointed out by Cooper in this volume, for incumbents, integrating "old" and "new" business, skillfully building upon innovation via start-ups or via vigor from inside the company, will be crucial, whereas newcomers need fast growth and access to customers. Obviously, cooperation is the supreme discipline toward sustainable entrepreneurial solutions in this fragmented and decentralized energy prosumer environment.

REFERENCES

Adolf, J., Schabla, U., Lücke, A., Breidenbach, L., Bräuninger, M., Leschus, L., Ehrlich, L., Otto, A., Oschatz, B., Mailach, B., 2013. Shell BDH Hauswärme-Studie: klimaschutz im Wohnungssektor—wie heizen wir morgen? Fakten, Trends und Perspektiven für Heiztechniken bis 2030. May, Hamburg, Cologne, Germany.

Al-Saleh, Y., Mahroum, S., 2015. A critical review of the interplay between policy instruments and business models: greening the built environment a case in point. J. Clean. Prod. 109, 260–270.

Ammon, M., 2015. Dezentrale stationäre Stromspeicher: Marktanalyse, Marktüberblick. Presentation at 8. Fachmesse und Kongress für Energieeffiziente Gebäude und Dezentrale Energieerzeugung, 22 May, Stuttgart, Germany.

Ascarza, E., Lambrecht, A., Vilcassim, N., 2012. When talk is "free": the effect of tariff structure on usage under two- and three-part tariffs. J. Market. Res. 49, 882–889.

Balck, B., Cracau, D., 2015. Empirical analysis in customer motives in shareconomy: a cross-sectional comparison. Working Paper No. 2/2015. Otto-von-Guericke-Universität Magdeburg, Magdeburg, Germany.

Bauwens, T., 2016. Explaining the diversity of motivations behind community renewable energy. Energy Policy 93, 278–290.

BDEW, 2016. BDEW-Strompreisanalyse Mai 2016 Haushalte und Industrie. 24 May, Bundesverband der Energie- und Wasserwirtschaft e.V, Berlin, Germany.

BMUB/BfN, 2015. Naturbewusstsein 2015—Bevölkerungsumfrage zu Natur und biologischer Vielfalt. April, Bundesministerium für Umwelt, Naturschutz, Bau und Reaktorsicherheit (BMUB) and Bundesamt für Naturschutz (BfN), Berlin, Bonn, Germany.

BMWi, 2014. Smart Energy made in Germany—Erkenntnisse zum Aufbau und zur Nutzung intelligenter Energiesysteme im Rahmen der Energiewende. May, Bundesministerium für Wirtschaft und Energie, Berlin, Germany.

BMWi, 2015. An electricity market for Germany's energy transition. White Paper, July, Federal Ministry for Economic Affairs and Energy (BMWi), Berlin, Germany.

BMWi, 2016a. Fragen und Antworten zum EEG 2017. Available from Bundesministerium für Wirtschaft und Energie: http://www.bmwi.de/DE/Themen/Energie/Erneuerbare-Energien/faq-eeg-2017,did=773140.html

BMWi, 2016b. Strommarkt der Zukunft: Zahlen und Fakten. Available from: Bundesministerium für Wirtschaft und Energie: http://www.bmwi.de/DE/Themen/Energie/Strommarkt-der-Zukunft/zahlen-fakten.html

BMWi, 2016c. Netze und Netzausbau: Die Digitalisierung der Energiewende. Available from Bundesministerium für Wirtschaft und Energie: http://www.bmwi.de/EN/Topics/Energy/Grids-and-grid-expansion/digitisation-energy-transition,did=769220.html

Bräutigam, A., 2016. Batteries for stationary energy storage in Germany: market status and outlook. Presentation at Intersolar Europe 2016, 22 June, Munich, Germany.

BSW, 2016. Statistische Zahlen der deutschen Solarstrombranche (Photovoltaik). March, Bundesverband Solarwirtschaft e.V., Berlin, Germany.

Bundesnetzagentur, 2016. Installierte EE-Leistung zum 31.12.2015 (vorläufig). Available from Bundesnetzagentur: https://www.bundesnetzagentur.de/cln_1412/DE/Sachgebiete/ElektrizitaetundGas/Unternehmen_Institutionen/ErneuerbareEnergien/ZahlenDatenInformationen/zahlenunddaten-node.html;jsessionid=E11FBCB6045A46DD42F33958218BC75F#doc40453 2bodyText3

Bundesnetzagentur/Bundeskartellamt, 2016. Monitoring Report 2015. 21 March 2016, Bonn, Germany.

Burger, C., Kuhlmann, A., Richard, P., Weinmann, J., 2016. Blockchain in der Energiewende: Eine Umfrage unter Führungskräften der deutschen Energiewirtschaft. Deutsche Energie-Agentur GmbH (dena) and ESMT European School of Management and Technology GmbH, November, Berlin, Germany.

Buzzn, 2016. Die Gemeinschaft. Available from Buzzn: http://www.buzzn.net/karte

BWP, 2016. BWP-Branchenstudie 2015—Szenarien und Politische Handlungsempfehlungen. Bundesverband Wärmepumpe e.V, Berlin, Germany.

Capital-Redaktion, 2016. Airbnb vs. hotels. Available from Capital: http://www.capital.de/dasmagazin/Airbnb-vs-hotels.html

Chafkin, M., 2016. Airbnb opens up the world. Fast Company 202, 76–95.

Crowdfunding.de, 2016. Crowdinvesting Deutschland: Marktreport 2015—Marktdaten: Volumen, Wachstum und Marktanteile. 1st Update 29 February 2016, Berlin, Germany.

Destatis, 2013. Wirtschaftsrechnungen: Einkommens- und Verbrauchsstichprobe—Wohnverhältnisse privater Haushalte. 28 November 2013, Statistisches Bundesamt, Wiesbaden, Germany.

DGRV, 2016. Genossenschaften in Deutschland. Available from Deutscher Genossenschafts- und Raiffeisenverband e.V.: http://www.genossenschaften.de/bundesgesch-ftsstelle-energiegenossenschaften

Dóci, G., Vasileiadou, E., 2015. Let's do it ourselves: individual motivations for investing in renewables at community level. Renew. Sustainable Energy Rev. 49, 41–50.

E&M, 2016. Ökostrom Umfrage 2016. Energie and Management, Herrsching, Germany.

Eble, G., 2016. Fremder, als nächster darfst Du laden. Zeitung für kommunale Wirtschaft. 23 October 2016.

EuPD Research, 2016. Über Tesla spricht der Markt, deutsche Speicheranbieter führen bei Marktanteilen. Available from EuPD Research: http://www.eupd-research.com/index.php?id=38&tx_news_pi1%5Bnews%5D=34&tx_news_pi1%5Bcontroller%5D=News&tx_news_pi1%5Baction%5D=detail&cHash=12870871d37aea2796544f95bdac31d8

Eurostat, 2016a. Energy price statistics. Available from Eurostat: http://ec.europa.cu/eurostat/statistics-explained/index.php/Energy_price_statistics

Eurostat, 2016b. Housing statistics. Available from Eurostat: http://ec.europa.eu/eurostat/statistics-explained/index.php/Housing_statistics

Forsa, 2015. Akzeptanz von variablen Stromtarifen: Ergebnisse einer qualitativen Vorstufe und einer bevölkerungsrepräsentativen Umfrage. Study for Verbraucherzentrale Bundesverband e.V. November, Berlin, Germany.

Fuhs, M., 2016. Batteriespeicher: der Sonnenstrum geht weiter. Available from PV Magazine: http://www.pv-magazine.de/nachrichten/details/beitrag/batteriespeicher--der-sonnensturm-geht-weiter_100024731/?L=1%25252F&cHash=f3ea2c14a57e9a9170663f3fe9c01fca

Gamel, J., Menrad, K., Decker, T., 2016. Is it really all about the return on investment? Exploring private wind energy investors' preferences. Energy Res. Social Sci. 14, 22–32.

Gangale, F., Mengolini, A., Onyeji, I., 2013. Consumer engagement: an insight from smart grid projects in Europe. Energy Policy 60, 621–628.

GIK, 2015. Best4Planning 2015. Gesellschaft für integrierte Kommunikationsforschung, Munich, Germany.

Gossen, M., Henseling, C., Bätzing, M., Flick, C., 2016. Peer-to-Peer Sharing: Einschätzungen und Erfahrungen—Ergebnisse einer qualitativen Befragung. PeerSharing Arbeitsbericht 3, February, Berlin, Germany.

Hagedorn, S., Piepenbrink, A., 2016. Absatzpotenzial für stationäre Batteriespeicher im privaten und gewerblichen Einsatz in Deutschland. Market analysis, 29 May 2016, E3/DC GmbH, Osnabrück, Germany.

Hake, J.F., Fischer, W., Venghaus, S., Weckenbrock, C., 2015. The German Energiewende: history and status quo. Energy 92, 532–546.

Hamari, J., Sjöklint, M., Ukkonen, A., 2015. The sharing economy: why people participate in collaborative consumption. J. Assoc. Info. Sci. Technol. 67 (9), 2047–2059.

Handelsblatt, 2016. GE steigt bei Allgäuer Sonnen GmbH ein. Available from Handelsblatt: http://www.handelsblatt.com/unternehmen/industrie/energiespeicher-ge-steigt-bei-allgaeuer-sonnen-gmbh-ein/13695352.html

Hannen, P., 2016. Sonnen Gruppe erhält 76 Millionen Euro Wachstumskapital. Available from PV Magazine: http://www.pv-magazine.de/nachrichten/details/beitrag/sonnen-gruppe-erhlt-76-millionen-euro-wachstumskapital_100024726/#ixzz4Or1qN5FC

Herbes, C., Ramme, I., 2014. Online marketing of green electricity in Germany: a content analysis of providers' websites. Energy Policy 66, 257–266.

Kairies, K.P., Haberschusz, D., van Ouwerkerk, J., Strebel, J., Wessels, O., Magnor, D., Badeda, J., Sauer, D.U., 2016. Wissenschaftliches Mess- und Evaluierungsprogramm Solarstromspeicher: Jahresbericht 2016. Institut für Stromrichtertechnik und Elektrische Antriebe, RWTH Aachen University, Aachen, Germany.

Kalkbrenner, B.J., Roosen, J., 2016. Citizens' willingness to participate in local renewable energy projects: the role of community and trust in Germany. Energy Res. Social Sci. 13, 60–70.

Korosec, K., 2015. Airbnb, Tesla partner to make road trips easier for Model S owners. Available from Fortune: http://fortune.com/2015/08/20/airbnb-tesla-partner.

Kostakis, V., Roos, A., Bauwens, M., 2016. Towards a political ecology of the digital economy: socio-environmental implications of two competing value models. Environ. Innov. Societal Trans. 18, 82–100.

Kurose, J.F., Ross, K.W., 2010. Computer Networking: A Top-Down Approach, fifth ed. Pearson, Boston, USA.

Lacey, S., 2016. The energy blockchain: how Bitcoin could be a catalyst for the distributed grid. Available from Greentechmedia: https://www.greentechmedia.com/articles/read/the-energy-blockchain-could-bitcoin-be-a-catalyst-for-the-distributed-grid

Lambrecht, A., Skiera, B., 2006. Paying too much and being happy about it: existence, causes, and consequences of tariff-choice biases. J. Market. Res. 43, 212–223.

Nakamoto, S., 2008. Bitocin: a peer-to-peer electronic system. Available from bitcoin: https://bitcoin.org/bitcoin.pdf

Netztransparenz.de, 2016. EEG-Umlage 2017. Available from Netztransparenz.de: https://www.netztransparenz.de/EEG/EEG-Umlage

Oerlemans, L.A., Chan, K.-Y., Volschenk, J., 2016. Willingness to pay for green electricity: a review of the contingent valuation literature and its sources of error. Renew. Sustainable Energy Rev. 66, 875–885.

Prognos, ifeu Institut, Hochschule Ruhr-West, 2013. Marktanalyse und Marktbewertung sowie Erstellung eines Konzeptes zur Marktbeobachtung für ausgewählte Dienstleistungen im Bereich Energieeffizienz. Final report, 5 July 2013, Berlin, Heidelberg, Mülheim a. d. Ruhr, Germany.

Prognos, Reutlinger Energiezentrum, Energetic Solutions, 2016. Entwicklung eines Portfolios von Energieeffizienzdienstleistungen für kommunale EVU. Unpublished final report for Verband Kommunaler Unternehmen, Berlin, Reutlingen, Germany, Graz, Austria.

Radtke, J., 2016. Bürgerenergie in Deutschland: Partizipation Zwischen Gemeinwohl und Rendite. Springer, Wiesbaden, Germany.

RAP, 2015. Report on the German power system. Version 1.0. Study commissioned by Agora Energiewende, Berlin, Germany.

Reichmuth, M., Lorenz, C., Beestermöller, C., Nabe, C., Markgraf, C., Schließer, J., Gerstenberg, J., Kramer, A., Megyesi, A., Neumann, R., 2014. Marktanalyse Ökostrom. Final report for Bundesministeriums für Umwelt, Naturschutz, Bau und Reaktorsicherheit, March, Dessau-Roßlau, Germany.

Rommel, J., Sagebiel, J., Müller, J.R., 2016. Quality uncertainty and the market for renewable energy: evidence from German consumers. Renew. Energy 94, 106–113.

Rothacher, 2016. Onlinevertrieb für Heimspeicher: auswahl des richtigen Speichers. Presentation at 8. Solbat Anwenderforum Areal- und Gewerbespeicher mit Elektromobilität, 23 September 2016, Berlin, Germany.

Ryberg, T., 2016. Smart metering in Europe. M2M Research Series, Berg Innsight, Stockholm, Sweden.

Sagebiel, J., Müller, J.R., Rommel, J., 2014. Are consumers willing to pay more for electricity from cooperatives? Results from an online choice experiment in Germany. Energy Res. Social Sci. 2, 90–101.

Shelly, C., 2014. Residential solar electricity adoption: what motivates, and what matters? A case study of early adopters. Energy Res. Social Sci. 2, 183–191.

Sieverding, U., 2016. Blockchain—Chance für Energieverbraucher? Potenziale und Herausforderungen. Report for Verbraucherzentrale Nordrhein-Westfalen e.V., 26 July 2016, Düsseldorf, Germany.

Sinus, 2015. Information on Sinus-Milieus 2015/16. SINUS Markt- und Sozialforschung GmbH, Heidelberg, Germany.

Soskin, M., Squires, H., 2013. Homeowner willingness to pay for rooftop solar electricity generation. Environ. Econ. 4 (1), 102–111.

Tabi, A., Hille, S.L., Wüstenhagen, R., 2014. What makes people seal the green power deal? Customer segmentation based on choice experiment in Germany. Ecol. Econ. 107, 206–215.

Thompson, R., 2013. The grid edge: how will utilities, vendors and energy service providers adapt? Available from Greentechmedia: https://www.greentechmedia.com/articles/read/the-grid-edge-how-will-utilities-vendors-regulators-and-energy-service-prov

TrendResearch, 2013. Anteile einzelner Marktakteure an Erneuerbare Energien-Anlagen in Deutschland (2. Auflage). Trend:Research, Institut für Trend- und Marktforschung, Bremen, Germany.

TrendResearch, 2015. Mieterstrom Kundenakquise und -bindung im Wohnungsmarkt: Geschäftsmodelle, Kooperationsmöglichkeiten und Marktpotenziale bis 2020. Unpublished study, Trend:Research, Institut für Trend- und Marktforschung, Bremen, Germany.

UBA, 2016. Erneuerbare Energien in Zahlen. Available from Umweltbundesamt: https://www.umweltbundesamt.de/themen/klima-energie/erneuerbare-energien/erneuerbare-energien-in-zahlen#textpart-1

USCB, 2016. Quarterly residential vacancies and homeownership: Third quarter 2016. US Census Bureau, Release Number: CB16-172, 27 October 2016, Suitland, US.

Wainstein, M.E., Bumpus, A.G., 2016. Business models as drivers of the low carbon power system transition: a multi-level perspective. J. Clean. Prod. 126, 572–585.

Chapter 16

Peer-to-Peer Energy Matching: Transparency, Choice, and Locational Grid Pricing

James Johnston
Open Utility, London, United Kingdom

1 INTRODUCTION

As other chapters of this volume have already explained, the electricity industry is rapidly transforming away from an analog, centralized, and fossil-fuel powered model toward a digital, decentralized, and renewable-powered one.

Traditionally, power flowed unidirectionally from a small number of large thermal power stations through an elaborate transmission and distribution network to end users, who were passive and uninterested in what happened upstream of the meter. Most users paid regulated tariffs and had minimal and infrequent interactions with their suppliers, if at all, beyond paying the utility bills.

This picture is rapidly changing. Businesses and households are taking control and installing on-site renewable generation (plus energy storage in the near future). These distributed energy resources (DERs) mean that power flows are no longer guaranteed to be unidirectional, complicating the traditional roles of balancing, transmission, and distribution. Furthermore, energy customers are no longer passive and uninterested; they are evolving into a new sophisticated type of user with the capability, interest, and financial motivation to participate actively in energy markets.

In the future, many experts see the needs of customers and generators being served by peer-to-peer (P2P) models, where they can exchange energy with each other and participate in wider energy markets through online platforms. With flexible generation, storage, and smart devices, customers can better manage their consumption, shift loads, and provide balancing services

Innovation and Disruption at the Grid's Edge. http://dx.doi.org/10.1016/B978-0-12-811758-3.00016-4

to local and national grids. Over the coming decades a smart, flexible, and decentralized energy system could deliver trillions of dollars of savings globally (Lovins, 2011).

While many experts agree on the long-term vision of a decentralized energy system, there are fewer practical guidelines on how the existing centralized energy system can migrate toward this vision. The existing energy industry is built on centralized principles, which during a century of development have been baked into regulations, software systems, and business models and all serve to maintain the status quo. Customer attitudes may be changing, but the industry so far has been slow to catch up.

Real innovation lies at the interface between the current centralized energy system and the proposed future decentralized energy system. This chapter will focus on these practical solutions—that can both function in today's world, but also act as stepping stones toward the decentralized vision.

Section 2 focuses on present day commercial opportunities for P2P energy, and illustrates this with a case study on Open Utility, one of the first companies in the world to launch a commercial P2P offering. Section 3 looks into the near future and explore how P2P models and regulatory frameworks could evolve over the next few years to lay down the stepping stones toward the vision of a decentralized and democratized energy system followed by the chapter's conclusions.

2 TRANSPARENCY AND CHOICE

All across the world, individuals and businesses are seeking different ways to cut their carbon emissions and reduce their ecological footprint. As a first measure, they look to reduce their net demand through energy efficiency actions and as a secondary measure install on-site renewable generation. Most buildings cannot go fully off grid due to space limitations or leasing restrictions. Additionally, as renewable generation sources are intermittent, they might have surplus power at certain times of the year, while requiring top-up at others. Energy storage helps alleviate this on an hour-by-hour basis, but cannot be relied upon for interseasonal variations.

For these reasons, many people and businesses want to buy zero carbon electricity via the electricity network. But how is it possible to match customers with specific renewable generators? As everyone knows, electrons, once injected into the network, are mixed up with others and there is no physical way to route electrons from a specific generator to a specific customer.

However, you don't need to reroute electrons for matching to work. A good analogy is that the grid is just like an online bank account. When you digitally transfer money to a friend, there is no direct transfer of physical currency (e.g., dollar bills) between the two bank accounts. Instead, it's an accounting process facilitated by a trusted intermediary—the bank. One account is debited and the other credited by the same amount. In the same way, you can buy renewable electricity from the grid: customers' accounts are credited when they take energy off the grid and generators' accounts are debited when they put energy onto the grid. Similar to banking, you also need a trusted intermediary (a licensed electricity retailer) to facilitate the transactions.

Traditionally, there are two methods for tracking the buying and selling of renewable generation over the grid and providing authentication of supply. The first method is using a regulated Guarantee of Origin (GoO) certification process. According to the EU guideline 2009/28/EC, all EU member states are required to establish a national GoO registry for electricity generated from renewable sources. In the United Kingdom, these certificates [called Renewable Energy Guarantee of Origin (REGOs)] are managed by Ofgem E-serve who verifies that no double counting has occurred and administers the certificates to electricity retailers.[1]

A secondary method for achieving authentication of supply is setting up a power purchase agreement (PPA) bilaterally between a generation site and a customer—called corporate PPAs. This contractual approach provides a stronger linkage between the customer and generators, and also enables the price to be fixed over a longer time period.

However, there are problems with both of these traditional approaches. The GoO process is opaque and complicated. In the United Kingdom, generators need to manually submit their information to Ofgem every year. Customers don't necessarily trust that the process stops double counting. For example, in Europe many certificates originate from hydro sites in Norway, even though many consumers in Norway do not realize that they are giving away the "renewable benefit" from their local generators (Jensen et al., 2016). Studies have shown that many businesses do not trust their electricity retailers, as they do not provide enough granular data on the source of their energy (Aasena et al., 2010; Hast et al., 2015). Finally, as matching is done on an annual basis, it does not reflect the true nature of intermittent renewables—which could further exacerbate perceptions that buying renewable energy over the grid is "greenwash."[2]

Corporate PPAs are more effective at determining authenticity of supply, however they do so in a clunky way. Due to the bespoke nature of the contracts, they are very expensive and complex to set up and hence are only an option for the largest of global corporations (Labrador, 2015).

---

1. REGO, https://www.ofgem.gov.uk/environmental-programmes/rego
2. Taking the bank account analogy further, this is like having an overdraft to alleviate cashflow problems throughout the year, but not paying for it!

P2P energy markets are disrupting these traditional approaches. A perfect storm is brewing due to a number of far-reaching changes across the industry:

- Smart meters: over 1 billion smart meters will be deployed across the world by 2022 (Martin, 2013). Remote access to near-realtime data on consumption and generation will become the norm rather than the exception.
- Distributed generation: by the end of 2015, the global capacity of distributed solar PV (installations less than 4 MW) was around 160 GW.[3]
- Sharing economy: through the successes in other sectors (most prominently Uber and Airbnb in transportation and hospitality, respectively), customers are becoming comfortable with online "sharing economy" business models.
- Blockchain technology: the distributed ledger technology promises a new era of distributed business models across every sector, including energy. Regulators and incumbents are taking notice of the potential and setting up trials across the world (Burger et al., 2016).

As a result, there are now few barriers to the development of highly scalable online P2P transactive models, which can automatically match generation and consumption using near-realtime smart meter data. Provenance of supply could become a central feature of market design, rather than an inefficient afterthought.

P2P energy markets are still in their infancy and a number of start-ups developing solutions are emerging across the world.[4] One of the most established platforms is called Piclo and is run by Open Utility, a technology company based in London, United Kingdom.[5]

In October 2015, the Piclo Platform was trialed by UK electricity retailer Good Energy in a project funded by the UK government. Good Energy is a UK electricity retailer with around 75,000 residential customers and 4,000 business customers. They source their electricity from more than 1000 distributed renewable generators across the United Kingdom.[6] The trial was composed of 25 renewable generators (covering wind, solar, and hydro technologies) and 12 business customers (tourist sites, hotels, factories, and farms). Full results of the trial were published in the report "Piclo—A Glimpse Into the Future of Britain's Energy Economy."[7]

---

3. Total global solar PV capacity in 2015 was 227 GW (REN21, 2016). From the Wiki Solar site (http://wiki-solar.org/), operational utility scale (>4 MW installations) accounts for 60 GW. Taking into account that some data might be missing, this leaves an approximately 160 GW of distributed solar.
4. P2P startups (nonexhaustive list): Lumenaza (Germany), LO3 Energy (United States), Open Utility (United Kingdom), Powerpeers (the Netherlands), and Vandebron (the Netherlands).
5. The lead author of this chapter is CEO and cofounder of Open Utility.
6. Approximate numbers correct as of December 2016.
7. Full trial report can be accessed at: http://bit.ly/1SP0WQJ

In August 2016, Open Utility signed a multiyear commercial agreement with Good Energy to roll out the Piclo Platform to all their business customers and renewable generators. At time of writing in December 2016, the service— under the *Selectricity* brand, was the largest P2P energy matching platform in the United Kingdom.

More details on Open Utility's Piclo Platform are outlined in Box 16.1.

BOX 16.1 How Piclo Platform Works?

The Piclo Platform enables a virtual P2P connection between customers and generators and automatically allocates energy between them according to their smart meter data (Fig. 16.1).

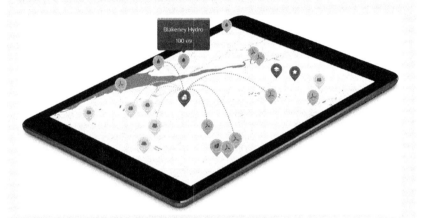

FIGURE 16.1   Connecting virtually with local generators.

Customers can control which set of generators they match with by setting preferences on the online service. Effectively, customers are creating a priority list of generators, which is analogous to a merit order, a concept that underpins traditional spot markets.[8] In P2P markets, every single customer can build their own unique merit order. The definition can also be relaxed to include multiple prioritization criteria, including cost, technology type, ownership, location, and other characteristics.

Once the merit order is set, the process is entirely automated. On behalf of each customer the matching algorithms try to match as much energy as possible with the highest priority generators. For example, a customer might prioritize a

---

8. Merit orders are traditionally a list of generation sources ordered by their marginal cost to run, used by centralized spot markets to determine the lowest cost set of generators needed to meet a certain demand level.

local community wind turbine but depending on the available wind resource it might only get a fraction of their needs from that source. The algorithms would then run through the merit order continually trying to match as much as possible.

As discussed earlier, no electrons are being rerouted in the matching process—it is simply a more granular way of debiting and crediting electricity being put on and taken off the grid. Matching is done on a half-hourly frequency (as meter data in the United Kingdom is available at that frequency) which means energy mixes made up of solar, wind, and hydro vary throughout the day according to the weather. Fig. 16.2 illustrates a typical week of matching data, where varying amounts of hydro, wind and solar generation sources are matched every half hour.

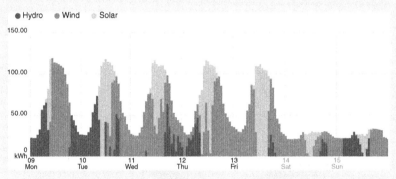

FIGURE 16.2   Typical week of matching data.

The Piclo Platform enables customers to engage with their energy supply in a unique way. Large business customers who have to spend millions each year on their electricity bills can receive more value: they can choose to buy from local community generators to visibly support the local economy and they get to tell positive marketing stories to better engage their staff and customers. They can, for example, say: "we buy from our local school solar array," or "our energy comes from the local wind farms," or "nearby hydro reservoir"— marketing messages that increasingly resonate with environmentally conscious consumers.

To access the Piclo Platform, customers and generators need to sign a contract with an electricity retailer that has bought a license to use the platform. Currently, they can only match with customers and generators who have signed up with the same electricity retailer, though this may change in the future.

The Piclo Platform is a cloud-hosted service. Smart meter is received from the electricity retailer via an API and both the electricity retailer and their customers get online accounts to access the matching data. The electricity retailer continues to manage customer contracts, customer service, billing, and payment as schematically illustrated in Fig. 16.3.

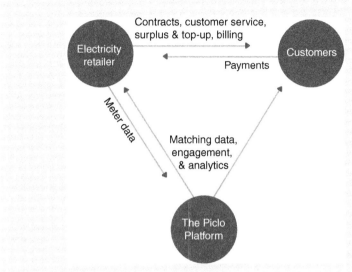

FIGURE 16.3   Piclo Platform business model.

Electricity retailers pay Open Utility to license the platform and in return they can offer customers a tangible renewable supply—a valuable differentiator in competitive markets. Those electricity retailers with an existing supply chain of contracted renewable generation can better leverage those assets and bring their renewable tariffs to life. They can develop new location-based sales strategies and improve retention metrics through better engagement.

P2P energy matching for provenance of supply could be seen by many as a "nice-to-have" for renewable-focused electricity retailers rather than mandatory function in the energy industry. However, it does offer several benefits over traditional renewable certification processes as outlined earlier, and regulation should recognize this.

Presently, there are no mandatory requirements for businesses, individuals, or electricity retailers to adopt it. Indeed, incumbent electricity retailers are resistant to change: they say that moving away from monthly or annual accounting of renewables would increase their administrative burden (Ofgem, 2016). Possibly, the real reason is that they are worried about what the extra transparency would reveal: renewables are commonly thought to be locally generated, when in many cases renewable tariffs are backed by certificates bought from overseas.

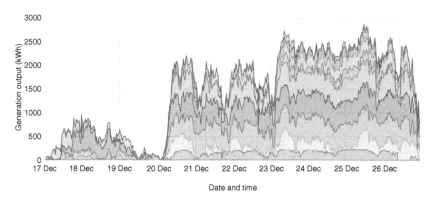

FIGURE 16.4   Generation output from 10 wind turbines across the United Kingdom in December 2016.

3   LOCATIONAL GRID PRICING

There is growing interest from DSOs, regulators, and policy experts in using locational grid pricing to incentivize local balancing to reduce network congestion and grid defection (MIT, 2016).

DSOs can no longer rely on the physical reinforcement of networks alone as costs would be prohibitive and networks oversized and underutilized. Renewables have low capacity factors (meaning they only reach their nameplate maximum capacity a fraction of the time) and peaks can often occur at the same time. This phenomenon is illustrated in Fig. 16.4, using data from 10 wind generators (situated across the breadth of the United Kingdom).[9]

Renewables are forcing DSOs to deploy more virtual means of managing congestion in their networks. One method is active curtailment of generators when local network limits are reached (Gill et al., 2014). Other direct control measures include deploying energy storage at substations to defer reinforcement; as of 2015 around 66 MW of energy storage was operational across UK distribution networks (REA, 2015).[10]

Alternatively, DSOs can create price-based incentives to promote certain behaviors (Eida et al., 2016). They are desirable for DSOs as it allows them to be agnostic to solutions and let the markets decide the most efficient method: thus driving competition and keeping costs down. As a downside, there is less certainty on the magnitude and timeliness of response as they

---

9. The generators were located across Scotland, England, and Wales. Storm Barbara passed over the United Kingdom on 23–24th December 2016 and Storm Connor on 25–26th December 2016. Output in excess of 2 MW was sustained during these 4 days, compared with a low of 3.6 kW at 2:00 a.m. on 20th December 2016.
10. Excludes pumped hydro storage.

rely on customers to act in their own best (economic) interest (Ruester et al., 2014).

In the United Kingdom, there is already a widespread adoption of price-based incentives. All large industrial and commercial energy customers in the country (accounting for around 50% of the UK demand) have time-of-use tariffs to encourage them to reduce or shift their demand at peak times (Moss and Buckley, 2014). These tariffs are split into red, amber, and green time periods.

The problem with these price incentives is that they only take into account temporal congestion factors. DSOs average out prices across all customers, regardless of where they are situated to create a "socialized" cost structure. This approach is attractive for regulators as it ensures nondiscrimination as customers are treated equally (MIT, 2016).

This system worked well for customers when most of their electricity came from big power stations and the distribution network was used more or less in full by everyone. However, with the rollout of distributed generation and community energy groups, an increasing number of people do not accept that they should be paying for parts of the network they are not using (Bennett, 2015). From a generator perspective, the lack of a financial incentive for local trading is driving them to install their own *private wires* to sell power direct to local customers and thus cutting out the DSO completely (Urquhart, 2016).[11]

Locational grid pricing that gives customers and generators a financial benefit for local trading could mediate both these problems. Understandably, regulators are hesitant in moving away from a low-risk socialized charging model. There are very few examples of locational pricing models deployed in the field and not much quantitative data on the value to distribution networks or impact on customers. For this reason, DSOs are keen to run trials to test out different approaches and build evidence for regulatory change (MIT, 2016).

P2P energy matching can facilitate new locational pricing models. It can be used to trace how much of the distribution network has been used in every energy transaction. In 2016, Open Utility in collaboration with economists Reckon LLP proposed a modification to UK DSO charging regulations to incorporate locational pricing. A summary of the model is shown in Box 16.2. A link to the full report (along with details on the industry change proposal) is in the footnote.[12]

---

11. Private wires is a legal definition as outlined in the Electricity (Class Exemptions from the Requirement for a Licence) Order 2001 (amended 2007). The class exemption enables anyone to distribute power directly to customers (up to 1 MW) without requiring a distribution license.

12. Report: Recognition of local electricity sourcing in DNOs' distribution use of system charging methodologies, available from: http://bit.ly/2iImhnB

BOX 16.2 Locational Pricing Case Study

Open Utility's proposed solution was to offer customers a lower DUoS rate for any energy sourced within the same 33 kV network group (Fig. 16.5). Effectively, if a customer matched with a local generator, they only pay the unit charges for the section of the network they used in those transactions. Any energy matched with generators via the parent 132 kV network is charged at normal DUoS rates.

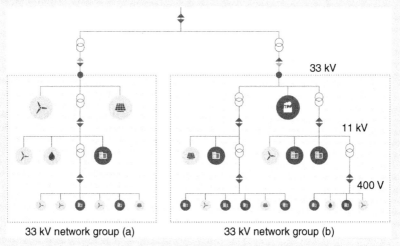

33 kV

11 kV

400 V

33 kV network group (a)   33 kV network group (b)

FIGURE 16.5 Schematic outline of Open Utility's locational pricing model.

Customers who modify their demand profiles to match more with local generation (e.g., undertake local balancing) would end up paying less DUoS charges than customers who act more passively. Therefore, a locational pricing incentive could reduce the amount of grid reinforcement required upstream which would offset the DSO's lost revenues (in offering the lower DUoS rates). In the example in Fig. 16.5, an upgrade on the 132 kV network could be avoided because of better local balancing within the two separate 33 kV network groups.

A desktop case study using the Eden Project, a large visitor attraction in the south-west of England was completed. Analyzing the potential for local matches, it was discovered that five solar farms and two generation wind turbine sites were connected to the same 33 kV network group (Fig. 16.6). With representative data on levels of local matching, it was estimated that the Eden Project could reduce its overall DUoS rates by 39% or around £20,000 per annum.

FIGURE 16.6 Distributed generation connected to the same 33 kV network group.

4 CONCLUSIONS

Much of the value of distributed renewable generation is locked up in the out-dated and centralized energy industry business models and regulations.

P2P energy matching introduces to the energy industry a simple concept of recording the source and destination of every energy transaction over the grid. The applications are widespread: from enabling electricity retailers to provide their customers with transparency and choice, to enabling DSOs to develop locational grid pricing.

A lack of transparency and choice is holding back the adoption of renewable tariffs. Some do not trust that buying renewable energy over the grid will have any meaningful impact. P2P energy matching is a relatively simple, yet powerful antidote to this and could help drive a strong customer demand for renewables into the future.

Finally, locational grid pricing could have a number of benefits for DSOs. By incentivizing local balancing it could help reduce losses on distribution networks, lower carbon emissions, and remove the need for network reinforcement to meet the same demand. Unlocking a financial benefit for local trading could support the development of a community-driven and democratized energy industry.

ACKNOWLEDGMENTS

The author would like to acknowledge the assistance of Fereidoon Sioshansi and the Open Utility team in finalizing this chapter.

REFERENCES

Aasena, M., Westskoga, H., Wilhiteb, H., Lindbergb, M., 2010. The EU electricity disclosure from the business perspective—a study from Norway. Energy Policy 38 (12), 7921–7928.

Bennett, P., 2015. Government must rethink local energy supplier rules, says 10:10. Clean Energy News. Available from: http://www.cleanenergynews.co.uk/news/renewable-heat/government-must-rethink-local-energy-supplier-rules-says-1010

Burger, C., Kuhlmann, A., Richard, P., Weinmann, J., 2016. Blockchain in the energy transition. A survey among decision-makers in the German energy industry, Deutsche Energie-Agentur GmbH (dena), Germany.

Eida, C., Codanib, P., Perezc, Y., Renesesd, J., Hakvoort, R., 2016. Managing electric flexibility from distributed energy resources: a review of incentives for market design. Renew. Sustain. Energy Rev. 64, 237–247.

Gill, S., Plecas, M., Kockar, I., 2014. Coupling demand and distributed generation to accelerate renewable connections. University of Strathclyde. Published online: https://pure.strath.ac.uk/portal/files/38906382/Gill_Plecas_Kockar_ARC_coupling_demand_and_distributed_generation_final.pdf

Hast, A., Syri, S., Jokiniemi, J., Huuskonen, M., Cross, S., 2015. Review of green electricity products in the United Kingdom, Germany and Finland. Renew. Sustain. Energy Rev. 42, 1370–1384.

Jensen, J., Drabik, E., Egenhofer, C., 2016. The disclosure of guarantees of origin: interactions with the 2030 climate and energy Framework. CEPS Special Report No. 149, November 10, 2016.

Labrador, D., 2015. Why corporate power purchasing is poised to be the next big thing in renewable energy. Rocky Mountain Institute. Published online: http://blog.rmi.org/blog_2015_04_15_corporate_power_purchasing_is_poised_to_be_the_next_big_thing_in_renewable_energy

Lovins, A., 2011. Reinventing Fire: Bold Business Solutions for the New Energy Era. Chelsea Green Publishing, Vermont.

Martin, R., 2013. The installed base of smart meters will surpass 1 billion by 2022. Navigant Research. Published online: https://www.navigantresearch.com/newsroom/the-installed-base-of-smart-meters-will-surpass-1-billion-by-2022

MIT, 2016. Utility of the Future report. MIT. Published online: http://energy.mit.edu/

Moss, A., Buckley, R., 2014. Competition in British business energy supply markets: an independent assessment for Energy UK. Cornwall Energy. Accessed online: https://www.energy-uk.org.uk/publication.html?task=file.download&id=3296

Ofgem, 2016. Consultation: proof of UK consumption of overseas electricity. Available from: https://www.ofgem.gov.uk/publications-and-updates/decision-proof-uk-consumption-overseas-electricity-consultation

REA, 2015. Renewable Energy Association, Energy storage in the UK: an overview. Published online: http://www.r-e-a.net/upload/rea_uk_energy_storage_report_november_2015_-_final.pdf

REN21, 2016. Renewables 2016 Global Status Report. Published online: http://www.ren21.net/wp-content/uploads/2016/06/GSR_2016_Full_Report.pdf

Ruester, S., Schwenen, S., Batlle, C., Pérez-Arriaga, I., 2014. From distribution networks to smart distribution systems: rethinking the regulation of European electricity DSOs. Utilities Policy 31 (2014), 229–237.

Urquhart, S., 2016. Private wire PPAs—why the focus now?. TLT. Published online: http://www.tltsolicitors.com/news-and-insights/insight/private-wire-ppas---why-the-focus-now/

Chapter 17

Virtual Power Plants: Bringing the Flexibility of Decentralized Loads and Generation to Power Markets

Helen Steiniger
Next Kraftwerke, Cologne, Germany

1   INTRODUCTION

Traditionally, grid operators dispatched thermal plants of various designs and capabilities in the so-called merit order to meet demand on the network at the lowest possible cost. In this paradigm, the load was assumed as a *given* and the supply was continuously adjusted to chase it from hour to hour, day to day, and season to season. Base load plants, usually large nuclear or coal/lignite-fired units, ran flat out to cover the minimum load throughout the year, since these units typically had the lowest per-unit costs.

To meet the hourly, daily, and seasonal variations in demand, midrange units were dispatched. Peaking units were sparingly dispatched only for limited duration to cover peak demand, such as on a hot summer day when air condition load would be high in hot regions of the world or on cold winter days when electric heating load would be high in cold regions of the world. The concept of traditional merit order is schematically illustrated in Fig. 17.1.

Historically, many networks had little renewable generation, with the exception of hydro units, which were used depending on the availability of water and the capacity of reservoirs to complement the output of thermal plants. Pumped hydro storage, when available, was also utilized to manage fluctuations in demand and to store excess generation.

Fast forward to 2017 in places like Germany, Denmark, California, Texas, or South Australia, where renewables now comprise more than half of total generation on many hours and many days, occasionally exceeding total demand on the network. In Germany, for example, the proportion of renewables has been on the rise and now dominates the energy mix on many days—a fact that

Innovation and Disruption at the Grid's Edge. http://dx.doi.org/10.1016/B978-0-12-811758-3.00017-6

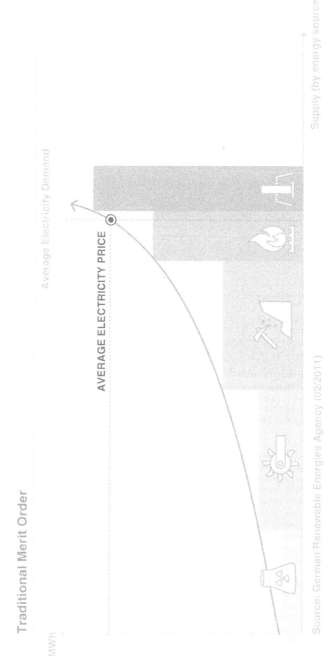

FIGURE 17.1   **Traditional Merit Order to match supply and demand in an electricity market without renewables.** (*Source: Next Kraftwerke, based on German Renewable Energies Agency 2016.*)

will become even more pronounced as the existing fleet of nuclear plants with almost 11 GW of installed capacity will be phased out by 2022 and lignite/coal capacities of more than 7 GW will leave the market in response to recently introduced carbon reduction policies (Fig. 17.2). These numbers are significant, considering Germany's peak load is about 80 GW.

In such renewables-dominated energy markets, the traditional notion of merit order dispatch no longer applies. Renewables such as solar and wind tend to be variable and they must be utilized when they are available—otherwise they would have to be curtailed, which means wasting low-cost, carbon-free electricity. Run-of-the-river hydro falls in the same category. If the grid operator does not use such resources when they are available, they are wasted.

Another big difference with the traditional dispatch paradigm, of course, is the fact that many renewable resources—with the exception of geothermal, hydro with massive reservoirs, biomass, and biogas with flexibility—are not dispatchable. They are available when they are available, and when they are not, the grid operator cannot order them to produce energy. This suggests that as more countries gradually move toward a future where the bulk of generation will be renewable, they must come up with alternative schemes to balance the load and generation and to keep the grid stable and reliable.

One fundamental way to think about this challenge is to think of a radically different operating and dispatching paradigm. Instead of taking load as a *given* and adjusting generation to chase it at all hours, why not devise a paradigm where both demand *and* supply are *flexible* and can be adjusted to dance together, rather than forcing one to follow the other.

As it turns out, this is precisely the approach that has been developed by a number of innovative enterprises who increasingly see the future of electricity markets and networks as a choreography of flexible load and variable generation—and in some cases distributed or centralized storage[1]—that work in unison to keep the grid secure and reliable while maximizing the use of variable generation resources.

This approach, sometimes referred to as price-responsive demand, demand response (DR), or demand-side management (DSM) is increasingly applied in Virtual Power Plants (VPPs) to intelligently shift demand based on wholesale power prices. Industries have long since been managing their electricity demand to reduce network charges. Now the upcoming VPPs show promise as a superior alternative to balance variable demand with variable generation, the emerging reality of the future of most electricity markets.

This chapter describes the fundamentals of this promising approach by focusing on the successful business model of Next Kraftwerke, a German digital utility and VPP operator that has grown substantially in the past few years, applying these principles in the context of Europe's challenged electricity market.

---

1. An example may be the large-scale VPP project under development by AGL in Australia, further described in the chapter by Orton et al.

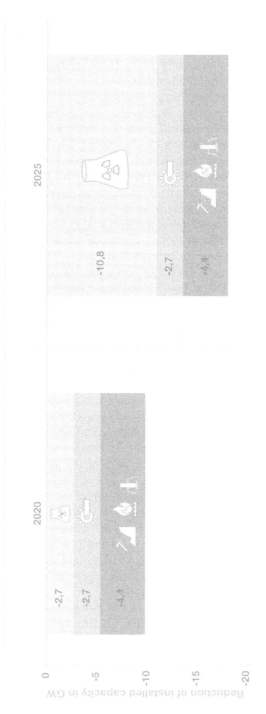

FIGURE 17.2 Scheduled reduction of German conventional generation capacity until 2025. (*Source: r2b energy consulting and Next Kraftwerke 2016.*)

By late-2016 they have aggregated more than 4000 electricity producing and consuming participants with a combined capacity of over 2700 MW—the equivalent of two large coal-fired power stations. The chapter argues that the future of electric systems will increasingly rest on aggregating flexible load *and* variable and flexible renewable generation—rather than dispatching thermal plants to meet an inflexible demand.

The chapter is organized as follows: Section 2 describes the rising significance of flexibility in the context of variable renewable generation. Section 3 introduces the role of aggregators and the concept of VPPs and describes how they work and what services they offer. Section 4 gives an outlook on the future evolution of VPPs followed by the chapter's conclusions.

2   FLEXIBILITY IN THE CONTEXT OF VARIABLE RENEWABLE GENERATION

As already described, the reality of many electricity markets around the world is the rising dominance of variable renewable generation which is frequently policy-driven, as in the case of Germany, and generally intended to reduce the carbon footprint of electricity generation over time. In Germany, roughly one-third of generation in 2016 was supplied from renewables with an overwhelming portion from wind and solar PVs, both of which are inherently variable (Fig. 17.3).

A related problem of renewable—especially wind—generation is the increasing frequency and costs of thermal redispatch and renewable curtailment. These measures are more and more often taken by the grid operators when on- and off-shore wind turbines in the windy North of Germany produce so much electricity that the grid transporting it to the industrial South gets congested. The problem is intensified by thermal power plants in the North and East that cannot be switched off, even when prices are negative. In these instances, the grid operators pay thermal or renewable plants in congested areas extra money to switch off; while at the same time, they pay thermal units in the South extra to ramp up.

This phenomenon has become common over the last few years and is expected to get worse in the future. In 2014, for example, the German grid operators—there are four of them—had to take action on 330 days to redispatch as described above resulting in €187 million cost compared to 232 days in 2013 and a cost of €132 million.[2]

The compensation costs for curtailing or switching off renewable plants rose from €43.7 million in 2013 to €82.7 million in 2014. While these costs are still quite small compared to the €24 billion that were paid out to renewable

---

2. http://www.bundesnetzagentur.de/SharedDocs/Downloads/DE/Allgemeines/Bundesnetzagentur/Publikationen/Berichte/2014/Monitoringbericht_2014_BF.pdf?__blob=publicationFile&v=4

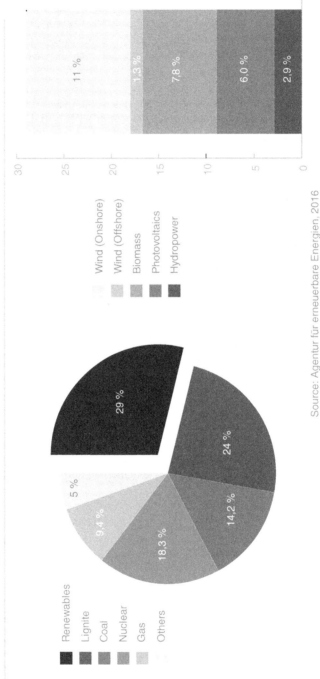

FIGURE 17.3 **Breakdown of renewables in German electricity generation mix in 2016.**
(*Source: Next Kraftwerke, based on German Renewable Energies Agency 2016.*)

generators in 2015 in total,[3] the relative cost increases in some years have been significant as illustrated in Fig. 17.4.

Germany, of course, is not alone in facing such a challenge. California and New York, for example, have 50% renewable targets by 2030 while Hawaii is striving to reach a 100% renewable goal by 2045. In South Australia, Denmark, and Texas, the transmission network is routinely overwhelmed during windy periods. A number of other countries including Norway and Iceland are virtually 100% renewable—mostly with hydro though which does not necessarily put stress on the network since water can be stored in reservoirs for generation as needed.

It is important to note that a high share of *variable* renewables does not necessarily mean less grid reliability—certainly not in Germany or Denmark, both countries with large and growing renewable supplies. In 2013 the lights went out for 12 min on average in Denmark and for 15 min in Germany; the corresponding figure for 2014 was a little over 12 min.[4] In nuclear-dominated France, by contrast, it was 68 min in 2013. In America and many other countries grid service interruptions exceed one or more hours per year, on average.

Moving forward, the most noticeable expected impact of so much variable renewable generation on the network is twofold:

- First, it tends to depress overall wholesale prices because wind, solar, and run-of-the-river hydro displace thermal generation with zero marginal cost while usually thermal capacity is shut down at a lower pace than new renewable capacity is built, leading to overcapacities in the market.
- Second, variability of renewable generation tends to make wholesale prices volatile because large swings in wind and solar generation impact overall market price spreads.

The first expected impact can easily be observed in the German wholesale market as illustrated in Fig. 17.5 where rising renewable generation—accompanied by excess conventional capacities with minimum loads—has depressed wholesale prices in recent years with grave consequences for the four major German power utilities.

The rise of renewables not only impacts overall prices in wholesale markets but changes the merit order of plants, which means that zero marginal cost renewables—principally solar and wind—displace more expensive base load plants including nuclear, lignite/coal, and gas, by pushing them to the right and eventually out of the picture entirely as illustrated in Fig. 17.6. This means that, all else being equal, thermal plants are dispatched less frequently, and for fewer hours. Many of them cannot be ramped down to zero so they run on minimum loads more frequently, further depressing prices. And when they do get dispatched, they receive lower prices on average. This explains why

---

3. https://www.cleanenergywire.org/factsheets/re-dispatch-costs-german-power-grid
4. http://www.renewablesinternational.net/german-grid-keeps-getting-more-reliable/150/537/89595/

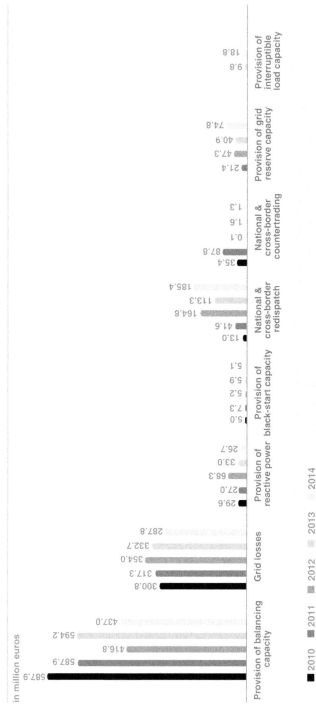

FIGURE 17.4 **Cost development of grid stabilization measures taken by German transmission system operators (TSOs).** *(Source: Next Kraftwerke, based on Monitoring Reports 2013, 2014, and 2015 of Federal Network Agency 2016.)*

Futures prices for electricity base load in Germany (Phelix Base Year Futures)

— 2013   — 2014   ⋯ 2015   2016

FIGURE 17.5 Development of electricity future prices in Germany since 2012. *(Source: Next Kraftwerke, based on data from the European Energy Exchange (EEX) 2016.)*

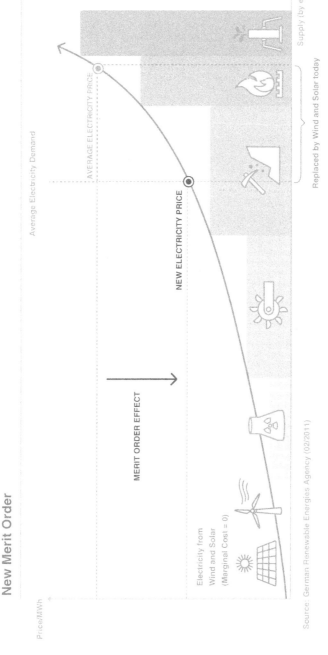

FIGURE 17.6   New Merit Order to match supply and demand in an electricity market with zero-margin renewables (solar and wind).
(Source: Next Kraftwerke, based on German Renewable Energies Agency 2016.)

thermal generators in Germany, for example, have had such a difficult time in recent years.

The trend is likely to get worse. Recent studies estimate that up to 230 GW of fossil capacity in Europe will remain unprofitable through 2020.[5] Due to the massive overcapacities of thermal generators, market prices cannot adequately reflect scarcity situations. Therefore, the value of flexible peak load assets is currently very low which reduces the incentive to invest in decentral, green flexibility. Still, a lot of nuclear and coal/lignite capacities are kept in the market through direct or indirect subsidies—a paradoxical outcome at odds with the EU's goal to decarbonize the energy sector by 2050.[6] If the EU goal is to be reached, investments in sufficient green flexibility options are needed soon to balance out the variable generation of wind and solar for the decades to come.

The second expected impact—wholesale prices becoming more volatile—is more difficult to detect in the market. It is intuitive to assume that with an increasing share of variable renewables the volatility of wholesale prices should increase, too. However in the case of Germany such a trend cannot be observed in wholesale price volatility—at least not if measured by the standard deviation of average prices on the day-ahead market where the bulk of wind and solar power is traded (Fig. 17.7). Rather, the overall trend suggests that the volatility of day-ahead prices has gone down since 2012. One of the reasons for this development is that the liquidity of the day-ahead market has increased significantly since the introduction of a market model for renewables—the so-called market premium model—in Germany in 2012. Since then, renewable electricity is regularly traded on day-ahead and intraday markets where forecasting errors can be corrected as forecasts improve shortly before delivery.

Consequently, in Germany at least, it is the case that electricity spot markets can very efficiently deal with the fluctuating nature of wind and solar so that variable renewables do not necessarily cause prices to become more volatile.

Nonetheless, renewables have immensely altered the price structures of wholesale markets. This is being felt, for example, in the collapse of midday prices when solar generation is most pronounced which has virtually eliminated the peak/off-peak paradigm of the old electricity world. As shown in Fig. 17.8, average day-ahead prices over one day have converged in the past five years— now peaks usually occur in the morning and evening hours with midday prices being closer to night prices. Another case in point is the well-known California duck curve, which refers to the daily dip in midday *net load* on the network caused by the rising solar generation, further described in Chapter 9 by Baak.

Variations of the California duck curve are now routinely experienced in other parts of the world including sunny Australia or windy Texas and Denmark where negative prices during periods of excess solar/wind generation are frequent.

---

5. McKinsey 2014: "Beyond the storm—value growth in the EU power sector."
6. https://ec.europa.eu/energy/en/topics/energy-strategy/2050-energy-strategy

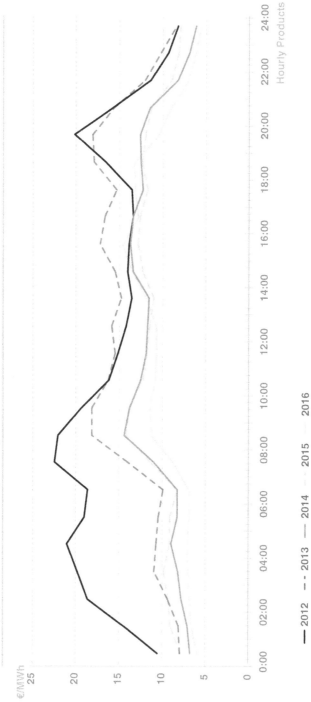

FIGURE 17.7   Development of day-ahead price volatility in Germany since 2012. Prices seem to become *less* volatile over time. *(Source: Next Kraftwerke, based on data from the European Power Exchange (EPEX SPOT) 2016.)*

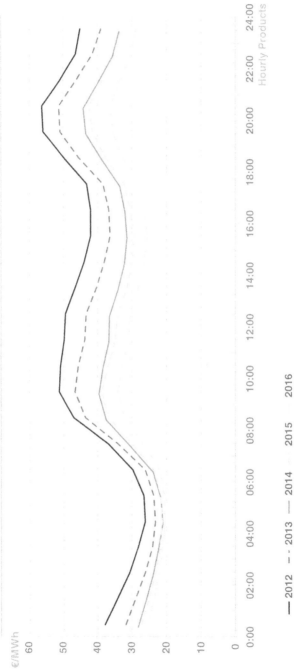

FIGURE 17.8 **Development of average day-ahead prices in Germany.** The peak/off-peak duality is vanishing. *(Source: Next Kraftwerke, based on data from the European Power Exchange (EPEX SPOT) 2016.)*

Combined, the net effect of these two phenomena—wholesale prices plummeting and the peak/off-peak differential eroding—is to:

- make thermal plants—particularly those with inflexible operational characteristics—less profitable; and
- make the grid operators' job more challenging as the task of keeping supply and demand in balance at all times becomes more complicated requiring new thinking, new tools, and new approaches.

The next section will primarily look into the second effect by offering a solution, namely how to balance variable renewable generation and variable demand on networks with increasing amounts of variable renewable generation, such as in Germany.

3   VPPS AND THE ROLE OF AGGREGATORS

The preceding discussion clearly demonstrates that business as usual is not likely to be sustainable, practical, or cost-effective in markets where variable generation is beginning to overwhelm the grid operator's traditional tool box and solutions.

A major problem caused by big swings in variable solar or wind generation is that thermal plants have to turn down to their minimum operating levels for most of the day, only to resume generation at full capacity in the late afternoon hours to meet the evening's peak demand. In the case of California and its Duck Curve, this up and down ramping is approaching 13–14 GW and has to be accomplished during a 3-h window. This not only puts a lot of wear and tear on the thermal plants but is hugely inefficient, expensive, and polluting.

Along with this vexing challenge faced by California and Germany alike comes the so-called minimum load problem, which refers to the fact that on many cool sunny days or windy nights, the grid operator simply runs out of thermal plant capacity that can be turned down during midday or night hours.[7] Many thermal units cannot be turned down for a variety of reasons, beyond a reasonable limit, so they have to run on minimum load during the day when they are not needed. This issue is likely to become more serious over time, certainly in California, Germany, and possibly elsewhere.

Which is why experts are looking beyond the grid operators' traditional tool box for practical, cost-effective, and non-polluting alternatives to balance variable generation and load.

One obvious solution, of course, is to:

- better manage the variable generation to the extent possible;
- foster flexible renewable generation alongside variable renewables; and
- make demand more flexible and price-responsive.

---

7. Refer to What does CAISO crave the most? Flexibility, EEnergy Informer, August 2016, p. 19.

Next Kraftwerke, hereafter referred to as NK for brevity, a German digital utility, is among innovative companies who are trying to do all of the above, and do it in a way that is *win-win-win* for the grid operator, for consumers with flexible demand, and for generators who can adjust their output.

The company's basic business model[8] is to create a portfolio of *flexible* generation and loads by aggregating a large number of participants who are willing and able to respond to price signals in their VPP.

As illustrated in Fig. 17.9, NK have developed a profitable business by bringing power producers, consumers, wholesale markets, and the grid operators together using a remote control unit called the "Next Box" that allows the parties to transact with one another in real time. By late-2016 they had networked more than 4000 electricity-generating and -consuming participants with a combined capacity of over 2700 MW.

The Next Box connects the parties, enabling supply and demand to respond to price signals through highly automated machine-to-machine (M2M) communications as schematically shown in Fig. 17.10.

This is particularly relevant for offering ancillary services or control reserves. Control reserves are tendered by transmission system operators (TSOs) to stabilize the grid between 49.8 and 50.2 Hz in Europe. In general, power generators and consumers can bid positive and negative capacity to the tender. When selected, this capacity must not be marketed elsewhere so the TSO can call it when the grid is in imbalance.

In times when power markets were dominated by a small number of large thermal power plants, this balancing function used to be a rather trivial task. With millions of decentral power plants in the system today, a much larger set of variables is involved, requiring a complex optimization problem.

To solve this problem, the Next Box[9] collects operating data of each decentral power producer and consumer in the VPP, for example data on the availability, current capacity level, and flexible capacity. These data are then encrypted and sent to the NK control system where it is decoded and processed.

The control system[10] validates the data sent in by all Next Boxes so it always knows the number of megawatts available in the pool. Along with weather data and price signals, the validated data from the Next Boxes are fed into an optimization scheme that determines the optimal output value for each unit. The control system then sends a signal with those numbers back to the Next Boxes and the decentral units adapt their output accordingly.

The value of NK's assembled portfolio of flexible resources comes from the diversity of its client base, which includes generators with flexible output, such as biomass, biogas, and hydro, customers with flexible onsite microgeneration,

---

8. Further details on NK may be found at company's website: www.nextkrafterke.com
9. https://www.next-kraftwerke.com/network-technology/next-box
10. https://www.next-kraftwerke.com/network-technology/control-system

How does a Virtual Power Plant Work?

We aggregate decentral power producers and consumers and act as their gateway to the power markets and the grid operators.

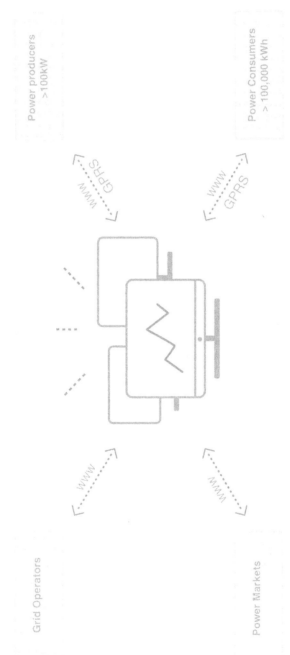

FIGURE 17.9   NK connect power generators and consumers to power exchanges and ancillary services markets organized by grid operators.
(Source: Next Kraftwerke.)

Technology (Next Box, Control System, & Trading)

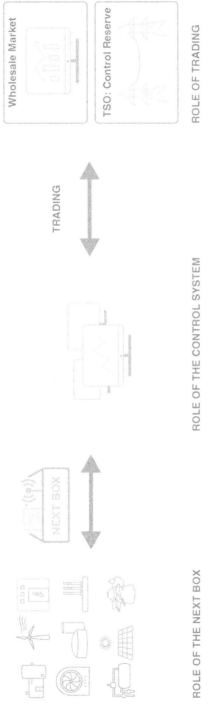

ROLE OF THE NEXT BOX

> Standardized interface between power plant & control system

> Follows IT security regulations of the transmission code by the TSOs

ROLE OF THE CONTROL SYSTEM

> Collects all information transmitted by the Next Boxes & the electricity system

> Through the central control system, the units are controlled, ramped up & down

MARKETS

Wholesale Market

TSO: Control Reserve

ROLE OF TRADING

> Trading finds highest value for electricity & flexibility

TRADING

FIGURE 17.10   NK network decentral units with their Next Box and the control system steers them automatically based on market signals.
(Source: Next Kraftwerke.)

as well as large commercial and industrial customers with flexible loads (DSM customers) as illustrated in Fig. 17.11. They also aggregate volatile solar and wind generators, forecast and market their electricity on power exchanges, and curtail their output if market conditions mandate it.

Over time, NK have expanded their service offerings to include power trading,[11] balancing services for utilities or volatile power plants, and (flexible) power supply to industrial and commercial power consumers. The basic idea behind NK's business model is simple: produce electricity when it is scarce and consume electricity when it is plentiful. Due to the increasing shares of solar and wind power in European electricity systems, scarcity and excess situations are more and more determined by the output of solar and wind plants, which makes weather forecasts a very relevant component of predicting price movements.

Based on its own weather forecasts, NK suggest ideal hourly schedules for the producers and consumers in their VPP one day before delivery. They then trade the corresponding electricity volumes for every hour on the day-ahead markets of European power exchanges. Deviations from feed-in forecasts due to forecasting errors or unexpected downtimes as well as within-the-hour gradients of wind and solar power are traded on continuous intraday markets for every quarter hour of the day. In turn, the price signals on day-ahead and intraday markets provide incentives for flexible power producers and consumers to shift their generation and consumption to those quarter hours with the best prices—for producers these are periods of power scarcity with high prices and for consumers these are periods of excess supply with low prices.

How NK manage the nuances of the interactions among the various participants in the scheme is complex. The fundamentals, however, are straightforward. When demand and prices are high:

- clients with flexible generation, such as CHP plants, biomass, or hydro are encouraged to produce *as much as they can*; and
- clients with flexible load are encouraged to do the opposite—namely *reduce consumption to a bare minimum.*

The incentives are reversed when prices and demand are low, namely:

- clients with flexible generation are encouraged to *reduce production* to the extent feasible; and
- clients with flexible load are encouraged to *increase consumption* to the extent practical.

This is accomplished automatically through M2M communication as clients respond to price signals provided by NK—who are constantly monitoring supply and demand conditions on the wholesale market and forecasting prices and a number of other critical variables in real time.

---

11. https://www.next-kraftwerke.com/network-technology/power-trading

FIGURE 17.11 Next Kraftwerke's customer portfolio. *(Source: Next Kraftwerke.)*

Many consumers have ample flexibility in *when* they use electricity, particularly customers with inert processes, such as pumping, heating, melting, crushing, processing, etc. These customers can adjust their heavy consumption periods and shift them to times of low cost electricity without compromising their operations. These are the types of customers that NK specifically recruit on the consumption side. And these customers learn how to better respond to price signals over time—it is a classic case of learning by doing. Participants get better at it over time, saving more and are less inconvenienced.

For NK, DSM means to shift demand from high-price to lower-price times of the day in a continuous optimization process—rather than only shaving off demand peaks during high-price periods as is the traditional approach to DSM. Consequently customers rarely if ever reduce the total amount of electricity consumed, while actively and dynamically optimizing *when* they use electricity as illustrated in Fig. 17.12.

One of NK's customers, for example, is an association for coastal management with large pumping loads to maintain water levels in dikes and embankments in Germany's low-lying coastal areas. So long as these customers maintain water levels by pumping rain back into the sea, the pumps need not run constantly. This offers valuable flexibility.

With its optimization scheme, NK determine those quarters of every day in which it is cheapest to pump given a predefined set of restrictions—for example, water levels must be maintained within certain limits to avoid flooding. Once these customers agree to change their pumping schedules to take advantage of lower prices, their pumping load is adjusted and their electricity bill is reduced without any adverse impact on the management of water levels. This particular product is called "Best of 96" because a day consists of 96 quarter hours and NK pick the best, that is the cheapest, for these consumers.

Fig. 17.13 illustrates how the scheme works where customers take advantage of low prices to schedule their heavy loads. In this example, the pumps used to be operated between 5:30 and 7:30 a.m. NK's optimization analysis, however, suggested that pumping should be done at 5:00 a.m., 12:30 p.m., 1:30 p.m., and 4:00 p.m. for half an hour each. By following this procedure the customer reduced his energy costs by 30% in 2015.

The same principle can be applied to most pumping loads virtually anywhere. These customers typically use large industrial pumps consuming large amounts of electricity. Any reduction in their power usage ends up as savings in the bottom line with no impact on their operations. While this is an easy-to-understand scheme, the same applies to many industrial and commercial loads where the exact time of when electricity is consumed is not critical.

For those electricity consumers who do not have so much flexibility that they can shift demand in quarter-hour periods, NK offer set price zones for a whole year in advance—a product called "Take your Time." Fig. 17.14 shows how a customer's required flexibility increases with the different options of Take your Time, peaking with Best of 96.

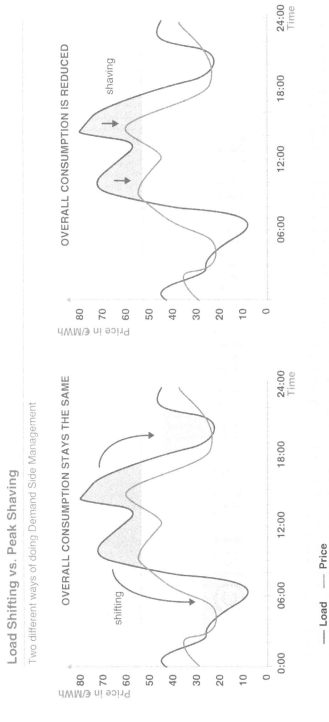

FIGURE 17.12 NK's principle of demand-side management (DSM): load-shifting rather than peak-shaving. *(Source: Next Kraftwerke.)*

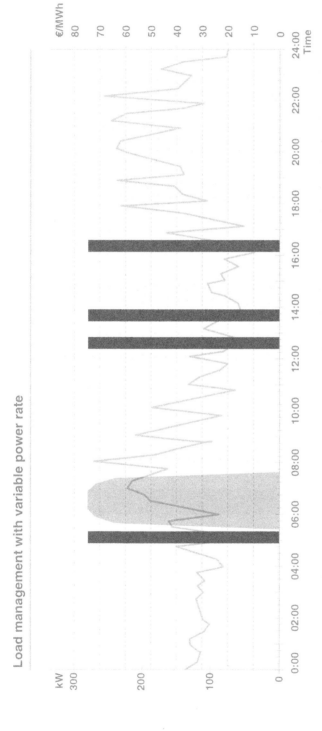

FIGURE 17.13 How customers with flexible demand can gain from shifting some of their demand into times with low prices (Best of 96).
(Source: Next Kraftwerke.)

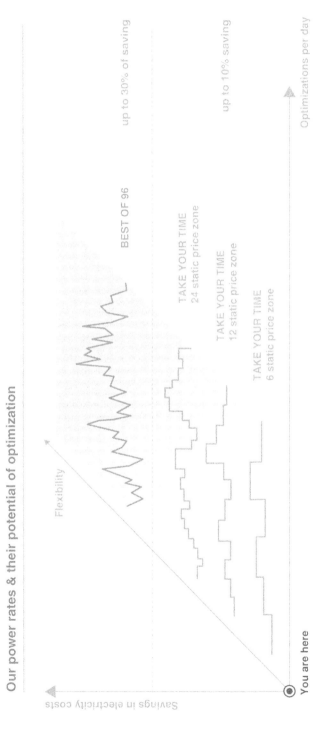

FIGURE 17.14 **Increasing flexibility means increasing cost savings.** The cascade of NK's flexible electricity rates. (*Source: Next Kraftwerke.*)

The same principle applies to clients with flexible generation, such as those with biomass or hydro plants. One customer on the supply side is a biogas plant whose operator optimized its design with respect to flexibility, a practice that is incentivized by recent revisions of the Renewable Energy Sources Act in Germany. For doing this, the plant owner receives a flexibility premium but in return only gets a feed-in premium for half a year's production at full throttle. As a consequence, on each day, the plant operator wants to choose those 12 hours, or 48 quarter hours, with the highest prices to switch on the generator. This is what NK do for this particular customer, as shown in Fig. 17.15.

Of course not every plant is able to do that, just as not every consumer is able to change her electricity-consuming processes on a quarter-hourly basis. Some biogas plants have contracts for heat supply that are linked to a minimum electricity production or their biogas storage tanks are relatively small. All of the processes involved in the plant's operation need to be analyzed to determine the range of flexibility of that particular power plant.

When bottlenecks or constraints are identified, they can frequently be overcome by investing in larger tanks or seamless operation of generators. For a plant operator, these investments need to pay off rather sooner than later. With a more volatile market, the payback periods generally become more difficult to estimate. But with the 18 GW of conventional nuclear and coal/lignite capacity leaving the market by 2025 in Germany alone, it is rather clear that price spikes will occur more often in the years to come. This is because even with more installations of wind and solar, there will be times in the year when there is hardly any wind or sun. And in these times flexible renewables that can ramp up and flexible consumers that can ramp down will be in high demand.

The business model of companies, such as NK, in other words, are robust today and are likely to become more valuable in the future.

4 WHAT FUTURE FOR VARIABLE DEMAND?

By now, it is generally accepted that the future of the power sector is shifting toward more decentralized generation, increasingly from variable renewable resources, and most likely augmented with better energy management and control systems on customers' premises with extra bells and whistles, such as distributed storage offering more flexibility, more autonomy, and less reliance on bulk commodity energy at undifferentiated prices from the grid.[12]

How such a future evolves, of course, depends on where you are and what sort of regulatory policies you operate under. What may happen in Germany or New York or California—say over the next decade—may not apply to Indonesia, Saudi Arabia, or Mali.

---

12. Refer to Sioshansi, F.P. (Ed.), Future of Utilities—Utilities of the Future. Academic Press, 2016.

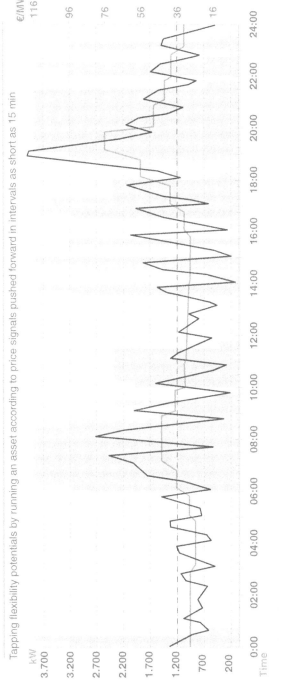

FIGURE 17.15 How a flexible biogas plant shifts its generation to those quarter hours of the day with high market prices. *(Source: Next Kraftwerke.)*

These realities, however, are already at the front and center of priorities to be addressed by regulators in New York,[13] California, and parts of Australia, to name a few. Other parts of the world are likely to confront similar issues in the near future.

As noted in an article in Energy Central,[14] the rapid rise of renewables in places like Germany and Denmark has already made the job of the grid operators more challenging than in the past. And as the example of NK and their cohorts suggests, this challenge has created opportunities to offer services that are sought by generators, consumers, and the grid operator. The question is, how will such new business models emerge, evolve, and how fast are they likely to grow? In this context, it is also important to ask what role will incumbent utilities play—whether generators, distribution utilities, or other stakeholders?[15]

Clearly, a growing number of companies, such as NK will emerge to aggregate decentralized generation capacity and loads to form VPPs while offering a myriad of products and services including ancillary services to TSOs (Fig. 17.16).

If the experience of NK is any indication, such enterprises are likely to grow exponentially, and as they do, offer enhanced services and expand into other markets.

Founded in 2009, Next Kraftwerke has enjoyed explosive growth. In 2013 they aggregated 1 GW of capacity from 2400 installations and traded 2.5 TWh, mostly from biogas, biomass, and CHP plants, up from 1 TWh and 400 installations only a year previously. According to the latest statistics, the company now has over 4000 decentral units of virtually every renewable technology and loads under management, trading 9 TWh (Fig. 17.17).

NK's rapid rise and commercial success can be traced to two valuable services increasingly sought after in future markets with high renewable penetration:

- Aggregating the output of its generators and selling it in various wholesale markets, such as the European Power Exchange (EPEX SPOT)
- Optimizing flexible capacity of both its consumers and producers on various balancing and spot markets

NK's highly automated central control room in Cologne, Germany, adjusts the aggregated generation and loads according to current market prices, grid congestion, and weather data using its proprietary Next Box system.

Both producers and consumers benefit from participating in NK's scheme—the typical savings depend on how flexible they are and how well they respond to price signals.

---

13. Refer to Introduction.
14. Are virtual power plants the future of European utilities?, November 20, 2014, Energy Central at http://community.energycentral.com/community/intelligent-utility/are-virtual-power-plants-future-european-utilities
15. Refer to the book's Foreword, Preface, and Epilogue.

FIGURE 17.16   The concept of NK's virtual power plant (VPP). Aggregate flexible load *and* generation to valorize their flexibility on power markets. (*Source: Next Kraftwerke.*)

Next Kraftwerke in a Nutshell

€ Revenue (2015): 273 million euros

Amount of traded power (2015): 9.1 TWh

Aggregated assets: 4,076

Aggregated power: 2,726 MW

Employees: 131

Operating Countries: Germany, Austria, Belgium, France, Netherlands, Poland

Prequalified Primary Reserve: 29, 7 MW

Prequalified Secondary Reserve: 657,7 MW

Prequalified Tertiary Reserve: 756,9 MW

FIGURE 17.17 Key facts and figures on Next Kraftwerke. (*Source: Next Kraftwerke.*)

For the balancing market, NK are active in secondary and tertiary control reserve, which pays for availability as well as supply. In Belgium and Germany, they recently started offering primary control reserve or frequency response as well.

The availability payments eventually come out of the consumer's wallet via grid charges, but the supply costs are covered by the market players who caused the imbalance—those who have not met their forecasts for generation or consumption.

NK's customers typically get to keep a share of the value of the balancing contract, depending on their flexibility.

When the company was founded in 2009, NK were expecting the volume of ancillary services requested by the TSOs to rise because the feed-in of volatile renewables increased so rapidly. Seven years later they were surprised to find that they had underestimated the exploding significance of the intraday power market, which did not even exist in 2009. In effect, through the doing of the likes of NK, the amount of control reserve called by the TSOs has decreased (dark gray bars in Fig. 17.18) while the amount of electricity traded on the intraday market of EPEX SPOT has increased over-proportionately [light gray bars (green bars in web version)].

The volatility brought into the electricity market by wind and solar is now balanced on the intraday market, reducing the overall imbalance of the grid and thus reducing control reserve calls. The company interprets the development as a sign that the (spot) market can most efficiently deal with volatile renewables while ensuring the high level of energy security Germany is used to.

What is the expected growth trajectory for companies such as NK? The company's co-founder, *Hendrik Samisch*, is confident in NK's business model. He envisions that VPPs, such as his, will enable Germany to reach its ultimate goal of "100% renewable energy at a reasonable price."

Highlighting his belief in market-oriented solutions for the integration of renewables into the energy system, Samisch says:

The continuously growing plant performance in our pool results in economies of scale that lead to size advantages in our trading position and, naturally, benefit the operator as well. Only those who achieve a reasonable price on the electricity exchanges and acclimatize to the free market early will emerge from the dependency on government funding in the medium term.

Next Kraftwerke's ambitions, of course, do not stop at Germany's borders. Samisch expects European electricity market designs to converge over the next five years, offering a huge opportunity to sell NK's grid services across the continent. As European Commission regulation mandates it, cooperation between several TSOs in Europe, for example, on secondary control reserve, is intensifying, so balancing regions are growing larger—and NK want to play their part in the new market designs.

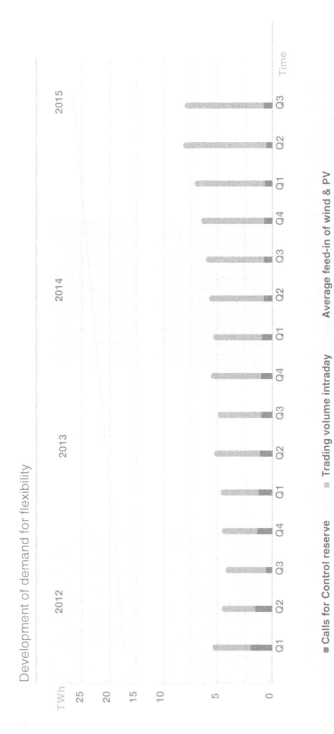

FIGURE 17.18 **Development of demand for flexibility in Germany.** Flexibility has moved away from control reserve markets toward intraday trading in recent years. (*Source: Next Kraftwerke.*)

This explains NK's recent expansions into neighboring markets in Austria, Belgium, France, Poland, the Netherlands, Switzerland, and most recently Italy. According to Samisch, TSOs are welcoming the opportunities of cooperation, such as lower redundancy requirements within each balancing area, and it gives NK a chance to grow. Primary reserve is already contracted transnationally in the ENTSO-E Regional Group Continental Europe.

What about the competition? Samisch expects increasing competition for grid services, not just from other VPPs, but also from Germany's municipal utilities and even its Big Four utilities, RWE/Innogy, E.ON/Uniper, EnBW, and Vattenfall.

How will this game play out? No one can be sure but NK's CEO says:

The role of the utility will change over the next decade. Virtual power plants are a new kind of utility, which perhaps provides a glimpse of what utilities might have to do differently in the next five to ten years. It's too early to say whether the likes of E.ON and RWE will become digital utilities like us, but they will put a great deal of effort into doing this. So far we have a head start.

What NK believe is that when nuclear and coal/lignite plants are gradually closed down and market prices will move freely, there will be sufficient market incentives for setting up decentral, green flexibility options of renewable generation and running loads more flexibly to absorb scarcity and excess situations and supply the short-term flexibility in-between.

Considering the volatility of wind and solar generation over time and space, there will be quite high uncertainties as to how often a specific technology will be needed. High fixed cost assets, such as conventional power plants will likely not get sufficient security over their amortization periods—at least without government support.

Instead, the assets with low fixed costs and low or high variable costs will be most efficient to provide the bulk of flexibility in the future: the plants that are already in place, that are amortized, or have other primary use-cases outside the energy market. These assets could be CHP or biogas plants, (small-scale) hydro plants, or batteries used in electric vehicles, homes and industrial and commercial applications, and of course DSM in general.

According to NK, the flexibility principle of the future energy market will be: first shift, then store.

5 CONCLUSIONS

Europe's electricity markets will gradually converge to form one electricity market in which hydro power plants in Norway, Austria, and Switzerland and solar farms in the Iberian Peninsula and Italy provide power to central Europe in times of scarcity caused by low wind production. At the same time, the flexible capacity of Europe's electricity consumers will be activated over the years such that the idea of peak demand will vanish over time: supply will no longer chase a fixed demand, but supply and demand will dance with each other to find a perfect balance at all times.

This complex and delicate dance of millions of decentral units will have to be choreographed by a digital utility in the center, ensuring the balance of supply and demand at all time while maintaining the grid's reliability.

Next Kraftwerke have sought out this role and developed a successful business model around it. It is thrilling to speculate how NK's green, decentralized vision might unfold and how the concept of variable demand responding to prices might evolve over time.

ACKNOWLEDGMENT

The author wishes to thank the editor for his encouragement and support in preparation of this chapter.

FURTHER READING

Hölemann, S., et al., 2016. Intelligent Grid—A Question of Timing and Standards. Engerati. 14.10.2016. Available from: http://insights.engerati.com/power-in-europe-2016#!/intelligent-grid

Next Kraftwerke Company Blog. OPX. Available from: https://www.next-kraftwerke.com/opx-energy-blog

Sämisch, H., 2016. Digitalisation: Where are the German Digital Utilities? EurActiv. 26.4.2016. Available from: https://www.euractiv.com/section/energy/opinion/digitalisation-where-are-the-german-digital-utilities/

Sämisch, H., Aengenvoort, J., 2016. The Illusion of the German Copper Plate Power Grid. Energy & Carbon. 2.3.2016. Available from: http://energyandcarbon.com/the-illusion-of-the-german-copper-plate-power-grid/

Chapter 18

Integrated Community-Based Energy Systems: Aligning Technology, Incentives, and Regulations

Binod Koirala and Rudi Hakvoort
TU Delft, Delft, The Netherlands

1 INTRODUCTION

As described in other chapters of this volume, technological advancement, falling costs, as well as support schemes for renewables, distributed self-generation, energy efficiency, and demand response has resulted in the rapid deployment of distributed energy resources (DERs) throughout the world, especially in Europe. DERs, by definition, not only include distributed generation but also energy storage, such as in the form of batteries, electric vehicles, and heat storage, as well as demand response. With increasing DERs penetration, the role of households and local communities is changing from passive consumers to active prosumers (van der Schoor and Scholtens, 2015). Accordingly, clusters of residential and community level DERs, in the form of local energy initiatives, such as integrated community energy systems (ICESs), capable of providing a viable alternative to the present centralized energy supply system are emerging. The future energy system is expected to be a combination of the centralized, large-scale system and the local distributed system. The interaction between central and local system will be determined by the ongoing innovation and the way energy market parties handle these developments.

In Europe, there are more than 2800 such initiatives in the form of energy cooperatives of which around 1000 are in Germany and around 350 are in the Netherlands, Fig. 18.1 (REN21, 2016; Morris and Pehnt, 2016; Hier Opgewekt, 2016). This has forced several energy utilities to develop new customer-centric business models for managing energy (Energy Post, 2013; E.ON, 2014; Burger and Weinmann, 2013, 2016). The important role of citizens

Innovation and Disruption at the Grid's Edge. http://dx.doi.org/10.1016/B978-0-12-811758-3.00018-8

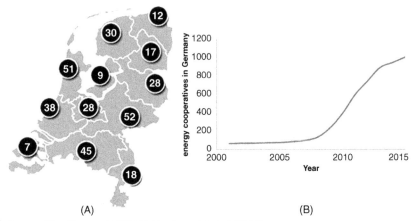

(A)   (B)

Source: Hier Opgewekt (2016); Morris and Pehnt (2016)

FIGURE 18.1   Energy cooperatives in the Netherlands (A) and in Germany (B).

and communities in the energy system has been highlighted also in the recent energy union package of the European union (Energy Union, 2015, 2016).

Although the local energy initiatives are rapidly emerging, the motivation so far has mainly been economic incentives. For example, the lucrative feed-in-tariffs in Germany attracted local investment in DERs through energy cooperatives. As a result, more than half of the renewables installed in Germany are now owned by local citizens and communities (Morris and Pehnt, 2016). However, the market conditions and support incentives in terms of feed-in tariffs have changed resulting in stagnation of the growth of energy cooperatives in Germany (DGRV, 2015). These cooperatives are now in dilemma on how to make most out of the locally generated energy. This means household and community generation have to compete with the centralized generation with economies of scale, highlighting further the need of higher local self-consumption of local generation. In addition, alternative business models, such as local balancing and ancillary services are needed to continue their growth.

A comprehensive and integrated approach for local energy systems where communities can take complete control of their energy system and capture all the benefits of energy system integration is still lacking. Many challenges, such as split-incentive problems, financing, operation, and complexity in decision-making remain for this new type of community energy organization (Koirala et al., 2016b). Would central community energy planning or market mechanisms better serve the objectives of ICESs? How can pricing and incentive schemes be structured to encourage DERs investments in ICESs? Which operational strategies lead to a reduction of peak demand? How can cost and revenue be fairly distributed to benefit the whole community and other stakeholders? In this chapter, we precisely address these issues for local energy systems, such as ICESs, which could also be applicable for broader solutions in the grid's edge.

Availability of numerous technologies, actors, institutions, as well as market mechanisms, further complicates the development of ICESs. Such complexity demands new mechanisms and institutional arrangements to optimally integrate generation and demand at a local level. New initiatives, such as reforming the energy vision in New York (NY REV) with goals to reduce 40% greenhouse gas emissions, to generate 50% electricity from renewables and to reduce energy consumptions of building by 23%, as well as its focus on sustainable and resilient communities can help further emergence of ICESs (NY REV, 2016), as further discussed in chapter by Baak in this volume.

This chapter consists of four sections in addition to the introduction. Section 2 provides new thinking for local energy systems and introduces the concept of ICESs. Section 3 covers the necessary institutional precursors of ICESs. Section 4 examines the institutional design of ICESs from technoeconomic perspective followed by the chapter's conclusions.

2 RETHINKING LOCAL ENERGY SYSTEMS

The technological and institutional changes in present energy system are rapid. In addition to aging infrastructures, these transformations have resulted in technical and economic changes in the power system as summarized in Table 18.1. The energy system is at the crossroad, providing a tremendous opportunity for the reorganization and transformation toward the more sustainable system. The key challenge of the future energy system is a seamless integration of increasing penetration of DERs. One of the prominent solutions lies in increasing self-consumption and matching supply and demand at the local level, such as in ICESs.

2.1 Integrated Community Energy Systems Concept and Definition

Local communities have started to respond to the challenges posed by unsustainable production and consumption practices in the energy sector. These communities are well-placed to identify local energy needs, take proper initiatives, and bring people together to achieve common goals, such as self-sufficiency, resiliency, and autonomy. Local energy projects are inclusive, democratic, and sustainable and might lead to job creation and economic growth (Lazaropoulos and Lazaropoulos, 2015). These initiatives can further the transition to a low-carbon energy system, help build consumer engagement and trust, as well as provide valuable flexibility in the market.

Bottom-up solutions are desired to capture all the benefits allotted by DERs. Recently, the interest of households and communities in generating, supplying, managing energy, as well as improving energy efficiency collectively has also increased and thereby local energy systems are being formed. Recent research also focuses on community energy system where citizens can jointly invest

TABLE 18.1 Technoeconomic Changes in the Energy Landscape

	Traditional power system	Future power system
Technical	Centralized	Centralized and decentralized
	Schedule supply to meet demand	Match both supply and demand
	Base load, off-peak, and peak power plants meet the demand	Decouple supply and demand with flexibility—grid expansion, demand-side management, storage and flexible back-up, low capacity factor for some technologies
	Passive network management	Active network management
	Flexibility from ramping-up and down, peak power plants, interruptible loads, interconnection	Flexibility market, demand response, storage, interconnection, curtailment
Economic	Centralized day-ahead, intraday, and balancing market	Centralized markets for energy and other services and decentralized market for local flexibility
	CO_2 emissions are external	CO_2 emission is internalized through carbon tax, carbon pricing
	Retail prices are in proportion to wholesale prices	Mismatch between wholesale and retail prices due to increasing fixed costs
	Volumetric network tariffs	Advanced network tariffs
	Price inelastic consumers	Price elastic consumers

and operate the local energy systems (Rogers et al., 2008; Walker et al., 2010; Bradley and Rae, 2012; Walker and Simcock, 2012; Wirth, 2014). In a liberalized market, it is possible to establish local producer/prosumer—consumer energy commons enabling them to cocreate commons-based smart energy system at the local level (Lambing, 2013).

In this context, ICESs are multifaceted smart energy system, which optimizes the use of all local DERs, dealing effectively with a changing local energy landscape. ICESs are capable of effectively integrating energy systems through a variety of local generation inclusive of heat and electricity, flexible demand, e-mobility, as well as energy storage. Such integrated approach at the local level helps in the efficient matching of local supply and demand, impacting the existing system architecture and influencing the way the energy systems evolve. The concept of ICESs, as building blocks for the smart grids, is further elaborated in detail (Koirala et al., 2016b; Mendes et al., 2011; Xu et al., 2015). ICESs also represent planning, design, implementation, and governance of energy systems at the community level to maximize energy performance while cutting costs and reducing environmental impacts (Harcourt et al., 2012).

ICESs should have defined system boundaries. Specifically, ICESs can integrate DERs at building and neighborhood scale. Typically, a cluster of households within a distribution transformer can be part of ICESs. The advantage of extending to multiple buildings lies in the variation of demand profiles and availability of multiple generation sources, increasing the flexibility of the system as well as economies of scale but is limited by the complexity of collective decision-making process.

2.2   ICESs as Sociotechnical System

ICESs are complex sociotechnical systems with a strong degree of complementarity enabled through physical and social network relationship, Fig. 18.2 (Künneke et al., 2010). The physical system consists of generation, distribution, storage, and energy management technologies to manage the commodities flow. The social system with different actors, such as consumers, prosumers, aggregators, energy suppliers, and system operators ensures efficient economic operation at minimum environmental effects at the same time providing consumers with different choice options. These systems are complex in the sense that they consist of different decision-making entities and technological artifacts that are governed by energy policy in a multilevel institutional space.

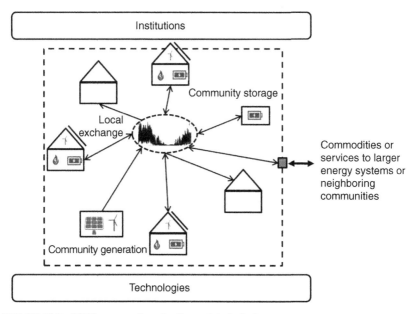

FIGURE 18.2   ICESs as complex adaptive sociotechnical systems.

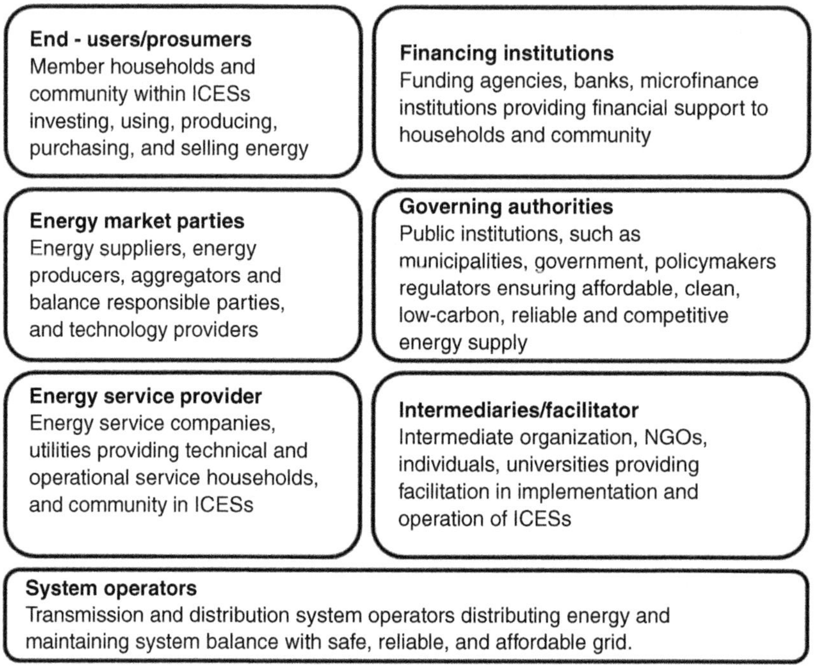

End - users/prosumers
Member households and community within ICESs investing, using, producing, purchasing, and selling energy

Financing institutions
Funding agencies, banks, microfinance institutions providing financial support to households and community

Energy market parties
Energy suppliers, energy producers, aggregators and balance responsible parties, and technology providers

Governing authorities
Public institutions, such as municipalities, government, policymakers regulators ensuring affordable, clean, low-carbon, reliable and competitive energy supply

Energy service provider
Energy service companies, utilities providing technical and operational service households, and community in ICESs

Intermediaries/facilitator
Intermediate organization, NGOs, individuals, universities providing facilitation in implementation and operation of ICESs

System operators
Transmission and distribution system operators distributing energy and maintaining system balance with safe, reliable, and affordable grid.

FIGURE 18.3   Various actors in ICESs.

2.2.1 Actors

The energy system comprises a great variety of public and private actors with different interest and functionalities within a specific institutional environment. The roles and responsibilities of these actors change in the context of ICESs as presented in Fig. 18.3. ICESs are community-based, providing more roles to them in investing, using, producing, selling, and purchasing energy. The complex technical operation in ICESs often needs the engagement of third-party actors, such as system operator or service provider.

2.2.2 Technologies

ICESs consist of households and community level DERs as shown in Fig. 18.4. Several DERs with flexible and intermittent generation, as well as demand- and supply-side management technologies are increasingly becoming available. The technology invested and topologies chosen by local communities is expected to substantially influence future energy system pathways. New services can be driven by information and communication technologies (ICTs) through advancement in the smart grids, for example, to align local demand and supply in time and location or to provide flexibility, as further discussed in the chapter by Knieps in this volume (Clastres, 2011; Järventausta et al., 2010). ICES can

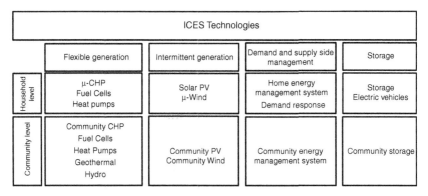

FIGURE 18.4   Few examples of available technologies for ICESs.

provide necessary local infrastructures for an efficient match between demand and supply complementing further development of smart grids. Development of smart-grid technologies and demand-side management technologies facilitate an increase in reliability and efficiency of such local energy systems.

2.3   Added Value of Integrated Approach

ICESs combine energy system integration and community engagement, Fig. 18.5. In this way, ICESs are capable of embracing technical and social innovation, cocreating sustainable and affordable local energy system.

The interactions and complementarities between the different energy carriers are increasing (Lund and Muenster, 2003, 2006; Lund and Kempton, 2014). Different energy carriers can work in synergies leading to a more sustainable and integrated energy system (Lund and Muenster, 2003, 2006; Lund et al., 2010, 2015; Orehounig et al., 2015). The advancement in ICTs as well as smart-grid technologies will further facilitate such integrated operation (Lobaccaro et al., 2016; Lund and Muenster, 2006; Orehounig et al., 2015). ICESs might provide cost-effective solutions to local congestions and help avoid or defer grid reinforcement foreseen with increasing penetration of local renewables.

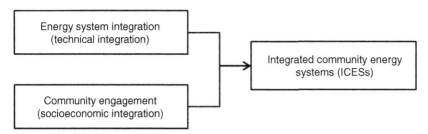

FIGURE 18.5   Technical and socioeconomic integration in ICESs.

ICES stand out from other energy system integration options due to engagement of the local communities. The engagement of citizens and communities increases the acceptance of new energy systems. ICESs also help to keep the local money for the local economy and help fight energy poverty. It not only creates more jobs at the local level but also increases values, such as trust, identity, and sense of community, helping to build stronger communities.

2.4 Benefits and Challenges of ICESs

2.4.1 Benefits of ICESs

The benefits of ICESs include reducing energy cost, CO_2 emissions, and dependence on the national grid, as well as (self-) governance. ICESs help to increase penetration of intermittent renewables and bring new roles for local communities, such as flexibility and ancillary services (Howard, 2014). ICESs provide opportunities for citizens and communities to decide about their energy future, thereby ensuring strong local support and social acceptance. Other benefits of ICESs include increased awareness, reduced energy poverty, affordable energy for all, as well as increased sense of community, pride, and achievement. The benefits of ICESs for communities, system operators, and policymakers are summarized in Table 18.2.

2.4.2 Challenges of ICESs

The main challenge for implementation of ICESs comes from the centralized design and regulation of present energy systems, which do not always provide level playing field for ICESs. In a centralized system, the energy and information flow are unidirectional. However, successful implementation of ICESs needs interaction among several actors of the energy system. For example,

TABLE 18.2 Benefits of ICESs

Community	System operators	Policymakers
Hedging against price fluctuations	Improved reliability of the energy system	Higher energy efficiency
Modular in development	Grid support— ancillary services and flexibility	Higher renewables penetration
Reliability		Local economic growth
Resiliency		Increased energy security
Economic benefits—savings and revenue generations	Occasional roles as service provider	Environmental benefits
Grid support within ICESs	Investment deferrals	Sustainability
Higher efficiency		
Integrated		
Improved power quality		
Sense of community		

TABLE 18.3 Challenges of ICESs

Challenges	Description
Operation	Need a service provider or expert companies for complex technical operation beyond its technical capabilities
Financing	Access to private finance, microfinance, and loans for ICESs
Cost–benefit sharing	Fair allocation of costs incurred and revenue generated among actors
Business case	New business model for flexibility and ancillary services
Monetization of services	Monetization of essential community as well as other ICESs services
Managing utility relations or grid issues	Network access and cost recovery of network investment especially when energy networks are a natural monopoly

selling electricity to neighbors is not allowed and affordable grid access for community generation can be long, complex, and costly.

Although the technologies for ICESs are ubiquitous, there are major challenges in its institutional organization, which must be satisfactorily resolved before they can be successfully deployed and integrated. As highlighted in Table 18.3, these challenges include financing, operation, revenue adequacy, community participation, as well as the fair allocation of costs and benefits.

3 INSTITUTIONAL PRECURSORS FOR ICESS

3.1 Regulation

Energy laws and policies around the world have been built to support centralized energy systems. Accordingly, there are legal barriers to the implementation of ICESs. One of the most prominent ones is EU energy market legislation, the third energy package (EU, 2009). According to this package, generation, distribution, and retail should be unbundled. With the engagement of citizens and community, ICESs are likely to control the local energy system and take over all these roles as a single entity, demanding rebundling.

In the Netherlands, there are similar obligations to both small and large producers in terms of the license of supply (Avelino et al., 2014). As also discussed in the chapter by Löbbe and Hackbarth in this volume, in Germany, after 2014 amendment to the renewable energy law (EEG), small- and medium-sized producers have to compete with large producers (BMWI, 2014).

As also discussed in chapters by Pelegry and by Haro et al. in this volume, recent self-consumption regulation in Spain discourages self-generation as well as ICESs (MIET, 2015). Moreover, administrative hurdles for renewable energy installation, legislative uncertainty, disincentive for self-consumption and production, as well as ineffective unbundling of integrated energy companies inhibit ICESs implementation in Spain. Similarly, in Portugal, the *Decreto Lei n. 153/2014,* a net-metering law despite allowing self-consumption and trading with 10% contribution going to network maintenance, still does not encourage local energy exchange. Similar issues in the United States and Australia are covered in chapters by Baak, Jones et al., and Mountain & Harris, respectively, in this volume.

For the emergence of ICESs, space for innovation, often introduced by new actors is a necessary precondition. As ICESs might take different forms based on local conditions, the legislation should keep open space for as much as possible options for the development of local models. Experiments should be encouraged so that the effects of different models can be assessed. Legal frameworks should promote a wide range of models for community ownership, participation, and investment in ICESs. Several countries in the world, such as Germany, Denmark, the Netherlands, the United Kingdom, and the USA already have policy incentives to promote community-based energy systems.

3.2 Support Incentives

As the focus is shifting to auction/tendering process to support future renewable energy development, the community participation should nevertheless be safeguarded. To speed up low-carbon transition, ICESs should also be given access to national support policies for renewable energy mainly designed for households and large investors, such as feed-in tariffs, tax incentives, grants, low-interest loans, grid access, guaranteed power purchase, and virtual net metering.

The implementation and success of these support incentives differ among countries, which again is affected by several institutional factors. Rather than a one-size-fits-all approach, support schemes designed and tailored to local conditions might prove beneficial in long-run. At the same time, support and mentoring of these local energy initiatives through dedicated intermediary organizations has been proven successful in the United Kingdom and Scotland (Seyfang and Smith, 2007; CES, 2016). At European level, European Federation for Renewable Energy Cooperatives (RESCOOP) is playing this role through networking and knowledge exchange among European renewable energy cooperatives (RESCOOP, 2016).

Few examples of the support incentives addressing community-based energy systems are postcode regulation for local energy exchange in the Netherlands; community net-metering in New York; priority access to the grid in Germany; government grants in Germany, the United Kingdom, and Scotland; as well as low-interest loans in Germany. In the United States, several states, such as New York through reforming the energy vision (REV), California through its community-based renewable energy self-generation program (SB 843), as well as several other states are pushing all sorts of opportunities for community energy (NY REV, 2016; Community Solar, 2012). Similarly, Australia is also expecting high shares of community solar (C4CE, 2016).

3.2.1   Postcode Regulation in The Netherlands

Since 2013, the Dutch postcode (*postcoderoosregeling*) regulation supports local generation and promotes DER penetration. Local entities, such as Energy cooperatives and housing corporation can jointly invest in community energy. Participants get a heavy discount in energy tax up to 10,000 kWh per members. For example, in 2016, the locally exchanged energy is exempted from energy tax. For details on Dutch postcode regulation, see Visbeek (2016).

3.2.2   Community Net Metering in New York

In July 2015, the New York Public Service Commission established a community net-metering in New York state (DOE, 2015). To qualify, the energy community should have a minimum of 10 members and maximum installed capacity of 2 MW. The energy community can have an individual member having a demand of more than 25 kW (with the generation from this member limited to 40% of the energy community output) whereas all other members should have less than 25 kW demand. Moreover, 60% of the generation from the energy community should be self-consumed. This policy enables renters, low-income citizens, and homeowners to engage in energy community. The sponsor of such energy community could be facility developers, energy services companies, municipal entities, and civic association who will be also responsible for building and operating such energy community.

3.3   Grid Access and Local Balancing

There can be resistance from the incumbent grid operator to transfer the owner-ship or lease the network to the community as seen in Feldheim and Schönau in Germany (EWS, 2015; NEFF, 2016). Feldheim had to build a parallel grid and Schönau had to buy back the local grid to realize the local energy system. As private utilities are often biased toward incumbent energy suppliers, increasing number of formally privatized distribution grids, including Hamburg, are remu-nicipalized and further 20% are planning such a step in Germany (Wagner and Berlo, 2015; Nikogosian and Veith, 2012).

Moreover, local energy exchange among ICESs members should be enabled and incentivized. The community can be connected directly to the national grid through a point of common coupling. As shown in Fig. 18.6, the local energy exchange might not always be straightforward. It might involve changing the point of delivery of energy, building a physical interconnection between house-holds across the street or utilizing higher level network infrastructure. In each case, the rules for access to technologies and networks should be well-defined to prevent the opportunistic behavior.

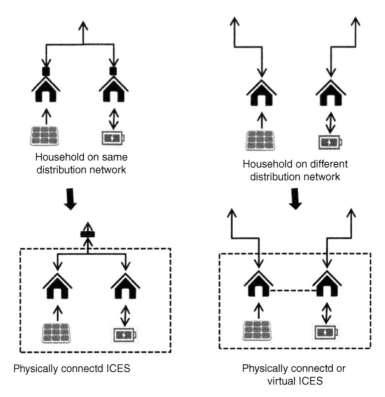

Household on same
distribution network

Household on different
distribution network

Physically connectd ICES

Physically connectd or
virtual ICES

FIGURE 18.6   **Grid access issues in ICESs.**

Increased incentive to follow load or to integrate renewables might improve local balancing and reduce stress on the grid during hours of peak generation. Moreover, although *time netting* of solar PV generation through net metering has proven beneficial for Dutch households, it might be counterproductive for the operation of energy storage. *Location-based netting* promotes cooperation among households through local exchange and might be beneficial for the emergence of ICESs. Moreover, ICESs should be provided with right incentives to collaborate with system operator on storage, energy management, and grid issues.

3.4   Aligning Institutions and Technology

Following the sociotechnical system perspective, ICESs should be seen as a combination of technical elements, characteristics, and links (Wolsink, 2012). Although technologies to realize such local energy system are widespread, the institutions to govern these energy systems are still lagging behind. "Institutions" are the systems of established and prevalent social rules that structure the social interaction (Hodgson, 2006). They are often considered to be the result of enduring interaction processes by which actors have developed ways to reconcile their conflicting interests (Klijn and Koppenjan, 2006).

The current centralized institutional arrangement does not always provide enough incentives for ICESs as the latter were not foreseen during the development of these institutions. New Institutional arrangements are needed to coordinate and shape collective action, thereby leading to further innovation through value-sensitive design and cocreation (Klijn and Koppenjan, 2006).

The institutions and technologies surrounding ICESs also need to be adapted and aligned to each other for optimal performance. The institutions should be established, (re-)designed or adapted, to enforce the necessary roles, responsibilities, control, and intervention. New models of partnerships between the energy distribution networks, utilities, private developers, and communities need to be allowed and examined. In addition, performance expectations, such as sustainability, flexibility, and cost minimization also play an important role in shaping technology and institutions in ICESs.

4 INSTITUTIONAL DESIGN OF ICES THROUGH TECHNOECONOMIC PERSPECTIVE

New energy systems, such as ICESs are not without technical and operational challenges. The technical design ensures commodity flow through reliable and robust system whereas market design ensures monetary flow through the efficient allocation of goods and services according to the community needs (Scholten et al., 2015). These two essential design approach although complementary may sometimes be in odds. A comprehensive design, called institutional design, is necessary which combines the technoeconomic perspectives in the design of institutions for ICESs. Moreover, Due to the involvement of multiple stakeholders and technologies, the institutional design of ICES is complex. For example, collective decisions have to be made to meet the individual needs. Different institutional arrangements for physical and financial administration are necessary for well functioning of these systems.

4.1 Technical Perspective

4.1.1 Flexibility

In recent years, technological change has enabled households and business to fine tune their energy consumption, as well as higher penetration of renewables and disruptive technologies, such as electric vehicles. The variations in electricity demand and supply can be forecasted but an unexpected mismatch might still occur and the system must ensure that supply and demand are always equal (Strbac et al., 2012). This feature of an energy system is called *flexibility* (Denholm and Hand, 2011).

ICESs are capable of decreasing or aligning the production and consumption depending on the requirement of the larger energy system. The technical

TABLE 18.4 Different Functionalities of Storage in ICESs

Functionalities	Community-level storage	Household-level storage
Balancing demand and supply	Seasonal/weekly/daily and hourly variations, peak shaving, integrated electricity and heat storage	Managing daily variations, peak shaving, integrated operation of electricity and heat storage
Grid management	Voltage and frequency regulation, ancillary services, participation in balancing markets	Aggregation of household storage for grid services
Energy efficiency	Demand-side management, better efficiency of ICESs minimize energy losses	Local production and consumption, behavior change, increase value of local generation, integrated operation

and socioeconomic integration make ICESs more flexible. For example, excess PV generation could be stored as heat through heat pumps or in the electricity storage. Similarly, when electricity demand is higher, combined heat and power units can continue to produce electricity storing the excess heat in thermal storage. At the same time, members of ICESs are more energy cautious allowing higher demand-side flexibility.

The flexibility provision, however, should be carefully designed to incorporate multiple households and communities as well as to avoid possible rebound effects. A clear, transparent, and reliable pricing mechanism, such as the one proposed by Universal Smart Energy Framework (USEF), for the trading of flexible energy might be beneficial (USEF, 2016). The energy markets itself should also be more flexible allowing trade close to the real time.

4.1.2 Storage

Energy storage is not only the great source of flexibility but also an enabler of integrated operation as illustrated in Table 18.4. Energy storage is vital to balance supply and demand at household and community level. Storage type and size differ based on seasonal, weekly, daily, or hourly demand to store energy. Long-term energy storage is still technologically challenging. Moreover, integrated operation of heat and electricity storage is desirable. The energy storage can enable location-based netting, ensuring local energy balance and overall higher energy system performance.

4.1.3 Energy Services

The increasing penetration of intermittent renewables and DERs in the energy systems is forcing a debate on new energy services as well as their pricing and provision (Perez-Arriaga et al., 2015). As presented in Table 18.5, these energy

TABLE 18.5 Energy Services Within ICESs and to the Larger Energy System

	Services	Description
Energy-related services	Electrical energy	Electricity sold or purchased at given location and time within the ICESs and to the system
	Flexibility	Upward and downward flexibility to the system
	Operating reserve	Primary: immediate, automatic, decentralized response to system imbalances stabilizing system frequency. For example, Feldheim energy community in Germany provides primary reserves for TSOs through its 10 MWh storage (NEFF, 2016) Secondary: up- or downregulation service to accommodate normal, random variations in system frequency, and normal variability and uncertainty of load and generation balance
	Firm capacity	A guaranteed amount of installed capacity that is committed to producing when called upon under system-stress conditions
	Black-start capability	The availability of resources to restore ICESs to normal conditions after black out
Network-related services	Network connection	Physical connection between the households, to the electricity distribution network and access to the associated services
	Voltage control	Maintenance of voltage within regulated limits throughout ICESs
	Power quality	Minimum voltage disturbance in delivered power
	Congestion management	Overcoming local congestion through network reconfiguration, redispatch/utilization of generators, modifications to load or generation, utilization of flexibility from ICES members
	Energy loss reduction	Local consumption reduces energy losses

services will also be relevant for ICESs; some of these services are internal to ICESs whereas others are system services. For more details on energy- and network-related services of the energy systems, see Perez-Arriaga et al. (2015).

4.1.4 Autarkic Design

The decreasing costs of DERs and the rising retail prices are creating an enabling environment for customers to optimize the planning and operation of the local energy system with the national grid or to get out of the national grid to manage their own local grid (Chaves-Ávila et al., 2016). ICESs might redefine the

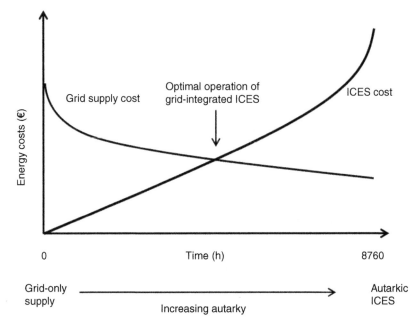

FIGURE 18.7 Trade-offs in autarkic design of an ICES.

relation between production and consumption as they enable resiliency through coproduction. This helps to reduce or substitute the industrial production of energy at centralized power plants by decentralized local production. The excess energy can be sold directly to the grid. The residual demand should be met by the industrial production until large-scale storage becomes financially viable.

Accordingly, ICESs can take different architecture: grid-integrated and grid-defected or autarkic ICES as demonstrated in Fig. 18.7. The most optimal solution is the hybrid system with a combination of the grid and the ICES. Such system can also be islanded during emergency situations to provide critical community functions. The driving forces for the grid defection are independence from the national grid, CO_2 emissions reduction at higher levels than the centralized system, self-governance, and other local preferences.

Under the current system of prices and charges and DER economics, grid-defected ICESs are not economically rationale (Koirala et al., 2016a). As also discussed in the chapters by Steiniger and Sioshansi in this volume, aggregation of the diversity of demand as well as generation profiles among the households within ICES might make community grid defection less expensive than individual grid defection. However, higher reliability needs lead to an oversized system with a very high unused energy to be curtailed and dumped. Nevertheless, it is important to identify the conditions under which such autarkic system can already be a policy option. The cost of an autarkic ICES should be estimated after considering the price of emergency service and the cost of avoided grid reinforcement.

4.2   Economic Perspective

4.2.1   Collective Financing

ICESs may require funding from a variety of sources, such as individuals, municipalities, local cooperatives, and banks. Each ICESs projects will require a customized approach for financing considering both costs and revenue streams. The willingness to invest in local energy initiatives again depends on several institutional factors and local conditions. For example, with long traditions of local energy and opposition to the nuclear energy, German citizens exhibit higher willingness to invest in local energy projects (Kalkbrenner and Roosen, 2016). ICESs can bring much-needed investment and financing to the local energy system through citizens' engagement.

4.2.2   Mismatch Between Wholesale and Retail Price

The current retail electricity price includes wholesale price, regulated costs, such as network costs and other surcharges, as well as taxes. Although the wholesale price might decrease with the high penetration of renewable, the retail price is expected to increase in future due to increasing fixed costs for network reenforcement, grid expansion, balancing costs, as well as other surcharges. ICESs enable local communities to hedge against fluctuating energy prices.

4.2.3   Business Case for ICESs

As the main technologies for ICESs, such as DERs, storage, and energy management systems are gaining maturity, the next step is to create the enabling environment for business model innovation through flexibility in regulation as well as energy policy. The success of ICESs depends on the business model adopted and its flexibility. These business model should reflect self-provision of energy, local exchange, as well as different energy services to the system.

4.3   Institutional Design of ICESs

In this section, institutional design recommendations are provided considering the technoeconomic perspective. ICESs should be managed in a flexible manner adapting to capabilities and interests of the community involved. This translates to efficient (self-) governance, lower transaction costs, fair cost–benefit allocation, and simplified legal requirements.

4.3.1   Roles and Responsibilities

In ICESs, the role and responsibilities of the actors will change. For example, the domestic consumers can have a more active role as prosumers. Local communities can have new roles as *flexibility providers*. The function of "aggregators" which is so far only exercised by the suppliers can also be performed by ICESs through aggregation of small consumers, similar to the virtual power plants concept elaborated in the chapters by Steiniger and Sioshansi in this volume.

Installers can finance as well as operate the installations themselves ensuring consumers "comfort" rather than just supplying equipment. The emergence of new roles and new interpretation of existing functions can ensure efficient development of ICESs.

4.3.2 Design and Coordination of Local Exchange

The local exchange can take several forms, such as peer–peer exchange and prosumer community groups (Giotitsas et al., 2015; Rathnayaka et al., 2015; Brooklyn Microgrid, 2016). Using a well-known blockchain technology, a new community microgrid project in Brooklyn, New York is providing a platform of peer–peer, transactive energy trading among neighbors in a local neighborhood (Brooklyn Microgrid, 2016).

Suitable institutional arrangements should be designed to prevent local energy exchange from being a monopolist. The commodities and suppliers should be well defined, ensuring efficiency, fair allocation of costs and benefits, right prices for participation, and preventing the opportunistic behavior. The local energy price should reflect all the capital costs, operation costs, as well as local network costs.

There is no single best organizational model applicable for ICESs but it should be based on the available sources, types of participants, as well as their needs and expertise. The technical and operational complexity of ICESs might require the involvement of the service provider. The service provider could be energy service companies (ESCOs), distribution system operators (DSOs), or private company with expertise in ICESs. The service providers not only provide assistance in ICESs planning and operation but also provide access to the financing resources.

Nevertheless, two models are outlined here to show how the ICESs could be operated namely service and cooperative model, Table 18.6. In the cooperative model, the actors jointly find the local planning and coordination site, operate the facility together, lifting the separation between production and use. The complex technical operation can be handled by the service provider, however, the ICESs remains in control of the local cooperative. In the service model, the social desire for local utility with a wide range of services is reflected with a great emphasis on the development of ESCOs.

There should be freedom to organize the energy to the local requirements. Laws and regulations should, as far as possible, create space for actors to actually try these or other models. In practice, it may turn out which models are viable and which are not. By analyzing the implications of these operation models in the short and long term, it will become clearer how the ICES can emerge and in what areas further legislation is possible or desirable.

4.3.3 Ownership and (Self-) Governance

Ownership refers to a source of control rights over a resource or property and power to exercise control when the contract is incomplete, such as excluding the

TABLE 18.6 Overview of Functions and Actors in the Service and the Cooperative Model

	Functions	Service model	Cooperative model
Final function	Energy use	Customers	Cooperative
System function	Production	Producers (decentralized)	Local production by the cooperative
	Storage	Customers, producers, or system operators	Cooperative
	Transport	Operators (capacity contracted by ESCO)	Within the community, the cooperative
	Balance responsibility	System administrator	System administrator
	Coordinator	None	On community scale, the cooperative
Marketing functions	Trade	ESCO	Within the cooperatives, local exchange and outside the cooperative through the national market parties
	Delivery	ESCO	
	Aggregation	ESCO	
	Program responsibility	ESCO	
Service functions	Installation	ESCO, installers	Installers in cooperation with cooperative
	Advising	ESCO	Cooperative
	Market coordination	ESCO (limited)	Cooperative
	Financing and insurance	ESCO (in terms of the project in the community)	Cooperative
	Metering	Metering responsible or ESCO	Metering responsible or cooperative
	Communication	Through public networks or desired by ESCO	Through public networks or desired by cooperative
	Switching	ESCO	Cooperative
	Billing	ESCO	Cooperative

ESCO, Energy service company.

TABLE 18.7 Ownership and Governance Model for ICESs

Ownership	(Self-) governance
Community	All costs and benefits are covered by ICESs. Cooperative structure for the management and operation can be outsourced to the service provider.
Utility (DSO)	Utilities remain relevant in ICESs as owner, service provider or grid connection enabler, or combination of these roles. ICESs can benefit from its technical and financial capability. The utility can decide independently and level of community engagement is subjected to the utility.
Private	Private expert companies own and operate ICESs. Incorporating social and economic objectives of the local communities requires negotiation and bargaining.
Public–private (hybrid)	Joint decision-making and planning through the engagement of local communities and private expert companies. Private expert companies can hedge against future uncertainty.

DSO, Distribution system operator.

nonowners from access, selling and transferring resources, as well as appropriately streaming the economic flows from use and investments (Grossman and Hart, 1986; Gui et al., 2016). The ownership in energy systems, such as ICESs is affected by the financing requirements, social welfare issues, as well as risk preferences (Haney and Pollitt, 2013; Walker, 2008). ICESs can have locally owned and controlled community ownership, utility ownership, private ownership, and public–private ownership, Table 18.7. Governance refers to a structure to practice economic and administrative authority, such as rules of collective decision-making among actors (Goldthau, 2014; Avelino et al., 2014).

In the context of ICESs, (Self-) governance refers a group of people that exercise the control over themselves by self-ruling or autonomy. Ostrom (2005) has demonstrated the robustness of self-governance in socioecological systems where government and markets could not do better. Cayford and Scholten (2014) has analyzed the viability of self-governance in community energy system and reported that it depends on communities' abilities to be adaptive to coordinate with different governance circles and may even take different forms according to the social and technical complexity.

4.3.4   Costs and Benefit Allocation

Local balancing reduces peak demand and volume of imported energy in ICESs. The energy losses of the centralized system are also reduced through local generation and exchange. The different energy and network services provided by ICESs avoid energy costs and generate revenues. Grid-defected ICESs provide ancillary services locally, saving on ancillary services. ICES can defer grid reinforcement required for accommodating increasing penetration of DERs or demand. These avoided costs and generated revenues are the benefits of ICESs.

ICESs costs involve capital costs for DERs and energy management system, fuel cost, operation and maintenance cost, as well as network costs for interconnection infrastructure. DER capital costs involve the cost of household- and community-level DERs and cost for corresponding energy management system. Operation and maintenance cost involves the cost of operating local energy exchange as well as the cost associated with operation and maintenance of DERs. Moreover transaction cost is associated with making contracts and billings. The cost of network interconnection and operation should also be considered.

The success of ICESs largely depends on the fair allocation of these costs and benefits. The cost must be paid by those who cause it and the benefits must accrue to those who previously made the investment. In ICESs, this is sometimes difficult to achieve because parts of the facility have the character of a public good.

4.4   Future-Proof Institutional Design

As discussed in Section 2, the complex sociotechnical system, such as ICES has to adapt and operate in changing energy landscape where new technologies will become available, new institutions will emerge, and role and responsibilities of the actors might also change. ICESs should be open for new interactions and experiments to allow further technological and social innovation.

Different actors of the ICESs will have important roles to steer and transform activities of ICESs. These activities namely consumption, storage, exchange, and collective purchasing are influenced by attributes of the technical world, such as available technologies, grids, as well as the environment, attributes of community in which actor and actions are embedded, and institutions which guide and govern actors behavior. This leads to patterns of interactions and outcomes, which could be judged by technical, economic, social, and environmental performance evaluation criteria. Policymakers should steer right transformation of ICESs through suitable policies, incentives, and support schemes.

5   CONCLUSIONS

ICESs are emerging in an environment that was designed for a centralized, top-down, unidirectional network with regulation assuming full reliance on the common network. Now prosumers have options that do not fit the old model and their aggregation in the form of ICESs need fertile support to get established in an otherwise hostile environment. It is important to create dedicated policy space for ICESs within climate and energy framework for the next decades. Policymakers should steer right transformation of ICESs through suitable policies, incentives, and support schemes.

ICESs offer strategic choices for households and communities to transform their energy system and become active prosumers. Households and communities need to understand the trade-offs between self-consumption and local energy balance, as well as to provide system services for the larger energy system. ICESs also address the desire of local communities to contribute toward sustainability and energy security locally.

This chapter has highlighted several institutional precursors for the emergence of ICESs, such as regulation, support incentives, grid access, and local balancing, as well as the alignment of technologies and institutions. Advancing ICESs requires supportive institutional environment for integrated operation as well as interactions among different actors. Unbundling should be relaxed for the long-term financial viability of ICES and partial rebundling is required for local ownership of energy supply infrastructure and energy grid. The local energy exchange platform should be developed to ensure further emergence of ICESs.

Several institutional design recommendations for ICES based on techno-economic perspectives are provided. Technical perspectives considered are flexibility, energy storage, energy services, as well as the autarkic design of grid-integrated and grid-defected ICESs. Collective financing and new business cases involving value of flexibility and ancillary services, as well as hedging against price fluctuations are important. The clear understanding of the changing roles and responsibilities, community ownership and self-governance, design and coordination of local energy exchange, as well as the fair allocation of cost and benefits are important institutional settings for the success of ICESs. These institutional settings need to adapt to the changing energy landscape.

REFERENCES

Avelino, F., Bosman, R., Frantzeskaki, N., Akerboom, S., Boontje, P., Hoffman, J., Paradies, G., Pel, B., Scholten, D., Wittmayer, J., 2014. The (Self-)Governance of Community Energy: Challenges & Prospects. DRIFT, Rotterdam.

BMWI, 2014. BMWi—Federal Ministry for Economic Affairs and Energy—2014 Renewable Energy Sources Act. Available from: http://www.bmwi.de/EN/Topics/Energy/Renewable-Energy/2014-renewable-energy-sources-act.html

Bradley, F., Rae, C., 2012. Energy autonomy in sustainable communities: a review of key issues. Renew. Sustain. Energy Rev. 16, 6497–6506.

Brooklyn Microgrid, 2016. Brooklyn Microgrid. Available from: http://brooklynmicrogrid.com/press/

Burger, C., Weinmann, J., 2013. The Decentralized Energy Revolution. Palgrave Macmillan, London.

Burger, C., Weinmann, J., 2016. European utilities: strategic choices and cultural prerequisites for the future A2. In: Sioshansi, F.P. (Ed.), Future of Utilities—Utilities of the Future. Academic Press, Boston, MA, pp. 303–322, (Chapter 16).

C4CE, 2016. Coalition for Community Energy. Available from: http://c4ce.net.au/

Cayford, T., Scholten, D., 2014. Viability of self-governance in community energy systems: structuring an approach for assessment. In: Proceeding of 5th Ostrom Workshop, Bloomington, IN, USA, pp. 1–28.

CES, 2016. Community Energy Scotland—Home. Available from: http://www.communityenergyscotland.org.uk/

Chaves-Ávila, J.P., Koirala, B.P., Gomez, T., 2016. Institute for Research of Technology Working Paper—Grid Defection: An Efficient Alternative? Universidad Pontificia Comillas.

Clastres, C., 2011. Smart grids: another step towards competition, energy security, and climate change objectives. Energy Policy 39 (9), 5399–5408.

Community Solar, 2012. Community Solar California. Available from: http://www.communitysolarca.org/

Denholm, P., Hand, M., 2011. Grid flexibility and storage required to achieve very high penetration of variable renewable electricity. Energy Policy 39 (3), 1817–1830.

DGRV, 2015. DGRV Annual Survey Reveals Sharp Fall in New Energy Cooperatives—Cooperatives in Germany—DGRV. Available from: https://www.dgrv.de/en/services/energycooperatives/annualsurveyenergycooperatives.html

DOE, 2015. Community Net Metering Department of Energy. Available from: http://energy.gov/savings/net-metering-23

Energy Post, 2013. Exclusive: RWE Sheds Old Business Model, Embraces Transition. Available from: http://www.energypost.eu/exclusive-rwe-sheds-old-business-model-embraces-energy-transition/

Energy Union, 2015. European Commission—PRESS RELEASES—Press Release—Transforming Europe's Energy System—Commission's Energy Summer Package Leads the Way. Available from: http://europa.eu/rapid/press-release_IP-15-5358_en.htm

Energy Union, 2016. Commission Proposes New Rules for Consumer-Centred Clean Energy Transition—Energy—European Commission. *Energy.* Available from: http://ec.europa.eu/energy/en/news/commission-proposes-new-rules-consumer-centred-clean-energy-transition

E.ON, 2014. Strategy—E.ON SE. Available from: http://www.eon.com/en/about-us/strategie.html

EU, 2009. Market Legislation—Energy—European Commission. Energy. Available from: https://ec.europa.eu/energy/en/topics/markets-and-consumers/market-legislation

EWS, 2015. EWS—Elektrizitätswerke Schönau: EWS—Elektrizitätswerke Schönau: Homepage. Available from: http://www.ews-schoenau.de/homepage.html

Giotitsas, C., Pazaitis, A., Kostakis, V., 2015. A peer-to-peer approach to energy production. Technol. Soc. 42 (August), 28–38.

Goldthau, A., 2014. Rethinking the governance of energy infrastructure: scale, decentralization, and polycentrism. Energy Res. Soc. Sci. 1 (March), 134–140.

Grossman, S.J., Hart, O.D., 1986. The costs and benefits of ownership: a theory of vertical and lateral integration. J. Polit. Econ. 94 (4), 691–719.

Gui, E.M., Diesendorf, M., MacGill, I., 2016. Distributed energy infrastructure paradigm: community microgrids in a new institutional economics context. Renew. Sustain. Energy Rev., doi:10.1016/j.rser.2016.10.047.

Haney, A.B., Pollitt, M.G., 2013. New models of public ownership in energy. Int. Rev. Appl. Econ. 27 (2), 174–192.

Harcourt, M., Ogilvie, K., Cleland, M., Campbell, E., Gilmour, B., Laszlo, R., Leach, T., 2012. Building Smart Energy Communities: Implementing Integrated Community Energy Solutions. ICES Literacy Series. Quality Urban Energy Systems for Tomorrow, Canada. Available from: http://questcanada.org/sites/default/files/publications/Building%20Smart%20Energy%20Communities%20-%20Implementing%20ICES.pdf

Hier Opgewekt, 2016. OVERZICHT Hier Opgewekt. Available from: https://www.hieropgewekt.nl/initiatieven

Hodgson, G.M., 2006. What are institutions? J. Econ. Issues 40, 1–25.

Howard, M.W., 2014. The Integrated Grid: Realizing the Full Value of Central and Distributed Energy Resources. The ICER Chronicle, second ed. Available from: http://www.icer-regulators. net/portal/page/portal/ICER_HOME/publications_press/ICER_Chronicle/Art2_9a

Järventausta, P., Repo, S., Rautiainen, A., Partanen, J., 2010. Smart grid power system control in distributed generation environment. Annu. Rev. Control 34 (2), 277–286.

Kalkbrenner, B.J., Roosen, J., 2016. Citizens' willingness to participate in local renewable energy projects: the role of community and trust in Germany. Energy Res. Soc. Sci. 13, 60–70.

Koirala, B., Chaves-Ávila, J., Gómez, T., Hakvoort, R., Herder, P., 2016a. Local alternative for energy supply: performance assessment of integrated community energy systems. Energies 9 (12), 981.

Klijn, E.-H., Koppenjan, J.F.M., 2006. Institutional design: changing institutional features of networks. Public Manag. Rev. 8 (1), 141–160.

Koirala, B.P., Koliou, E., Friege, J., Hakvoort, R.A., Herder, P.M., 2016b. Energetic communities for community energy: a review of key issues and trends shaping integrated community energy systems. Renew. Sustain. Energy Rev. 56 (April), 722–744.

Künneke, R., Groenewegen, J., Ménard, C., 2010. Aligning modes of organization with technology: critical transactions in the reform of infrastructures. J. Econ. Behav. Organ. 75, 494–505.

Lambing, J., 2013. Electricity commons—toward a new industrial society. The Wealth of the Commons: A World Beyond Market and StateLevellers Press, Amherst, MA.

Lazaropoulos, A., Lazaropoulos, P., 2015. Financially stimulating local economies by exploiting communities' microgrids: power trading and hybrid techno-economic (HTE) model. Trends Renew. Energy 1 (3), 131–184.

Lobaccaro, G., Carlucci, S., Löfström, E., 2016. A review of systems and technologies for smart homes and smart grids. Energies 9 (5), 348.

Lund, H., Kempton, W., 2014. Renewable Energy Systems: A Smart Energy Systems Approach to the Choice and Modeling of 100% Renewable Solutions, second ed. Academic Press/Elsevier, Amsterdam.

Lund, P.D., Mikkola, J., Ypyä, J., 2015. Smart energy system design for large clean power schemes in urban areas. J. Clean. Prod. 103 (September), 437–445.

Lund, H., Möller, B., Mathiesen, B.V., Dyrelund, A., 2010. The role of district heating in future renewable energy systems. Energy 35 (3), 1381–1390.

Lund, H., Muenster, E., 2003. Modeling of energy systems with a high percentage of CHP and Wind Power. Renewable Energy 28, 2179–2193.

Lund, H., Muenster, E., 2006. Integrated energy systems and local energy markets. Energy Policy 34 (10), 1152–1160.

Mendes, G., Loakimidis, C., Ferraro, P., 2011. On the planning and analysis of Integrated Community Energy Systems: a review and survey of available tools. Renew. Sustain. Energy Rev. 15, 4836–4854.

MIET, 2015. New Spanish Self-Consumption Regulation, Royal Decree 900/2015. Ministerio de Industria, Energia y Tourismo. Available from: http://www.omie.es/files/r.d._900-2015.pdf

Morris, C., Pehnt, M., 2016. Energy Transition: The German Energiewende. The Heinrich Boell Foundation. Available from: http://energytransition.de/wp-content/themes/boell/pdf/en/German-Energy-Transition_en.pdf

NEFF, 2016. New Energies Forum Feldheim. Available from: http://nef-feldheim.info/?lang=en

Nikogosian, V., Veith, T., 2012. The Impact of ownership on price-setting in retail-energy markets—the German case. Energy Policy 41 (February), 161–172.

NY REV, 2016. "REV." Reforming the Energy Vision (REV). Available from: http://rev.ny.gov/

Orehounig, K., Evins, R., Dorer, V., 2015. Integration of decentralized energy systems in neighbourhoods using the energy hub approach. Appl. Energy 154 (September), 277–289.

Ostrom, E., 2005. Understanding Institutional Diversity. Princeton University Press, Princeton, NJ.

Perez-Arriaga, I., Bharatkumar, A., Burger, S., Gomez, T., 2015. Towards a Comprehensive System of Prices and Charges for Electricity Services. Available from: https://mitei.mit.edu/research/utility-future-study

Rathnayaka, A.J.D., Potdar, V.M., Dillon, T., Kuruppu, S., 2015. Framework to manage multiple goals in community-based energy sharing network in smart grid. Int. J. Elec. Power Energy Syst. 73 (December), 615–624.

REN21, 2016. Renewables 2016 Global Status Report. Available from: http://www.ren21.net/wp-content/uploads/2016/06/GSR_2016_Full_Report_REN21.pdf

RESCOOP, 2016. REScoop.eu European Federation of Renewable Energy Cooperatives. Available from: https://rescoop.eu/

Rogers, J.C., Simmons, E.A., Convery, I., Weatherall, A., 2008. Public perceptions of opportunities for community-based renewable energy projects. Energy Policy 36, 4217–4226.

Scholten, D., Künneke, R., Groenewegen, J., Correljé, A., 2015. Towards the comprehensive institutional design of energy infrastructures. In: van den Hoven, J. (Ed.), Handbook of Ethics, Values, and Technological Design: Sources, Theory, Values and Application Domains. Springer, Dordrecht.

Seyfang, G., Smith, A., 2007. Grassroots innovations for sustainable development: towards a new research and policy agenda. Environ. Polit. 16 (4), 584–603.

Strbac, G., Aunedi, M., Pudjianto, D., Predrag, D., Gammons, S., Druce, R., 2012. Understanding the Balancing Challenge. London Imperial & NERA Consulting, London.

USEF, 2016. Universal Smart Energy Framework: A Solid Foundation for Smart Energy Futures. Available from: https://www.usef.energy/Home.aspx

van der Schoor, T., Scholtens, B., 2015. Power to the people: local community initiatives and the transition to sustainable energy. Renew. Sustain. Energy Rev. 43 (March), 666–675.

Visbeek, S., 2016. Post Code Regulation: Regulation for Reduced Energy Tax. Postcoderoosregeling/Regeling Verlaagd Tarief. Available from: http://www.postcoderoosregeling.nl/

Wagner, O., Berlo, K., 2015. The Wave of Remunicipalisation of Energy Networks and Supply in Germany: The Establishment of 72 New Municipal Power Utilities, June. Available from: http://epub.wupperinst.org/frontdoor/index/index/docId/5920

Walker, G., 2008. What are the barriers and incentives for community-owned means of energy production and use? Energy Policy 36, 4401–4405.

Walker, G., Devine-Wright, P., Hunter, S., High, H., Evans, B., 2010. Trust and community: exploring the meanings, contexts, and dynamics of community renewable energy. Energy Policy 38, 2655–2663.

Walker, G., Simcock, N., 2012. Community energy systems. International Encyclopedia of Housing and Homevol. 1Elsevier, Oxford, pp. 194–198.

Wirth, S., 2014. Communities matter: institutional preconditions for community renewable energy. Energy Policy 70, 236–246.

Wolsink, M., 2012. The research agenda on social acceptance of distributed generation in smart grids: renewable as common pool resources. Renew. Sustainable Energy Rev. 16, 822–835.

Xu, X., Jin, X., Jia, H., Yu, X., Li, K., 2015. Hierarchical management for integrated community energy systems. Appl. Energy 160 (December), 231–243.

Chapter 19

Solar Grid Parity and its Impact on the Grid

Jeremy Webb*, Clevo Wilson*, Theodore Steinberg* and Wes Stein**
*Queensland University of Technology, Brisbane, QLD, Australia; **CSIRO, Canberra, ACT, Australia

1 INTRODUCTION

With the closing of the unsubsidized price gap between solar, wind and conventional power sources, photovoltaic (PV) solar is set to attract an increasingly dominant share of renewable power uptake. For the first time that uptake will be market driven with commercial, manufacturing, and large-scale residential buildings likely to figure prominently in this uptake. There is equally the prospect of a growing number of small- and medium-scale hybrid PV, wind, and concentrated solar power (CSP) plants located on city fringes and beyond where high grid infrastructure costs make dependence on renewable energy and storage competitive.

This chapter describes how such developments will promote increased distributed power generation and in doing so change the nature of the interface with power grids and, consequently, the commercial relationship between business customers and power utilities. In this environment, location, size, and type of business will increasingly determine the mix of renewable energy used, and the way in which it is supplied. The chapter explains the likely nature of the interface of remote and city fringe microgrids with utilities. Over the medium-to-long term the effect of future ongoing technological improvements in batteries and CSP solar will play a key role in determining the extent to which commercial and remote communities will remain reliant on the grid or become largely or even wholly independent of it.

Section 2 described the relative cost reductions occurring in renewable power generation-PV, wind, and CSP. Described in more detail are the specific drivers of PV uptake in the United States and Australia. Section 3 shows how the uptake of PV is creating a greatly expanded class of commercial and industrial prosumers. Section 4 describes the way purchasing power agreements (PPAs) are being shaped by PV uptake. The nature of Australia's uptake of PV solar is

Innovation and Disruption at the Grid's Edge. http://dx.doi.org/10.1016/B978-0-12-811758-3.00019-X

outlined in Section 5 and that of California in Section 6. In Section 7 the financing and management of PV solar uptake by business clients is discussed, and in Section 8 the future of a largely PV/wind-based renewable energy system is examined. Section 9 looks at the need for community-based microgrids where PV solar is dominant in the power mix, followed by the chapter's conclusions in Section 10.

2   THE SOLAR ENERGY COST WATERSHED

The current increasingly rapid uptake of solar—albeit from a low base of a little over 1% of global power generation—is being driven by the continuing and substantial fall in the cost of PV and storage installations. This is illustrated by the comparative analysis from the recent MIT Energy Initiative (2015) study on the future of solar energy (Table 19.1).

Thus until very recently, renewable uptake has been almost entirely driven by legislative mandates, subsidies and feed-in tariffs, witnessed by the rapid adoption of highly subsidized residential PV power systems in countries and regions such as Australia, California, and Germany. "In Australia, where more than 1.6 million households are solar powered during the day—the highest globally—subsidies are generally no longer needed. In the US, the Consumer Energy Alliance (2016) is reporting that in some states the combination of

TABLE 19.1 Levelized Cost Of Electricity (LCOE) Estimates for Selected Generation Technologies (C/kWh)

Item	California (C/kWh)	Massachusetts (C/kWh)
Utility-scale PV	10.5	15.8
Residential-scale PV	19.2	28.7
Utility-scale CSP	14.1	33.1
Average over 22 US regions	**Maximum (C/kWh)**	**Minimum (C/kWh)**
Utility-scale PV	10.1	20.1
Utility-scale CSP	17.7	38.8
Onshore wind	7.1	9.0
Gas combined cycle	6.1	7.6
Conventional coal	8.7	11.4
Conventional gas turbine	10.6	14.9

Note: Numbers are current cost estimates for specified locations in Southern California and Central Massachusetts. The EIA numbers are maximum and minimum 2019 costs across 22 US regions. CSP, Concentrated solar power; PV, photovoltaic.
Source: MIT Energy Initiative (2015) and Energy Information Agency (2014).

incentives provided are now so significant that, they are approaching or exceeding a solar system's total cost". Now, with the continuing fall in renewable energy costs the drivers of PV uptake are increasingly market driven. However for different localities the extent to which there is parity with grid power varies. As Mountain & Harris note in Chapter 5, in Victoria, Australia, extremely high grid connection fees and the nature of peak and off-peak prices can push payback periods for rooftop solar out to an average of 12 years.

2.1 Solar Energy Grid Parity

The continuing speed of this cost reduction is illustrated by more recent US estimates (EnergyTrend, 2016), indicating that utility-scale PV costs had fallen 17% year on year for the third quarter of 2015 and are set to fall a further 15% during 2016 and 2017. That would put the Levelized Cost Of Electricity (LCOE) at just $0.07/kWh and, therefore, below coal and in many cases below that of natural gas power plants, as well as being marginally below the cost of wind (it is noted, however, that the LCOE does not adequately reflect capacity value). The narrowing of the LCOE solar/grid cost gap is also a global phenomenon. A Deutsche Bank study (Shah and Booream-Phelps, 2015) compared the retail price of electricity to that of PV solar, finding that in 30 markets grid parity had already been reached. Thus in Australia, USA, Germany, Spain, and China, they found that retail prices—$US0.63, 0.58, 0.50, 0.40, and 0.20 respectively—were below that of the LCOE of solar power, around $US0.50, 0.40, 0.35, 0.28, and 0.10, respectively.

The report which focuses on the potential of industrial solar points out: "In markets heavily dependent on coal for electricity generation, the ratio of coal-based wholesale electricity to solar electricity cost was 7:1 four years ago. This ratio is now less than 2:1 and could likely approach 1:1 over the next 12–18 months." The report goes on to note that part of the reason for the narrowing has been the increase in grid electricity costs accounted for by transmission and distribution (T and D) estimated to be around 40% of the average electricity bill in the United States. Other studies (Wolfe, 2015) project overall grid parity (wholesale and retail) being reached for most countries between 2017 and 2020.

2.2 Drivers of Solar Power Uptake: USA and Australia

With the critical juncture of PV parity with new nonrenewable energy being reached, nonsubsidized commercial drivers are being added to existing subsidies. Thus in the United States the 30% Federal Government's tax credit and the mandated Zero Net Energy (ZNE) for commercial buildings in California and a number of other states are creating a turning point in renewable uptake. That is already being reflected in a number of contracts in the United States (Fig. 19.1) following the estimated 73% decline in the cost of PV installations over the past decade.

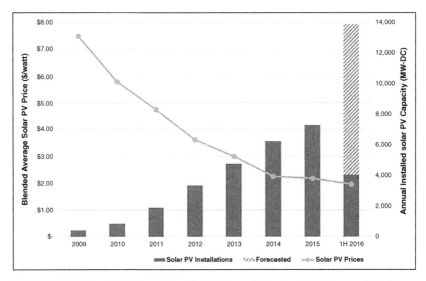

FIGURE 19.1   Solar PV price and installed PV capacity in the USA. *(Source: Solar Energy Industries Association/Gtm Research (2016).)*

This particularly sharp increase projected in the graph for 2016 does, however, seem to reflect the fact that the 30% tax break offered by the US Federal Government for solar installations was due to end in 2017 and, therefore, induced frontloading.

However, with the extension of the full tax break in late 2015 to 2020 and a gradated reduction thereafter, this critical incentive plus the projected continuing fall in the cost of PV solar creates the environment for a rapid and increasingly market-driven future uptake. While 2017 uptake levels may, therefore, indicate tax break–induced frontloading in 2016, the projected continuing fall in solar PV below grid prices will provide an ongoing and broad-based incentive for PV uptake. Thus estimates suggest a continuing rapid increase, as illustrated in Fig. 19.2.

The market-driven rise of solar is also a result of its new cost parity with wind. Moreover, given the speed at which countries are now committed to raising their renewable energy mix, the lead-in time for projects is also becoming a critical issue. This is a relevant issue for wind power whose lead time is generally twice that of solar PV. A further emerging problem for wind power is that in a number of countries its installation is being subjected to increasing resistance from sections of the community on environmental grounds.

The way in which the global shift to PV solar as a result of its recently achieved cost parity with wind is clearly illustrated in countries such as Australia, where renewable energy uptake is driven by a mandated increase of renewable energy to 20% by 2020. The Australian Renewable Energy Authority (ARENA) short

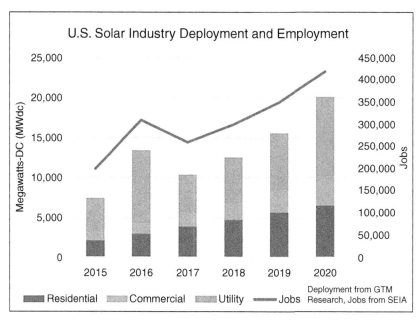

FIGURE 19.2 Projected new US solar uptake to 2020. Note: The left to right order of the legend for power-use sectors is replicated in the figure's bar charts from bottom to top. *(Source: Solar Energy Industries Association/Gtm Research (2016).)*

listed 20 projects for its $100 million large-scale funding round.[1] While this was expected to support around 200 MW of solar, because of the dramatic cost reductions that occurred over the past several years, the 10 successful projects will achieve a power output of 480 MW, with an average size of 40 MW AC.

Industry experts, such as the head of First Solar, describe the ARENA tender as a watershed for commercial-scale solar in Australia describing it as "... the biggest tipping point in the industry we've seen..." (PV Magazine, 2016). Moreover as Fig. 19.2 projects for the United States, as PV solar falls below the LCOE of grid power in an increasingly wide number of countries, the rise in solar uptake is expected to continue unabated.

3 THE RISE OF DISTRIBUTED COMMERCIAL/SOLAR

What has been less clearly articulated, however, is the likely nature of the accelerated uptake of PV-derived energy in terms of how companies interface with their power grids and power utilities. Bodies, such as the Solar Energy Industry Association (SEIA) in the United States, are betting that utilities will

---

1. This round is likely to be the last in which subsidies will be offered for mainstream renewables, such as photovoltaic (PV) and wind.

take the lion's share of the uptake, as indicated in their projections (Fig. 19.2). However, much depends on the nature and size of the commercial enterprises. A recent 2016 survey of large US-based corporations by PWC[2] showed some three-fourth were actively planning further PV renewable uptake and 85% intending to so in the next 18 months. Reasons for doing so were roughly evenly split between a desire to meet sustainability goals, reduce greenhouse gas emissions, generate an attractive return on investment (ROI) (three-fourth of recipients) and to a lesser, but still substantial extent, to limit exposure to energy price variability (almost two-thirds). Key attractions of PV for commercial uses are clearly its scalability, locational versatility, and usability. The latter advantage relates to the fact that the period of greatest power demand of corporates generally matches the period when PV's generating capacity is at its most productive.

4 THE SHAPING OF PPAS BY PV UPTAKE

With the rise of businesses' and particularly commercial establishments' uptake of PV solar, the range of grid edge agreements has inevitably grown. A PWC (2016) survey shows PPAs being chosen by two-third of those surveyed with slightly less than half choosing on-site PPAs.[3] Of the off-site PPAs, 30% are in the form of virtual PPAs.[4] However, it should be noted four-fifths of the companies surveyed are planning to build out their renewables portfolio with multiple types of transactions (e.g., an off-site PPA and an on-site financial investment). These choices indicate that large companies with high levels of power usage (in the PWC survey an average of $100 million/annum) clearly have an incentive to directly manage the on-site or off-site generation, given the advantage of being able to take advantage of the Federal Government's tax credit and lock-in long-term supply at an assured price. In this way, lower costs can be derived from avoiding the expected increase in grid-supplied power which, in the past decade, has been rising at an annual average of over 10% for commercial customers in the United States (Energy Information Administration, 2016).

Unsurprisingly those enterprises that occupy large surface areas have been the first to take advantage of on-site PPAs. The potential for a greatly expanded uptake is considerable. A recent study by the National Renewable Energy Laboratory (2016) indicated that if all suitable rooftops were utilized in the United

---

2. Around two-thirds of companies surveyed had turnovers of over $10 billion and energy bills in excess of $100 million annually.

3. Of those companies indicating they were not about to take up solar power in the foreseeable future, a little over half cited the perceived unattractive return on investment, indicating that a residual of companies may have considered the absence of a tax incentive as a sufficient drawback to not proceed.

4. Virtual purchasing power agreements (PPAs) involve the renewable power facility supplying power directly into the grid and the company deriving the contracted output of the facility from the grid.

States, almost 40% of its power needs could be derived from this source. Much of this would come from commercial-sized premises. A study by the US National Renewable Energy Laboratory (2016) puts the areas in the United States suitable for PV installation for small, medium, and large buildings at 4.92, 1.22, and 1.99 billion m^2, respectively. Under PV it is estimated the potential annual power generation would represent 25, 5.4, and 8.2% of national sales. Given an average house roof is around 1500 ft.^{2}, there are clearly a considerable proportion of the "small" segment, which represents small/medium-sized businesses in addition to residential.

The attractions of on-site solar for large companies with substantial power needs and large roof areas are set to drive a radical increase in distributed generation for such enterprises is in the pipeline. That is already being led by large single-site retail establishments in the United States, where on-site PV solar avoids the overhead costs which come from being supplied from a distant facility through an established grid.

For the two top commercial direct users of renewable energy in the United States—Target and Wal-Mart—economies of scale of rooftop solar are clearly the key drivers. Wal-Mart generates 105 MW of power from solar panels installed on 327 stores and distribution centers. However, this is only about 6% of all their locations, a percentage it intends to double by 2020. Underpinning this investment is a solid commercial advantage: a 9% reduction in power cost per square foot of retail space is claimed. Target (currently the largest commercial user of solar in the United States) plans to add rooftop solar to 500 more of its stores by 2020. Underlining the market-driven potential of commercial solar uptake, Environment America (2016) in a recent study estimates that if all 96,000 large retail establishments in the United States adopt rooftop solar, it will triple the nation's solar capacity.

5   COMMERCIAL SOLAR UPTAKE: AUSTRALIA

In countries such as Australia, while residential solar panels have in the past been heavily subsidized, such subsidies have not generally been extended to commercial apartment block complexes. Equally the uptake of commercial-scale solar has been constrained by the absence of ZNE requirements prescribed for commercial buildings and the lack of a tax credit for renewable energy generation of the type available in the United States. Thus the uptake of PV by Australian commercial power consumers—now that PV-derived solar prices are reaching and are set to fall below grid prices— is likely to become a major part of the renewable power uptake.

The spread of commercial-scale PV in Australia cities has particularly high potential where there is a high percentage of (reasonably direct) sunlight hours. The Western Australian Minister for Energy, Mike Nahan notes that the state would not need additional base load capacity during the day over the next decade given the projected increase in rooftop solar (Parkinson, 2016a,b) that

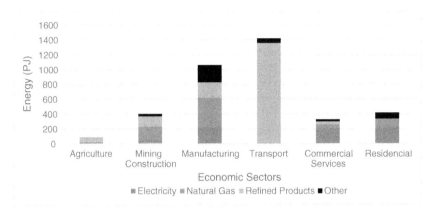

FIGURE 19.3   Australia's annual energy consumption by sector 2009–10. Note: The left to right order of the types of energy listed in the figure's legend are replicated on chart's bars from bottom to top. *(Source: Zero Carbon Australia Buildings Plan (2013).)*

would largely come by way of extending PV installations to commercial properties, particularly premises such as supermarkets and large retail stores.

As shown in Fig. 19.3, the annual energy consumption of the commercial services sector in Australia (overwhelmingly in the form of electricity) accounts for around 8% of total power consumption while manufacturing and mining absorb 56%, of which around 40% is in the form of electricity.

Figs. 19.4 and 19.5 show that in Australia, nonresidential buildings, which typically inhabit cities and which are major power consumers, also typically have large surface areas suitable for PV installations. Moreover, if the retail sector is examined, shopping centers—which have a particularly high potential for extensive use of cost-effective solar PV power generation—account for over half of all retail energy.

6   COMMERCIAL PV UPTAKE: CALIFORNIA

In California the commercial sector is considerably larger and the mining sector considerably smaller although their combined energy consumption is very similar to that of Australia (Figs. 19.6 and 19.7). Given the high utilization of electricity by these sectors, there is a corresponding wider scope for PV installations.

An emission-driven approach is very much a part of Californian and the EU[5] strategies. The Californian Energy Commission announced in 2016 that

---

5. The 2010 EU Energy Performance of Buildings Directive and the 2012 Energy Efficiency Directive put in place measures that required EU members to mandate that all new buildings be near-zero energy buildings by December 31, 2020 (public buildings by December, 31 2018). For existing buildings, member states are required to draw up national plans to increase the number of nearly zero energy buildings, though no specific targets have been set.

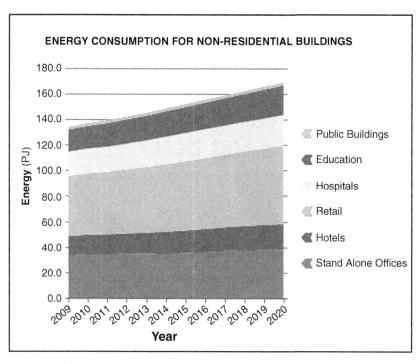

FIGURE 19.4 Energy consumption nonresidential buildings. Note: The bottom to top order of the figure's legend showing building types is replicated in the figure's levels from the bottom up. *(Source: Zero Carbon Australia Buildings Plan (2013).)*

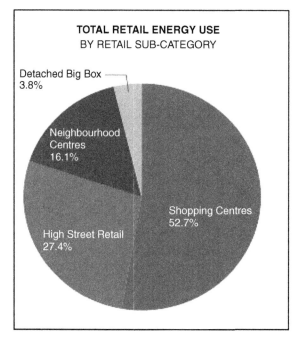

FIGURE 19.5 Total retail energy use by retail category.

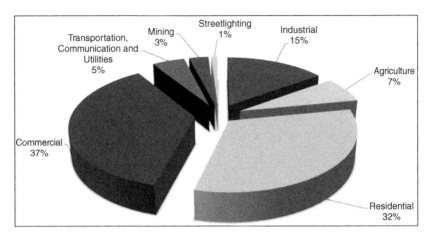

FIGURE 19.6   California electricity consumption by sector. *(Source: California Energy Commission (2008).)*

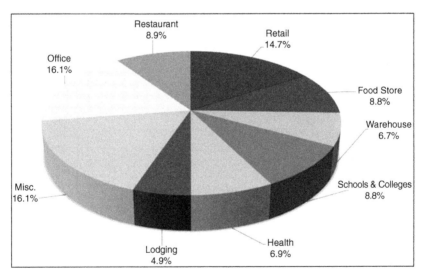

FIGURE 19.7   Californian commercial buildings by type. *(Source: California Energy Commission (2013).)*

all new commercial buildings and 50% of existing commercial buildings have to be ZNE by 2020 (residential by 2030). As 75% of the existing housing stock and 5.25 billion ft.^{2} of commercial space were built before these standards were introduced, their conversion will represent a substantial proportion of power savings.

Indeed, a 2013 study commissioned by the California Public Utilities Commission (2015) found that the commercial sector had the greatest potential for

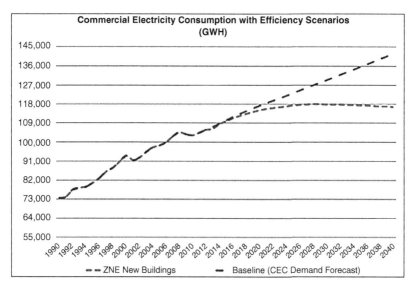

FIGURE 19.8   Californian Energy Savings Potential: Commercial Zero Net Energy (ZNE) buildings. Note: This represents an approximate assessment of the effect of ZNE for all projected future new construction, which is projected to require some six new 500-MW power plants to meet load. *(Source: Fogel (2015).)*

growth and the lowest market barriers (Fig. 19.8). As part of the 2020 target for new buildings they now must provide roof space for the addition of future solar collectors.

The cost benefit to do so for some, however, will depend on the way in which the ZNE concept is defined. It is simply expressed as the net of the amount of energy produced by on-site renewable energy resources, which is equal to the value of the energy consumed annually by the building. The measurement of power consumption, however, uses a time-dependent valuation metric, which means that energy generated/used in peak hours has a considerably higher metric than energy generated/used in off-peak times. While such a metric may pose issues for domestic houses to which it also applies, it is obviously more favorable for commercial and industrial premises, a substantial proportion of which, has peak power usage during periods of peak PV power generation.

7   FINANCING AND MANAGEMENT OF LARGE-SCALE CORPORATE UPTAKE OF PV SOLAR

As the uptake of PV-derived power has gained momentum, the changing grid interface is creating a greater variety of financing and management instruments. Large corporations, such as Wal-Mart and Target, with exceptional on-site PV potential have chosen to outsource their financing to third parties. Thus in Wal-Mart's case a third party, SolarCity, has provided the finance and responsibility for the installation of PV store solar. Wal-Mart then purchases back from Solar-City the power generated at an agreed prices on a long-term contract. Target has similar arrangements in place. In this way large retail enterprises indirectly derive benefits from the US Federal Government's tax breakthrough, lowering its effect on the contracted power supply price with the third party. Importantly the retailer derives the key benefit from this arrangement in the form of a long-term assured price for power. Such an arrangement will, therefore, be an increasingly attractive proposition for enterprises with economically large enough adjacent space or rooftops to justify PV installation. In addition the relatively new variation in the form of virtual PPAs provides a means of price hedging the cost of renewables.

However, many large organizations do not have a PV-friendly physical profile. Thus Starbucks with a large number of small establishments has taken the alternative route of claiming to be 70% carbon free through the purchase of carbon credits, thereby retaining its existing linkages with power utilities. For Starbucks this has evidently made good commercial sense in terms of branding. However, while this represents a rapid, administratively and financially simple method of greening the company, as PV energy prices fall and grid prices continue to rise, an effective increase in the price of electricity will be imposed.

With a rapidly falling price of solar (and to a lesser extent of wind), the incentive is now for corporates to create distributed forms of generation in which they capture both green credentials and the expected lower power bills in the future. Thus the trend for large corporations in accessing renewables has been away from the use of carbon credits and toward an uptake of PPAs. Companies, such as Google, that claimed carbon neutrality in 2007 through carbon credit purchases have since moved to create "true" carbon neutrality though PPAs (as indicated in Fig. 19.9). For Google that was achieved largely through power from wind farms, which, over the past decade, were the most cost-effective means. Google says it will achieve PPA-derived carbon neutrality in 2017.

Other major corporations have followed (Fig. 19.9). These companies have either used off-site PPAs (in which the renewable power generation is fed into the grid and an offtake taken by the enterprise at an agreed price and contract period). The relatively new instrument—the virtual PPA—is also being used to hedge against price variations over time.

There are, therefore, good reasons why various forms of PPAs will increasingly be the corporate path to green renewable energy. It is reasonable to assume that as greenhouse gas reduction targets become increasingly tight, green credentials will become a more widely desired corporate profile. It is also likely that use of PPAs will become more attractive to smaller medium-sized corporations given the hard and soft costs of PV solar and to a lesser extent wind, will continue to fall (particularly soft costs in the United States where they are substantially higher than in Australian and Europe).

In doing so utilities will clearly need to be developing new business models in which their grid focus becomes one of managing a complex of distributed

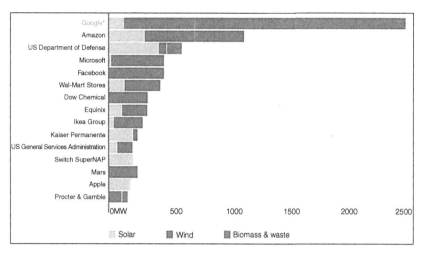

FIGURE 19.9   Cumulative corporate renewable energy purchases United States, Europe, and Mexico to November 2016. Note: The left to right order of the figure's legend listing power sources is replicated in the figure's bars from left to right. Biomass and waste are represented by the bar's third segment for the US Department of Defense and Proctor and Gamble. *(Source: Bloomberg New Energy Finance (2015).)*

facilities involving both off- and on-grid instruments. While grid feedback issues are less relevant to commercial customers whose usage more generally matches solar generation cycles, this is not the case for condominiums that choose to collectively develop renewable solar power and for some industrial plants. Structuring of PPA's may, therefore, need considerable creativity in their construction.

As noted in the Californian Energy Commission's (2016) Draft "Existing Buildings Energy Efficiency Action Plan," this issue of feedback tariffs is still to be resolved and clearly remains a controversial one. As the draft notes: "Procurement-based energy efficiency may be helpful for reaching the Governor's objective to double efficiency gains in existing buildings. In a procurement setting, many of the details around delivery of energy and related grid services would be contained in procurement contracts, rather than emerging from the energy efficiency program portfolio." That is, long-term contracts for power supply with utilities could be made in return for concessions relating to power feedback into the grid. This outcome may be a (albeit complex) formula (similar to that which has been offered to a range of residential clients in California), where long-term contracts for power supply with commercial enterprises are made in return for concessions relating to power fed back into the grid.

California's work in progress on feed-in tariffs reflects, as pointed out by Baak in his chapter, that California's power management system has focused more on the technical enabling aspects of distributed energy resources (DERs), rather than the market mechanisms needed to accelerate deployment.

8 THE FUTURE OF A PV/WIND-DOMINATED POWER SUPPLY

For most countries that are in the process of rapid uptake of renewables and in particular PV, a critical point will be reached where the associated process of intermittent and increasingly distributed power generation creates a base load capacity problem. Analysts indicate that will be the case for Australia and California around 2030 when renewables will account for 30%–40% of generated power. Such problems are already showing up in South Australia, where statewide blackouts have occurred due to a coincidence of storm damage to renewable installations and failure of grid backup. Similarly there are signs of strain already in the Californian energy network.

At this point, the "duck curve" becomes an issue, as the lack of base power in periods of peak residential use becomes a critical issue. For utilities the key issue is whether renewable sources will be able to handle this demand and in what form, as this will very much determine what happens at the gird edge between consumers and suppliers. The capacity of the grid to handle the DERs of commercial customers will depend crucially on the mix of clients that choose to uptake PV solar (or wind if cost effective). Those that operate during the day and have on-site solar may have only a limited interaction with the grid, given they would typically use all that they produce. Others with less suitably matched energy-use profiles and/or use the grid to draw from off-site power installations, may well, when in large numbers, lead to grid stress, as the renewable proportion increases.

In part base power may be supplied by minigrids, which cost effectively aggregate direct feedback from surplus distributed solar and battery sources. This is what is described in a landmark study by Energy Networks Australia (2016) (the national industry association representing Australian electricity networks and gas distribution businesses) and CSIRO (Australia's lead government-funded scientific research organization). The study suggests that by 2050, a zero-emissions grid for Australia could viably have several million customer–owner prosumers supplying between 30% and 50% of the grid needs.

However if a renewable grid's integrity is to be assured, there are those who argue that *both* distributed battery- and CSP-based storage will need to play major roles in providing a substantial sources of dispatchable and flexible renewable power. The US Department of Energy projects that CSP's costs will continue to fall substantially (Fig. 19.10) through to 2020, as improved technology is applied.[6] If so, the speed at which renewable (wind and PV) uptake is achieved is likely to accelerate (Fig. 19.11) given CSP's key advantage as a beneficial enabler of PV and wind, where their generation is curtailed (with a

---

6. For concentrated solar power (CSP) to successfully provide flexibility to a grid with an increasing penetration of variable renewables, the future power block will need to be more advanced than today's subcritical steam turbines, such as closed turbine cycles based on supercritical CO_2 as the working fluid. The resulting higher temperatures will also mandate new storage materials, such as particles or phase change materials.

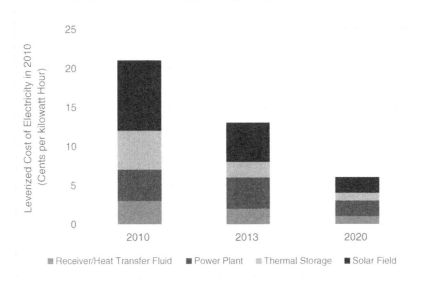

FIGURE 19.10 Projected fall in the cost of CSP. Note: The left to right order of the legend's listing of different types of power generation systems are replicated on chart's bars from bottom to top. *(Source: US Department of Energy (2016).)*

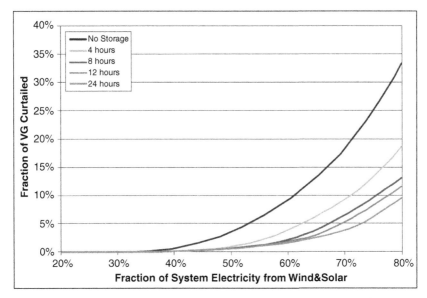

FIGURE 19.11 Effect of CSP with storage on uptake of PV and wind. Note: The legend's top to down order for storage hours is replicated in the same order by trend lines on the figure from left to right. *(Source: Denholm and Hand (2011).)*

subsequent loss of revenue). That is, CSP with its flexible combined generation and storage can ensure that renewable energy continues to be used as illustrated in Fig. 19.11.

9 COMMUNITY-BASED MICROGRIDS

A further evolving development from the falling cost of renewable energy is its increasing capability to provide reliable energy for the edge of urban and remote communities, where extension of and additions to grid-supplied power has become excessive.

The economics of such distributed, largely or wholly off-grid solar installations can be substantively affected by, first, the location (e.g., local irradiation levels and weather conditions) given solar power generation can vary substantially according to latitude. Second, as noted, there is the uneconomic cost of grid extension to locations on city fringes and remote communities. Third, in the future the spread of remote off-grid solar/wind power systems will rely on further cost reductions flowing from innovation and economies of scale.

Such drivers are particularly relevant to countries, such as Australia, which has high levels of solar irradiation, many highly remote communities, and extensive city fringe regions. Spanning over 5000 km, the Australian electricity grid is the largest interconnected power system in the world and, therefore, has a high per capita cost. There is, consequently, a very real prospect of highly distributed urban-scale solar systems outside major urban agglomerations and which provide distinct economic advantages.

However, very remote areas typically mean relatively high levels of dispatchability will be needed if grid support is limited or absent. A declining cost curve for CSP is, therefore, likely to be a critical element in the spread of distributed rural power generation. That is particularly so given that, as illustrated in Fig. 19.12, thermal energy storage, of the kind offered in conjunction with

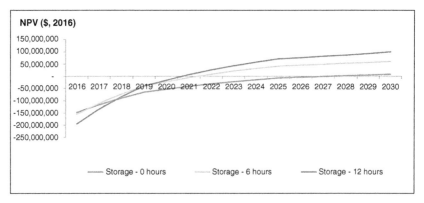

FIGURE 19.12   CSP with thermal energy storage compared to PV plus batteries when more than 3 h of storage is required; net present value basis. *(Source: Stein (2015).)*

CSP, remains more economic than PV/battery storage for periods of over 3 h (calculated over a 20-year payback period).

However, this is not viable if the energy has to be transported long distances. In such cases PV/battery storage becomes viable for those in rural communities.

The key then to the rise of community-based (and possibly owned) remote community microsolar grids is further substantial technological breakthroughs in CSP and, in the more immediate term, further substantial falls in the cost of battery storage. In the meantime in the Australian Capital Territory, where the local government's aim is a 100% renewable grid, storage is being based on the use of battery storage. Currently underway is the deployment of around 5000 battery storage systems in homes and businesses in the Capital Territory. This will be equivalent to around 36 MW of battery storage capacity, which is estimated to deliver around $220 million in network savings to the local grid.

It should nevertheless be noted that CSP is not a viable option if the energy it stores has to be transported long distances. In such cases PV/battery storage becomes viable for those in rural communities. In addition, in the medium term there is the option of hybridized storage, involving the use of gas as a booster for CSP thermal storage.

Just how a fully independent off-grid community facility is to be financed and managed is likely to become a critical issue for utilities. While CSP is locationally inflexible Jones et al. (Chapter 4) argue that given up to a half of the US population are not able to participate in solar net metering (all roofs are not suitable for solar array or residents are in rented premises), community solar arrangements are logical developments to assist them in benefiting from renewables. Thus they note regional aggregation of solar arrays linked to renewable installation off-site (which could include CSP and/or use of a gas booster) for at least a portion of the customers' needs make a great deal of sense. In such inclusive community microgrids, all community members would pay according to their net contribution/usage of power.

A future model for remote communities is being pioneered in the United States by the Vermont utility Green Mountain Power, which is trialing the financing of customers into completely off-grid power system–based solar panels and backup batteries. This "Off-Grid Package" marks the first time that a US utility has offered customers the option of getting utility financing and technical assistance to generate their own power independent of the power grid.

Under such an agreement, customers will pay the utility a monthly fee. In the case of the Vermont trial, community residents would pay between $400 and $850 for their power supply for an average monthly energy consumption between 400 and 800 kWh. That turns out to be less than installing their own power systems or paying for the utility to extend its power lines to reach them. As part of the deal Green Mountain Power recovers its fixed costs, plus a small margin that flows back to the rest of its customer base.

10 CONCLUSIONS

There is sufficient evidence to indicate a watershed has arrived in terms of pricing of renewable energy. The unprecedented recent and continuing decline in the costs of PV—and to a lesser extent, wind and CSP—are, for the first time in history, making renewables cheaper than carbon-based power. These are rates and speed of price declines, which very few advocates of solar, and certainly not policy decision-makers, anticipated.

This watershed in pricing relativities now means that renewables—and in particular solar—uptake will be market driven, self-sustaining, and far more broadly based. Thus while subsidy withdrawal is likely in the short term to slow residential PV uptake, a substantial uptake of PV power by private and public enterprises and from urban fringe and remote communities is in the market-driven pipeline. At the same time, as PV solar costs fall below wind, it should be expected that the mix of renewable power generated will be increasingly solar based—and in particular PV based—and increasingly distributed.

Such developments, and in particular the scalability of PV and its locational flexibility in urban areas, will provide the environment in which a considerable proportion of commercial consumers will have the opportunity to generate a substantial proportion of their power. This trend is likely to be reinforced by the expected lowering of its cost compared to wind. This development is already being led by large retailers with high power consumption of large solar panel–suitable roof areas. In light of projections of continuing price declines for PV many corporates will be able to better capture the cost-reduction benefits through direct and indirect ownership of power generation. However, as the MIT Energy Initiative (2016) study points out, that has the potential, if commercial solar arrays are located in close proximity, to stress the local distribution network.

On the other hand, given that most enterprises' power consumption largely matches daylight hours, power from solar arrays would generally be directly and fully consumed with minimal feedback to the grid. Utilities will, therefore, face some difficult choices as large-scale grid demand shrinks. As Sioshansi (2015) notes, US utility revenues may experience a 13% fall in revenue by 2025 due to the dual effect of increased solar PV self-generation coupled with energy efficiency gains. Moreover with the rise in DERs, their capacity to directly manage power prices with commercial customers will commensurately recede.

One of their keys to retaining commercial customers will, therefore, be an early recognition that different locational profiles of commercial and industrial enterprises will demand different and often creative types of agreements for financing and managing increasingly distributed power generation. Needed, therefore, will be the capacity to offer a range of PPAs, both on-site and off-site.

A further challenge and opportunity will occur when grids become overburdened with solar and wind renewables and base load power becomes a critical issue: based on present trends this is projected for Australia and California around 2030. The challenge for utilities is the prospect that this demand for base load may well be met in part by distributed battery storage, as their costs continues their long-term decline.

However, it can be argued that ultimately, if base load is to be renewable, it would need to be substantially and cost effectively derived from CSP. Importantly, the likely emergence of cost-competitive CSP over the next decade will allow large enterprises and remote/urban-edge communities to create wholly off-grid power generation facilities. In such cases utilities, if they are to retain their corporate clientele, will need to move their business reach to financing, construction, and managing such facilities. Clearly, in such an environment, the capacity of utilities to transition from management of relatively few stand-alone plants to managing a large number of small plants is likely to define their success or failure to survive and prosper.

REFERENCES

Californian Energy Commission, 2016. Draft Existing Buildings Energy Efficiency Action Plan. Available from: http://www.energy.ca.gov/ab758/

California Public Utilities Commission, 2015. Report and Workshops. Available from: http://www.cpuc.ca.gov/General.aspx?id=10740

Consumer Energy Alliance, 2016. Incentivizing Solar Energy: An In-Depth Analysis of U.S. Solar Incentives. Consumer Energy Alliance 2211 Norfolk St. Suite 410 Houston, Texas 77098713.337.8800 www.consumerenergyalliance.org.

Denholm, P., Hand, M., 2011. Grid flexibility and storage required to achieve very high penetration of variable renewable electricity. Energy Policy 39 (3), 1817–1830.

Energy Information Administration, (2016). Electricity Data Browser. Available from: https://www.eia.gov/electricity/data/browser/#/topic/7?agg=2,0,1&geo=g&freq=M

Energy Networks Australia, CSIRO, 2016. Electricity network transformation roadmap: key concepts report 2017-2027. Available from: http://www.energynetworks.com.au/sites/default/files/key_concepts_report_2016.pdf

EnergyTrend, 2016. Prices in PV Supply Chain Have Peaked and Expected to Drop in Late November. Available from: http://pv.energytrend.com/price/Prices_in_PV_Supply_Chain_Have_Peaked_and_Expected_to_Drop_in_Late_November.html

Environment America, 2016. Solar on Superstores: How the Roofs of Big Box Stores Can Help America Shift to Clean Energy. Environment America Research & Policy Center, 2016. http://www.environmentamerica.org/sites/environment/files/reports/AME%20Solar%20Stores%20Feb16.pdf.

Fogel, C., 2015. California's Zero Net Energy building goals how local governments can get involved. SEEC 2015, Conference paper, June 18, 2015.

MIT Energy Initiative, 2015. The Future of Solar Energy: An Interdisciplinary MIT Study. MITEI, Cambridge.

MIT Energy Initiative, 2016. Utilities of the Future. Available from: http://energy.mit.edu/wp-content/uploads/2016/12/Utility-of-the-Future-Executive-Summary.pdf

National Renewable Energy Laboratory, 2016. Rooftop Solar Photovoltaic Technical Potential in the United States: A Detailed Assessment. Technical Report NREL/TP-6A20-65298.

Parkinson, J., 2016a. Australia readies for big solar boom as PV costs continue fall. RENEWecon-omy. Available from: http://reneweconomy.com.au/australia-readies-big-solar-boom-pv-costs-continue-fall-66123/

Parkinson, R., 2016b. Corbell says solar farms competitive with wind energy on pricing. RenewEn-ergy. Available from: http://reneweconomy.com.au/corbell-says-solar-farms-competitive-with-wind-energy-on-pricing-42705/

PV Magazine, 2016. Australia readies for big solar boom as PV costs continue fall. Available from: http://www.pv-magazine.com/news/details/beitrag/australia-readies-for-big-solar-boom-as-pv-costs-continue-fall_100025931/#axzz4QtxN7jlu

PWC, 2016. Corporate renewable energy survey insights. Available from: www.pwc.com/us/renewables

Shah, V., Booeram-Phelps, J., 2015. Industrial Solar. Deutsche Bank Market Research.

Sioshansi, F., 2015. Electricity utility business not as usual. Econ. Anal. Policy 48, 1–11.

Stein, W., 2015. Commonwealth Scientific and Industrial Research Organisation (CSIRO). CSIRO National Solar Research Centre. Available from: www.csiro.au

Wolfe, P. 2015. Utility Scale soar growth: economically viable? Available from: http://www.renewableenergyworld.com/articles/2015/06/is-utility-scale-solar-growth-economically-viable.html

FURTHER READING

ABC Rural Report, 2016. Electricity Networks Australia and CSIRO outlines route to zero emis-sions electricity by 2050. Available from:http://www.abc.net.au/news/2016-12-06/electricity-network-transformation-roadmap-released/8096240

Bhavnagri, K., 2015. Growth of renewable energy in NSW. TransGrid Insight Series, Bloomberg New Energy Finance. Available from: https://www.transgrid.com.au/news-views/blog/Docu-ments/2015_09_28%20_BNEF_Kobad%20Bhavnagri_TransGrid_FINAL_EXTERNAL.pdf

California Energy Commission, 2015. New Residential Zero Net Energy Action Plan 2015–2020. Efficiency Division.

Chung, D., Davidson, C., Ran, F., Ardani, K., Margolis, R., 2015. US photovoltaic prices and cost breakdowns: Q1 2105. National Energy Renewable Energy Authority, Technical Report, NREL/TDP-6A20-64746.

Electric Power Research Institute, 2015. Australian Power Generation Technology report. Available from: http://old.co2crc.com.au/dls/brochures/LCOE_Executive_Summary.pdf

Green, J., Newman, P., 2015. The emerging power paradigm. Curtin University Sustainability Poli-cy Institute (CUSP) Citizen Utilities: The Emerging Power Paradigm.

Mehos, M.S., 2016. Beyond LCOE: the value of CSP with Thermal energy storage. CSP Energy Summit. Sunshot, US Department of Energy. Available from: http://www.astri.org.au/wp-content/uploads/2014/11/MARK-CSP-SunShot-Summit-Plenary-Presentation-Mehos-April-2016-Final.pdf

Moné, C., Stehly, T., Maples, B., Settle, B., 2015. 2014 Cost of wind energy review. National Re-newable Energy Laboratory Technical Report (NREL/TP-6A20-64281).

Norwood, Z., Nyholm, E., Otanicar, T., Johnsson, F., 2014. A geospatial comparison of distributed solar heat and power in Europe and the US. PloS One 9 (12), p.e.112442.

Pitchumani, R., 2015. Towards Ubiquitous, cost-competitive solar power. Presented at the ASTRI Annual Workshop, Brisbane, Australia, February 11, 2015.

Schmalensee, R., 2015. The future of solar energy: a personal assessment. Energy Econ. 52, S142–S148.

SMA Solar Technology AG, 2013. Commercial consumption of solar power. Available from: http://www.sma.de/en/partners/knowledgebase/commercial-self-consumption-of-solar-power.html

Epilogue

This volume is a timely contribution to the topic of the innovation and disruption, which is increasingly dominating the discussion within the energy industry, and is coming to the fore at academia. As the first contributions in this volume demonstrate, innovation is not limited to the use of new technical devices, but entails a new organization of the business and is centered at the grid's edge, that is, at the border of the industry and regulation. The traditional utilities, as well as regulators, are being equally challenged by this development.

The European Union prides itself of being at the forefront of the transition process in the electricity system. The COP 21 commitments are yet another sign that policy is firmly set to a low-carbon path until 2050. The midterm target of the European Union is to cut CO_2 emissions by 40%, and to achieve a share of 27% of renewables in total final energy consumption by 2030.

However, energy policy concerns itself with different things from what the energy industry is discussing nowadays, at least so it seems. Political discussions are about changing the energy mix, about capacity markets, and whether they are in line with competition law or just another means to subsidize incumbent producers; they are about near to real-time markets, which allow renewable generation to participate in wholesale markets, and how to integrate those markets across national borders; all in all more of what has been discussed for the last 15 years, when competitive markets were created by virtue of EU-level legislative acts. The industry, however, has been pointing toward a disruption in their business model for a number of years. A wave of restructuring of big energy companies ensued after the drop in prices due to the economic crisis in 2008. For at least 5 years the industry has been looking for fundamentally new ways to add value, and new competitors with new business models have been entering the market.

Only recently a clearer picture of what we may expect in the future has emerged. The transition is not restricted to substituting big fossil fuel energy production infrastructure with mainly small, low-carbon entities. It may not be confined to the normal network environment, but instead entail changes along the whole value chain and beyond. The direction of the transformation is more likely bottom–up than top–down, and value generation may well be based on services, such as efficiency or information services, and new bundles of services, more so than on the provision of energy itself.

The different chapters in this book deal with such phenomena, new sources of electricity demand, such as electric vehicles; new business models, such as those based on blockchain technology; the Internet of Things development; the integration of flexibility in generation and demand; and finally, the probably changing role of the public supply system, just to name some of the developments explored in this volume.

European regulators have been discussing the future role of distribution operators as a consequence of the present transformation for many years. Finally, in November 2016, the European Commission published a new legislative package for discussion.[1] Apart from traditional considerations, such as security of supply mechanisms or solidarity issues, the topics dealt with in this book, such as prosumers, local energy communities, microgrids, aggregators, and demand response are also essential elements of this package. Some of the proposed mechanisms may not be supported by lawmakers in the end, but the discussion is valuable in and by itself. The core of the package clearly is the transformation of the role final customers play in future electricity systems. They are to become active participants in the market and the package aims at facilitating this transformation.

From a regulatory perspective developments at the grid's edge are challenging the traditional objectives and mechanisms of network regulation, which are described in this book in detail. The main pillars of regulation are separation of monopoly business from competitive activities, incentive regulation for efficient investment, and operation of the grid, as well as nondiscrimination by the monopolistic operators. In the European Union, as the additional dimension is cross-border market integration, these elements have been enriched by setting up cross-border mechanisms for establishing market rules and defining priority investment projects. The aim of these measures has been to establish efficient wholesale markets and guarantee sufficiently effective competition for final customers to get their fair share from increased efficiency in the industry.

After many years of development, the market now quite efficiently organizes dispatch in the European Union, inefficient assets had to leave the market, and consumers are benefiting from efficiency increases. This is the beaten track of market liberalization.

However, can our regulatory system withstand a phase of Schumpeterian destruction? Former certainties are being challenged, and regulation has to follow suit. One of the tenets of regulating the distribution industry is its monopoly status, which is called into question by the idea of local energy communities. These can separate themselves from the public distribution grid, consumers can opt in and out of the network: difficult times for a style of regulation that distributes network cost over a stable customer and/or demand base. This has economic consequences, such as potential cherry picking or even the construction of parallel infrastructure, as further discussed in this volume, but may also have

---

1. Clean Energy for all Europeans (COM 2016 860 final).

more fundamental consequences for the entire inherent solidarity mechanism in the long run.

Where should innovation happen? Expenditure for innovation is part of regulatory cost accounting, as is labor cost or depreciation of assets, as long as cost is incurred efficiently. The cost is covered by the community of all customers and such a system works well, as long as cost structures are comparable. In the Austrian regulatory system, network operators also have the possibility to invest outside the regulated framework, taking into account principles of unbundling and competition law. This may not be the case in all regulatory systems, though.

The Austrian regulator E-Control launched a consultation process on new principles for network tariffs in 2016. The aim of the consultation was to gauge different means of accommodating foreseeable changes in demand and supply conditions, such as flexibility providers or prosumers. More fundamental changes to the system have not been proposed as yet; the potential developments that we may see are still too vague to prepare for them in concrete terms. So far, network regulation has proven flexible enough to tackle or integrate new aspects of network business. This book certainly helps to get a better understanding of potential future paths of development and allow regulators, as well as the industry itself, to accommodate for these changes ahead.

Regulation must keep up with the evolving roles of market participants. This may not be restricted to network regulation, but could extend to the whole set of legal requirements defining roles and responsibilities, regulating entry into the market as noted by top regulators in Australia, California, and New York in this volume.

A process of open debate is vital to securing the confidence of all involved parties in the energy system's ability to solve the issues ahead. This volume provides a selection of articles covering developments in the industry and regulation around the world. Regulators are informed about new ways of organizing energy supply or energy markets, international examples and trends are presented, along with aspects that might not be desirable from the perspective of society as a whole. This publication tries to grasp early signs of a changing environment, and regulators are well advised to follow this discussion.

Johannes Mayer
Head of Competition and Regulation
E-Control Austria

Index

A

Access economy, 44
Access rights, 280
Account reliability constraints, 228
Affordable electric service, 213
Aggregators, 10–11, 367
  function of, 379
AGL Energy, 241, 271
AI. *See* Artificial intelligence (AI)
All-IP Internet, 252, 253
Alternative business models, 364
Annual Energy Outlook, 208
Appliance purchases, 50
Artificial intelligence (AI), 3
Asset-neutral approach, 30
Asset ownership, 30
Australia
  Australian Consumer Energy Alliance, 390
  Australian Consumer Law (ACL), 279
  Australian Energy Regulator (AER),
      92, 279
  Australian National Electricity Market
      (NEM), 51
  Australian Renewable Energy Authority
      (ARENA), 392
  Australian State and Federal Energy
      Ministers, 261
  Australian System of National Accounts
      Information Technology
    net capital stock, 58
  behind-the-meter storage forecasts to 2030, 266
  composition of retail price, of electricity, 191
  electrical loads (customers), 188
  energy market, challenges, 283
  essential energy services, 13
  evolution of feed-in tariffs 2008–2015, 192
  household expenditures, 55
  national accounts, 57
  net consumption, 195
  price paid
    for output of embedded generation, 195
    by retail load, 195
  retail load in, 194
    tariffs for, 191
  self-produce electricity, 187
    status quo in, 195
    tariffs for embedded generation in, 192
Authentication, 250
Auto configuration, 248
Automated Metering Infrastructure (AMI), 252
Automotive battery technology, 109–110
  cost of, 109
Autonomous cars, 103
  companies, Tesla, 108
Autonomous vehicles (AV), 103, 104
AV. *See* Autonomous vehicles (AV)
Average residential load profile, 273, 274, 276–278
Average SA residential load profile, 272, 275

B

Barcelona metropolitan zone, 230
Battery-powered vehicles (BEV), 102
  projected uptake of, 110
Battery storage capacity, 112
Battery storage devices, 291
B2C. *See* Business-to-Customer (B2C)
Berkshire Hathaway Energy (BHE), 45
Best of 96 product, 350
  flexible demand, 352
BEV. *See* Battery-powered vehicles (BEV)
Bias investment, 236
Big Bang Disruption, 149
Big Data, 34, 250
  methodologies, 35
Billing code, 50
Bitcoin, 4
Black box, 235
Blockchain technology, 4, 287, 308, 322
Breaking up microgrid/embedded network, 194
Building design, 53
Business models, 149, 150, 210
  digitalized, typology of competitors in, 313
  paradigm shifts, 60
  transformation process, 310–313
    success factors, 310–312
      digitalization, customization and
        size, 311
      entrepreneurship, 312
      marketing, 311
      realization and adaptation, 312

Business practices, 150
Business-to-customer (B2C), 288
  potentials and major game changers, 291–301
    competition, 296
    energy communities, 296–301
      limitations, 298
      peer-to-peer business, lessons
        learnt, 300–301
    market potential, 294–296
    target groups and customer benefit, 292
Business transformation, 150

C

California
  CAISO duck curve, 172
  California Independent System Operator's
    (CAISO) markets, 171
  Californian Energy Commission, 396
  California Public Utilities Code, 69
  California Public Utilities Commission
    (CPUC), 68, 215, 396
  California Solar Initiative (CSI) program, 171
  commercial buildings by type, 403
  duck curve, 341
  electricity consumption by sector, 401
  energy efficiency programs, 169
  energy savings potential, 403
  heat maps, 178
  impacts of climate change, 169
  locational benefits of DER, 179
  Purchased Power Agreements (PPAs), 182
  regulatory approaches, 174
    California Public Utilities Commission
      (CPUC), 174
    DER planning and deployment, 174
    potential benefits and challenges, 174
    technical aspects, of identifying ideal
      locations for DER, 174
  virtual net metering, 200
Canada
  energy efficiency programs, 213
  limits on volume of embedded generation, 201
Capital-intensive assets, 227
Carbon emission, 227
Carbon neutral energy, 73
Carbon pricing, 51
Car ownership rate, 112
Car sharing, 248
CCA. *See* Community choice aggregation
    (CCA)
Clean-energy policies, 68
Climate change, 41, 150

Cloud computing, 250
Cloud-hosted service, 324
Combined heat and power (CHP), 208
Commercial enterprise, load profile, 397
Commercial solar uptake
  Australia, 395–399
  California, 396–398
Commons-based smart energy system, 365
Communication networks, 252
Community-based microgrids, 389, 404–405
Community choice aggregation (CCA), 65, 67
  in California, 68
  comparison of cases, 78
  energy services, 70–71
  governance and structure, 69
  grid resilience and storage, 71
  in the United States, 65, 67
Community energy system, 365
Community solar, 65
  challenges, 67
  development, 66
  growth, 66
Companies
  business models, 44
Competitive markets, 283
Concentrated solar power (CSP), 389
  projected fall in cost, 403
  with thermal energy storage, 404
Connected economy, 44
Consolidated Edison of New York, 74
Consumer behavior, 34
Consumer demand, for electricity, 208
Consumer empowerment, 123
Consumer price index, 125
Consumers, 367
  investments, 221
  protections, 279
    in distributed energy marketplace, 279
    for use of solar energy, 281
  regulators role, 5
  surplus, 59
  vs. prosumers, 8
Consumption profiles, 222
Continuous load growth, 44
Controlled load circuit, 282
Cooktops, 41
Cooling, 227
Cost development of grid stabilization measures
    by German transmission system
    operators, 338
Cost-reflective pricing, 262, 282
Cross-subsidization, 234
Cryptography, 251

CSP. *See* Concentrated solar power (CSP)
Cumulative corporate renewable energy
    purchases, 404
Customer assets
    beyond meter, 54–59
Customer-centric business models, 363
Customer-centric energy system, 261
Customer energy management service, 242
Customer internal electricity infrastructure, 53
Customers bifurcation, 8–10
Customer service, 35
Customer-side value chain, 53
Customers' individual loads, 44
Customer value, 49, 50
Cybersecurity, 250
    programs, 251

D

Data acquisition, 156
Data analytics, 154
Data processing technologies, 34
Data protection, 250
    issues, 250
Day-ahead market prices, 228
Decentralized decision-making, 36
Decentralized energy resources (DER), 148
Decentralized energy system, 293, 319
Decentralized industry, 26
Decentralized storage, 236
Deelectrification, 61
Demand- and supply-side management
    technologies, 368
Demand management benefit, 60
Demand response (DR), 333
Demand–response management service, 242
Demand-side management (DSM), 213, 333
    energy efficiency and Demand Response, 213
    programs and activities, 213
Demand-side response, 237
Department of Energy, 66
DER assets, 271
DERs. *See* Distributed energy
    resources (DERs)
Developing countries
    grid infrastructure, 6
Diesel *vs.* electric cars, 227
Digital and customer-centric
    transformation, 26
Digital signatures, 250
Dispute-resolution services, 284
Distributed automation, 26
Distributed commercial/solar rise, 393–397

Distributed energy resources (DERs), 3, 4, 41,
    66, 78, 123, 207, 261, 319, 363, 400
    benefits, 365
    CAISO duck curve, 172
    capacity analysis, hosting, 177
    challenges and opportunities of high levels
        of, 169
        demand response programs, 169
        drivers spurring, current evolution of, 170
        intersection of distribution network and
            customers' premises, 169
    consumer pyramid, 8
    critical factors, 17
        economics, 17
        innovation and disruptions, 17
        regulations, 17
    customer load and demographic data, 177
    economics *vs.* traditional bundled service
        economics, 5–8
    financial motives with, 183–184
    forms of, 182
    household prosumers and PV utility scale,
        134–140
        economic assessment of domestic PV
            installation, 137–139
        utility-scale PVs, 139–140
    Integration Capacity Analysis (ICA) map, 178
    levels of penetration, 182
    Locational Net Benefits Analysis
        (LNBA), 179
    market drivers, 172
    need for granular geographical and
        temporal data, 175
    prosumer energy space, 168
    public policy drivers, 170
    sample pacific gas and electric company
        ICA circuit data, 179
    sourcing, 180
    technologies, 272
    utility/grid needs drivers, 171–172
Distributed energy systems
    business models for, 301–309
        based on blockchain technology, 308
        innovative contracting, 305–308
            flat rates, 307
            local green energy for tenants, 307
            PV leasing and contracting, 307
            storage cloud, 306
        overview, 301
        peer-to-peer energy delivery, 303–304
            national, 303
            regional, 304
    future market structure for, 312

Distributed generation, 322, 329
Distributed intelligence, 26
Distributed power generation, 389
Distributed self generation, 5
Distribution capacity expansion, 216
Distribution network
  augmentation, 60
  regulators role, 5
  service provider (DNSP), 280
Distribution system operators (DSOs), 380
Distribution System Platform Providers
        (DSPPs), 174
Divergent customer requirements, 276
Diverting load, 194
Domestic scale batteries, 30
DR. *See* Demand response (DR)
DSM. *See* Demand-side management (DSM)
DSOs. *See* Distribution system operators (DSOs)
Dynamic Innovation Cycle, 151
Dynamic network tariffs, 9

E
Economic savings, 73
Economic theory, 227
EEG law, 289
Efficient tariffs, for generation and load in
        theory, 189
  cost of production, of electricity, 189
  delivery unit of electricity, 189
  locational marginal prices, 190
  lower-voltage distribution network level, 190
  marginal generator/load, 189
  peer-to-peer trading in electricity, 189
  public policy issues, 189
  time-averaged retail prices, 190
  virtual net metering, 189
e-Health, 248
Electrical heating equipment, 36
Electrical resistance storage hot water
        systems, 41
Electrical storage, 137
Electric charging infrastructure, 14
Electric grid, 300
Electric heating load, 331
Electricity
  bill, 224
  consumers, 187, 294
  consumptions, 124, 138, 231
  customers, 58
    long term interest, 58
  demand, 129, 208, 375
  distribution system, 208
  economics, 83

generation, 391
grids, 105, 111, 140
and household appliances
  relative prices, 56
industry, 188, 319
markets, 333
prices, 125
production, 83
retailer, 321
sector, 41
service, 15
storage, 33
supply industry, 221
tariff, 124, 139
Electricity-delivery system
  factors affecting growth and decline
        in demand for, 209
  modernization of, 208
Electricity service
  greater comfort and convenience, 42–46
Electricity utilities investments
  business model for, 45
Electricity value chain, 41, 50–59
  cross-subsidies, 59
  DER additions, 54
  DER benefits, effect of, 59
  DER costs, effect of, 59
  DER, role of, 59–61
  economy, role in, 59
  grid impacts, effect of, 59
  hot shower and cold beer approach, 59
  Lancaster characteristics theory, 59
Electric meters, 15
Electric power sector, 3
Electric power trains, 106
Electric Service Agreement, 73
Electric utilities, 65, 147, 288
  benefits, 154
  business model, 154
Electric vehicle–charging stations, 249
Electric vehicles (EVs), 101, 141–143, 168,
        237, 261, 375
  battery technology, 114–116
  commercial drivers of uptake, 108
  comparative cost of, 106
  cost advantages, 105
  fueling infrastructure, 106
  global and country uptake of, 102
  government and regulatory drives of
        uptake, 116–118
  international comparison of policy support, 114
  investment in, 108
  issues needed to be addressed, 141
  peak car and, 103

power outlets, 107
  government subsidies for, 107
  projections for electricity demand, 113
  refueling infrastructure, 111–114
  sales, 101
  storage capacity of, 112
  subsidization of, 116
Electrification, 104
Encryption, 250
End-customers
  differences in prices for generation and load
    affects incentives, 195
  incentive, to install embedded generation, 199
End-to-end security, 250
End-use customers, 42
Energiewende, 287, 293
Energy, 223
  alternatives, 149
  application, 208
  conservation law, 124
  consumption, 123, 222, 224
  distribution networks, 375
  economy stakeholders, 148
  ecosystem, 150
  efficiency, 249
  imports, 275
  internet, 149
  market volatility, 33
  modernization process, 35
  policies, 123
  prices, 150, 228
  resilience, 65
  service costs, 3
  storage, 261
    technologies, 221
  suppliers, 367
  supply agreement, 37
  supply arrangements, 280
  supply chain, decarbonization of, 148
  systems technoeconomic changes, 366
  transformation, 223
  transition, 223
  utilities, 363
Energy Independence and Security Act of
    2007, 208
Energy innovation market, 160–163
  marketing data, 161
  rising prosumers, 160–161
Energy management, 9
  technologies, 367
Energy Networks Association (ENA), 270
Energy sector transformation, 29
  challenges associated, 26–29
Energy service companies (ESCOs), 68, 380

Energy trading, 32
  bilateral model, 33
e-Privacy, 250
Ergon Energy, 271
ESCOs. *See* Energy service companies (ESCOs)
Estonia, 35
Ethernet, 248
European Commission
  clean energy package, 32
European countries
  feed-in-tariffs (FiTs), 6
European Federation for Renewable Energy
    Cooperatives (RESCOOP), 372
European Power Exchange, 348
European Union
  energy market legislation, 371
  greenhouse gas emissions, reduction of, 25
  ongoing market design changes, 32
  utility stocks performance
    vs. EU stocks, 28
EVs. *See* Electric vehicles (EVs)
Exchange energy, 319
Expensive reinforcements, 235
Experimental ZNE house, 7
Export energy, 277
Exporting solar generation, 275

F

Factory line shaft power drive, 61
Federal Energy Regulatory Commission
    (FERC), 15
Feed-in management cost, 335
Feed-in tariffs, 194, 364
FERC. *See* Federal Energy Regulatory
    Commission (FERC)
Flexibility providers, 379
Flexible biogas plant, 355
Fossil fuels, 41
  generation, 140
Fuel-oil boilers *vs.* heat pumps heati, 227
Fully automated home, 50
Future energy company
  characteristics, 30–37
    automation tools, 35–37
    big data control, 34–35
    customers demands, 37
    flexible demand, access to, 30
    optimization, 32–33
    portfolio generation, access to, 30
    risk management, 32–33
    storage, access to, 30
    trading, 32–33
    user-friendly applications, 35–37

G

Garbage collection service, 15
Generalized DiffServ architecture, 254
Generation output, from wind turbines, 326
Germany
    average day-ahead prices, 343
    development
        of demand for flexibility, 360
        of price volatility since 2012, 342
    electricity future prices, development since
        2012, 339
    energy cooperatives, 364
    energy market, 288–291
        liberalization of, 288
        market structure, 288
        smart metering, 291
        support schemes for renewable
            energy, 289
    German grid operators, 335
    German power utilities, 337
    market premium model, 341
    priority access to grid, 372
    renewable energy law, 371
    scheduled reduction of conventional
        generation capacity, until 2022, 334
    thermal generators, 337
GHG emissions, 116
    Paris agreement, 116
Government bonds, 125
Granular tariffs, 13
Great rebalancing act, 9
Green electricity, 296
Green-energy suppliers, 296
Green Mountain Power, 405
Grid
    boundary, 52
    capacity, 44, 277
    defection, 5
    demand, 273
    depend on technology choices, customer
        requirements, 278
    edge, 41, 250, 364
    electricity value proposition, 60
    elements pricing models, 61
    expansion, 379
    feeding net energy metering (NEM) laws,
        role of, 7
    investment in, 51
    operators, 331
    parity, 83
    redesign, 150
    sources, 279
    stability, 271

supplied
    electricity, 6, 83
    power, 394
GridCredits product, 270
Grief Cycle, 152
Gross metering, 194, 196
    definition of, 197
    increase in on-site generation, 198
    on-site inspection, 199
    policy director, of clean energy council, 198
    requirement, for separate metering, 198
    summary, of key policy issues, 197
Guarantee of Origin certification, 321

H

Hawaii, impacts of climate change, 169
Heating, 227
    pumps, 237
Heterogeneous virtual service networks, 254
High-priced energy, 234
High-voltage network, 208, 232
Home energy storage system, 273
Home Plug, 248
Hourly
    changing energy prices, 238
    energy price, 225
    final demand, 228
    payments, 224
Household
    electricity asset, 54
        asset imbalance, 54
    energy demand, 262
Hydrogen-powered car, 116
Hydro storage, 331

I

IBIS world revenue estimates, 57
ICESs. *See* Integrated community energy
    systems (ICESs)
ICTs. *See* Information and communication
    technologies (ICTs)
ICV. *See* Internal combustion engine
    automobile (ICV)
Ideal network tariff curve, 229
Import
    energy, 277
    power supply, 275
Incentive-compatible pricing, 255
Independent System Operators (ISOs), 174
Industrial revolution, 60
Industry maturity, 50
Industry transformations, 29

Information and communication technologies
(ICTs), 3, 241, 368
innovations and standards as drivers for
microgrids, 242
Innovation, 147, 150, 155
Integrated community energy systems
(ICESs), 363
actors, role and responsibilities of, 379
benefits
affordable energy, 370
increased awareness, 370
reduced CO_2 emissions, 370
reduced energy cost, 370
reduced energy poverty, 370
(self-) governance, 370
challenges, 364
community participation, 371
complexity in decision-making, 364
financing, 364, 371
operation, 364, 371
revenue adequacy, 371
split-incentive problems, 364
citizens' engagement, role of, 379
energy costs, effect on, 382
energy services, 377
grid access issues, 374
grid-defected or autarkic, 378
grid-integrated, 378
institutional design, 379–383
costs and benefit allocation, 382–383
design and coordination of local
exchange, 380
future-proof, 383
ownership and (self-) governance, 380–382
role and responsibilities, 379
institutional design through technoeconomic
perspective, 375–383
economic perspective, 379
business case, 379
collective financing, 379
wholesale and retail price, mismatch
between, 379
technical perspective, 375–378
autarkic design, 377
energy services, 376
flexibility, 375
storage, 376
institutional precursors, 371–375
aligning institutions and technology,
374–375
grid access and local balancing, 373–374
regulation, 371–372
support incentives, 372–373

Netherlands, postcode regulation, 373
New York, community net metering, 373
ownership and governance model, 382
revenue generation, 382
service and cooperative model
functions and actors, overview of, 381
smart grids, role in, 366
social network relationship, role in, 367
storage, different functionalities of, 376
technical and social innovation, role in, 369
technical and socioeconomic
integration, 369
trade-offs in autarkic design, 378
Integrated energy services model, 11
Integrated product/service combination, 11
Integration capacity analysis (ICA), 177
Integrators, 10–11
Intermediaries, 10–11
Internal combustion engine automobile
(ICV), 101
Internal rate of return (IRR), 137
International climate change policy, 51
International Organization for Standardization
(ISO), 242
International Weather for Energy
Calculations, 92
Internet Engineering Task Force (IETF), 242
Internet of everything, 41
Internet of things, 35, 249, 251, 254
Internet of things-electric (IOT-E), 216
Investment planning, 25
Investments customers, 61
Investor, 46
Investor-owned utilities (IOUs), 10, 172, 212
IOUs. *See* Investor-owned utilities (IOUs)
IP-based sensor networks, 248
IP-based standards for virtual microgrids,
evolution of, 247
IP-based wired home networks, 249
IPv4 protocol, 250
IPv6 protocol, 248, 250, 252
IRR. *See* Internal rate of return (IRR)

J

Joint consumption, 236

L

Larger energy system
energy services, 377
Last leg link, 249
LCOE. *See* Levelized cost of electricity
(LCOE)

Levelized cost of electricity (LCOE), 137, 390, 391
    estimates for selected generation technologies, 390
    solar/grid cost gap, 391
Li-ion batteries, 109, 111, 141
Load diversity
    critical advantages, 44
Load profiles, 272
Local distributed system, 363
Local energy creation, 42
Local energy initiatives, 364
Local energy projects, 365
    economic growth, role in, 365
    job creation, role in, 365
Local energy systems, 364–371
    integrated approach, added value of, 369
    integrated community energy systems (ICESs), 365–368
        actors, 368
        benefits, 370
        challenges, 370–371
        technologies, 368, 369
Local exchange
    forms, 380
        peer–peer exchange, 380
        prosumer community groups, 380
        transactive energy trading, 380
Local generators, 323
Locally generated energy, 364
Local sustainability, 65
Locational grid pricing, 326–329
Locational marginal prices, 190
Locational net benefits analysis (LNBA), 179
    consolidated components for PG&E's, 181
Locational pricing
    case study, 328
    open utility's model, schematic outline, 328
Location-based netting, 374
Location-specific pricing, 270
Logically isolated network partitions (LINPs), 245
Long-term marginal network cost, 236
Low-carbon energy system, 365
Low-carbon generation penetration, 234
Lowell, Massachusetts community choice power plan
    case study, 72–74
    energy procurement, 74
    goals, 73
    governance and structure, 72
Low energy consumption appliances, 55
Low Power Wireless Personal Area Networks (6 LoWWPANs), 248
Low valley network tariff, 232
Low-voltage electricity network, 244
Low-voltage microgrids, 242
Low-voltage network, 232

M

Machine-to-machine (M2M) communication, 3, 241
Management cost, 221, 228
    drawbacks, 228
Marin Clean Energy (MCE)
    case study, 68–71
    energy procurement options, 70
        Deep Green program, 70
        local solar option, 70
        standard, 70
    feed-in tariff program, 71
    formation, 70
    grid resilience benefits, 71
    net metering program, 71
Marin Energy Authority, 69
Markets
    competitive landscape, 28
    mechanisms, 61
    nature of risks, 32
    participants, 33
    price risk, 32
    values, 51
Massachusetts Department of Energy Resources, 73
Massachusetts Department of Public Utilities, 72
Massachusetts Municipal Aggregation Statute, 72
Mass-market business models, 37
Mastery Cycle, 152
MCE. See Marin Clean Energy (MCE)
Meter
    identifier, 50
    mystery beyond, 50–59
Microenergy grid, 242
Microgrids, 4, 241, 242, 249, 389
    characterization of, 243–245
    ICT traffic management, 254
    internet protocol–based virtual microgrids, 246–247
    nodes of distribution networks, 252
    and relation to next generation networks, 251
        outside communications, 251
    and virtual networks, 245
Minimum load problem, 344
MIT (Massachusetts Inst Tech), 5, 223, 326, 327, 390, 406

Mobile phone
  industry, 14
  network, 14
Mobile technologies
  services, effect on, 28
Modern power-delivery infrastructure, 208
Monotonic
  curves of power flowing, 231
  network flow curves, 231
  network tariff curve, 232
Multifaceted smart energy system, 366
Multiple generation sources, 367

N

National Association of Regulatory
      Utility Commissioners (NARUC),
      13, 212
National electricity grid, 279
National electricity law (NEL), 261
National electricity market (NEM), 262
  consumer market developments in, 262–264
  distributed technologies, outlook for, 264
    electric vehicles, 268–269
    energy storage, 265–267
    Solar PV, 264
    virtual power plants, 269–271
    virtual power plants
National electricity rules, 281
National Energy Consumer Framework
      (NECF), 279
National grids, 319
National Institute of Standards and Technology
      (NIST), 242, 250
National Regulatory Authority, 224
Natural gas, 150
Net electricity demand, 208
Net energy consumption, 235
Net energy metering (NEM), 216
Netherlands, energy cooperatives, 364
Net hourly energy consumption, 235
Net metering
  average retail utility energy rate, 200
  definition, 197, 200
  extended forms of, 199
    failure of, 202
    integration of generation and load, 199
    local distribution network, 199
    problems, 201
    solar panels, 199
  scope of, 201
Network acknowledged costs, 230
Network capacity, 222, 227, 275
  driven costs, 222

Network costs, 222
Network fee methodologies, 228
Networking physical microgrids, 244
Network investments, 222
Network layer security (IP security/IPsec), 250
Network tariffs, 224, 229, 234, 236
  map, 233
Network virtualization, 245
New energy company, 37–38
  consumers' needs, 38
  conventional generation investments, 38
  relationships with customers, 37
  trading and risk management tools,
        invest in, 38
New York
  community net-metering, 372
  reforming the energy vision (REV), 372
  regulatory approaches, 174
    DER planning and deployment, 174
    New York Public Service Commission
        (NYPSC), 174
    potential benefits and challenges, 174
    Reforming the Energy Vision
        (REV), 174
New York ISO (NYISO), 174
New York Public Service Commission,
      68, 78
New York State Electric & Gas Corporation
      (NYSEG), 74
New York State Energy Research and
      Development Authority
      (NYSERDA), 77
Next Box, 345, 347
Next Generation Networks (NGNs), 242,
      252–255
  incentive-compatible QoS differentiation
        within, 253
Next Kraftwerke
  customer portfolio, 349
  key facts and figures, 358
  principle of demand-side management, 351
  virtual power plant concept, 357
Nontree networks, 234
Nuclear generation, 41
NYSERDA. *See* New York State Energy
      Research and Development
      Authority (NYSERDA)

O

Off-grid customer, 53
Ofgem E-serve, 321
Old grid model, 45
Ombudsman, 284

On-site renewable generation, 319
Open utility, 35
Output curtailment, 281

P

Pacific Gas and Electric Company (PG&E), 71
Paid through capacity, 237
Paris Agreement, 150
Peak car, socioeconomic drivers of, 103
Peak demand days, 275
Peak/valley price ratio, 232
Pecan Street project, 215
Peer economy, 44
Peer-to-peer (P2P)
    business models, 298
    energy networks, 291, 296, 300
    models, 319
        transparency and choice, 320–325
    trading, 4, 188, 250
Personal energy, 150
PG&E. *See* Pacific Gas and Electric Company
        (PG&E)
PHEV. *See* Plug-in hybrids (PHEV)
Phone bills, 14
Photovoltaic (PV), 123
    drivers of low prices, 140
    rooftop PV, 140
    solar, 389
        in Australia, 389
        dominated power supply, future, 402–404
        financing and management of corporate
            uptake, 399–400
        price and installed capacity in the USA, 396
        and storage installations, 390
        in the United States, 389
    systems, 294, 295
    technical potential, estimated suitable area
        and rooftop, 390
Piclo Platform, 323
    business model, 325
Platform economy, 44
Plug-in electric vehicles (PEVs), 208
Plug-in hybrids (PHEV), 106
    impact on distribution networks, 106
    projected uptake of, 110
Policymakers, 284
Portugal, net-metering law, 371
Power charge, 227
Power generation, tax on, 125
Power industry
    aggregators, integrators, and
        intermediaries, role of, 10

Power purchase agreement (PPA), 279, 321
Power sector, 221
    technological innovations in, 101
Power system
    different networks of, 234
    economic incentives, 125
Power transmission and distribution, 124
    remuneration, 125
Power utilities, 124
P2P. *See* Peer-to-peer (P2P)
PPA. *See* Power purchase agreement (PPA)
PPAs. *See* Purchasing power agreements
        (PPAs)
Precio Voluntario para el Pequeño Consumidor.
        See PVPC
Premium feed-in tariff, 194
Pretax investment, 125
Price parity, 5
Privatized distribution grids, 373
Prosumagers, 3, 58
Prosumers, 3, 58, 102, 148, 150, 154, 287, 292,
        298, 309, 367
    energy space, 168
    regulators role, 20
Public Service Commission, 74
Purchasing power agreements (PPAs), 389
    shaping by PV uptake, 394
PV. *See* Photovoltaic (PV)
PVPC (fall-back tariff), 226
PVPC market energy, 227
PVPC tariff, 223, 224, 233

Q

QoS-enabled transport technologies, 252
QoS parameters, 255
Quality of Service, 253
Quality of Service (QoS), 242, 249, 252
Quasitax, 235
Quasitax-free energy
    self-generation of, 235
Quintessential innovation (Q^2i), 154–160
    data analytics and disruption, 154–156
    for electric utilities, 158–160
    iterative business transformation cycle,
        156–158
    iterative data innovation cycle, 156

R

Radio frequency identification (RFID)
        networks, 246
Rate of return regulation, 211
    advantages, 212

comparison with transactive energy (TE)
and, 218
disadvantages, 212
fair and reasonable prices, for electricity, 211
power plants and power-delivery systems, 211
regulatory compact, 211
Real price deflation, 54
Real time–based generation, 241
Real-time communication, 242, 245
Reckon LLP, 327
Reference retailers, 224
Reforming energy vision (REV), 13, 174
in New York (NY REV), 365
Regional power systems, 221
Regional transmission operators (RTOs), 174
Regulator, 210, 233
adoption of DERs, 210
development and adoption of new uses, of
electricity, 210
focus on consumer, 210
full-service strategy development, 210
great rebalancing, 210
innovative and dynamic rate design
options, 210
role, 12–17
fundamental guiding principle, 12
Remote control, 241
via smartphone, 36
Renewable energy, 167, 389
generation, 289, 296
guarantee of origin, 321
policies, 250
potential for, 168
retailers, 270
Renewable Energy Sources Act (EEG), 290
Renewable generation plants, 31
economic incentives for, 125
Renewable Portfolio Standard (RPS), 171
Renewable power, 389
generation, 389
Renewables, 76, 123
electricity demand from, 129
payment rates for, 124
regulatory promotion of, 124
resources, electricity produced from, 287
Repost power, 270
Residential electricity bill breakdown, 225
Retail tariff, 53
for generation and load in practice, 191
existing retail tariffs for loads, 191
composition of retail price, of
electricity, 191
fixed costs, types of, 191

presence of substantial fixed costs, 191
ubiquitous practice of time, 191
Right for customers, 282
Risk management services, 31
Rooftop PV installations, in Australia, 84
Rooftop solar, 112, 261, 395
PVs, 5
systems, 275
Royal Decree Law (RDL), 124, 125
Rural customers, 6

S

Salt River Project (SRP), 214
Scale economies, 44
benefits, 44
Scotland, dedicated intermediary
organizations, 372
SEAV. *See* Shared electric autonomous
vehicles (SEAV)
Security standards, 250
Self-produce electricity, 187
Self-sufficiency ratio, 138
Semiautonomous microgrids, 6
Service cost
components, 12
electric energy price, 12
energy-related services price, 12
network-related services price, 13
policy-related objectives costs of, 13
SETS. *See* Smart electrical thermal storage (SETS)
Shared cars (SV), 103
Shared electric autonomous vehicles (SEAV),
103, 105
Share economy, 44
business models, 322
Smart charging strategies, 112, 113
Smart electrical thermal storage (SETS), 36
Smart energy technology, 36
Smart grid connection, 249
Smart-grid technologies, 368
Smart inverters, Hawaii's Grid, 170
Smart meters, 59, 177, 221, 227, 241, 250,
291, 322
Smart thermostats, 36
Smoke/CO detectors, 36
Solar customers, 6
Solar electricity, 391
Solar energy, 67, 154
cost watershed, 390–393
grid parity, 391
for selected countries, 392
timeline, 393

Solar Energy Industries Association
(SEIA), 214, 393
Solar energy products, 280
Solar exist fees, 9
Solar exports, 275
Solar generation, 271
Solar net metering, 66
Solar panels, investment in, 112
Solar photovoltaics (PV), 65, 83
analytical methodology, 91–94
bill reduction, 92
capital cost, 89, 90
and average prices, 90
daily load profile before and after, 91–93
installation, summary statistics on difference
in bills, 94
payback period, summary statistics, 95
power, 207
sensitivities, 91
fixed charges into variable charges
convertion, 92
increase feed-in tariff by 5 cents
per kwh, 92
Solar power, 271
plants, 139
projected US solar uptake to 2020, 397
uptake in the USA and Australia, 391–393
Solar system, 273
Solar thermoelectric
rates, 125
Solar thermoelectric plants, 125
Solar *vs.* nonsolar customers, 8
Sophisticated decision-making, 36
Space heaters, 41
Spain
development of renewables and costs,
129–134
electricity generation in, 124, 127
electricity system
regulated costs of, 132
renewable energy produced, 135
photovoltaic power development, 133
regulation/legislation on renewables,
124–126
RD 6/2009, 125, 126
RD 13/2012, 125
RD 436/2004, 125, 126, 129
RD 647/2011, 126
RD 661/2007, 125, 126, 129
RD 900/2015, 126, 136
RD 1578/2008, 125, 126
RD 1614/2010, 125, 126
RD 1699/2011, 126, 136

RD 2366/1994, 124, 126
RD 2818/1998, 124, 126, 129
RDL 1/2012, 125, 126
RDL 2/2013, 125, 126
RDL 9/2013, 125, 126
RDL 15/2012, 125, 126
RDL 24/2013, 126
renewable energy plans, 127, 128
National Renewable Energy Action
Plan, 127
Plan Energético Nacional (PEN)
1991–2000, 127
Renewable Energy Plan 1986–1988, 127
Renewable Energy Promotion Plan
2000–2010, 127
renewable energy sources in, 127
renewable power and cogeneration,
130, 134
development of accumulated, 131
Spain
case study, 230–233
government, 224
network costs, 235
PV developments, 235
regulations, 223
residential electricity bills, 223
Spinning reserve, 270
State budget, 227
State territorial policy, 227
Status quo, 195
in Australia, 195
material public policy issues, 195
Stock market, 108
Storage, 236
Stored electricity, 188
Store energy, 236
Subscription product, 277
Sun's power, harnessing, 167
Supercapacitors, 115
graphene, advantages, 115, 116
Supergrid, 4
Surcharge, 289
Sustainable Westchester, 77
SV. *See* Shared cars (SV)
Switch-rated power, 222
System-size restrictions, 281

T

Tariff cost stack, 42, 50–59
generation pricing, role of, 51
Tariff for generation, 192
differences in price paid for, 192

price paid, for output of embedded generation, 194
Tariff for load, 192
  differences in price paid for, 192
Tariff network, 234
Tariffs on incentives
  impact of differences in
  differences in price paid for, 193
TCO. *See* Total Cost of Ownership (TCO)
TCP/IP Internet, 242, 252
TCP/IP protocol, 252
Technology
  advancements, 221
  change, 65
  connections, 280
  development, 123
  innovation, 207
  neutral framework for grid connections, 282
Telecommunications, 254
  Standardization Sector of ITU (ITU-T), 242
Thermal generation, 271, 337
  assets, 26
Thermal plants, 331, 337
  projected load factors for, 27
Thermal power
  plants, 335
  stations, 319
Third energy package, 371
Top-down grid management, 149
Total cost of ownership (TCO), 141
Total energy flow, 230
Total low-voltage demand, 232
Total tariff cost stack value, 52
Trading hedges, risk mitigation, 33
Trading instruments, 33
Traditional centralized utility network model
  advantages, 43
Traditional consumers, 235
Traditional electricity grid value chain, 52
Traditional grid, 4
Traditional merit order, in electricity market, 332
Transactional energy, 4, 250
  comparison of Rate of Return Regulation and, 218
  distribution system operators (DSOs), 216
  regulatory model, 207
Transmission network, 229
Transmission system operators (TSOs), 338
Transportation, 227
TSO forecasts, 233
TSOs. *See* Transmission system operators (TSOs)
Two-way flow of electricity, 208

U
United Kingdom
  dedicated intermediary organizations, 372
United States
  energy efficiency programs, 213
  limits on volume of embedded generation, 201
  net energy metering (NEM) laws, 6, 9
  net metering, 200
  residential customers by numbers, 16
  solar market, 214
  US Federal Energy Regulatory Commission (FERC), 213
Universal Smart Energy Framework (USEF), 376
User-friendly apps
  with energy applications, 37
Utility business models, 9
  future of, 14
Utility public relations, 42
Utility-scale generation, 10
Utility sector applications, 34
Utility value chain, 10

V
Value chain, 46–50
  institutional and financial elements, 43
Value of customer reliability (VCR), 57
Variable renewable generation
  breakdown in German electricity generation mix, 336
  demand future, 354–361
  flexibility in context, 335–344
Victoria
  consumption, export, and load profile data pre- and post-PV installation, 94
  count of retail offers, by distribution zone, 86
  electricity market
    residential rooftop PV market, 98
  electricity tariffs, 85
    annual charges and fixed charges, 87
    summary statistics, October 2016, 88
  reduction in annual bill *vs.* average price, 93
  residential electricity bill components, 88
  residential rooftop PV market
    annual bill, 93
    investment, 95–97
Victorian Essential Services Commission, 84
Virtualization, 254
Virtual microgrids, 244, 252, 253
  and next generation networks, 252
Virtual net metering, 188
Virtual networks, 254

Virtual power plants (VPPs), 252, 305, 333
  and role of aggregators, 344–354
Virtual resource management, 253
Virtual solar/renewable customers, 10
VPPs. *See* Virtual power plants (VPPs)

W

Westchester, New York
  case study, 74–78
    governance and structure, 75
  Westchester Power program, 74
    energy procurement, 75
      services/community solar/demand
        response/microgrids, 76
    governance and structure, 75
    structure, 75

Wholesale markets, 26
  energy prices, 229
Wholesale+network+retail tariff
    cost stack, 51
WiFi, 248
Wind energy, 124, 237
Wireless networks, 249

Z

Zero electric bills, 6
Zero net energy (ZNE), 391
  buildings, 3, 4, 15
ZigBee IP, 248
  Smart Energy Protocol V2.0, 248
  standard version IEEE 2030.5, 248
ZNE. *See* Zero net energy (ZNE)

Printed in the United States
By Bookmasters